Solid Mechanics and Its Applications

Founding Editor

G. M. L. Gladwell

Volume 274

Series Editors

J. R. Barber, Department of Mechanical Engineering, University of Michigan, Ann Arbor, MI, USA

Anders Klarbring, Mechanical Engineering, Linköping University, Linköping, Sweden

The fundamental questions arising in mechanics are: Why?, How?, and How much? The aim of this series is to provide lucid accounts written by authoritative researchers giving vision and insight in answering these questions on the subject of mechanics as it relates to solids. The scope of the series covers the entire spectrum of solid mechanics. Thus it includes the foundation of mechanics; variational formulations; computational mechanics; statics, kinematics and dynamics of rigid and elastic bodies; vibrations of solids and structures; dynamical systems and chaos; the theories of elasticity, plasticity and viscoelasticity; composite materials; rods, beams, shells and membranes; structural control and stability; soils, rocks and geomechanics; fracture; tribology; experimental mechanics; biomechanics and machine design. The median level of presentation is the first year graduate student. Some texts are monographs defining the current state of the field; others are accessible to final year undergraduates; but essentially the emphasis is on readability and clarity.

Springer and Professors Barber and Klarbring welcome book ideas from authors. Potential authors who wish to submit a book proposal should contact Dr. Mayra Castro, Senior Editor, Springer Heidelberg, Germany, email: mayra.castro@springer.com

Indexed by SCOPUS, Ei Compendex, EBSCO Discovery Service, OCLC, ProQuest Summon, Google Scholar and SpringerLink.

David J. Steigmann · Mircea Bîrsan · Milad Shirani

Lecture Notes on the Theory of Plates and Shells

Classical and Modern Developments

Springer

David J. Steigmann
Department of Mechanical Engineering
University of California
Berkeley, CA, USA

Milad Shirani
Department of Mechanical Engineering
University of California
Berkeley, CA, USA

Mircea Bîrsan
Faculty of Mathematics
University Duisburg-Essen
Essen, Germany

Department of Mathematics
University "Alexandru Ioan Cuza" of Iaşi
Iaşi, Romania

ISSN 0925-0042 ISSN 2214-7764 (electronic)
Solid Mechanics and Its Applications
ISBN 978-3-031-25676-9 ISBN 978-3-031-25674-5 (eBook)
https://doi.org/10.1007/978-3-031-25674-5

This Springer imprint is published by the registered company Springer Nature Switzerland AG
The registered company address is: Gewerbestrasse 11, 6330 Cham, Switzerland

Preface

The theory of plates and shells originated in the need on the part of engineers for tractable models to analyze and predict the response of thin structures to various kinds of loading. The basic idea is to exploit the thinness of the structure to represent the mechanics of the actual thin three-dimensional body under consideration by a more tractable two-dimensional theory associated with an interior surface. In this way, the relatively complex three-dimensional continuum mechanics of the thin body is replaced by a far more tractable two-dimensional theory. To ensure that the resulting model is predictive, it is necessary to compensate for this 'dimension reduction' by assigning additional kinematical and dynamical descriptors to the surface whose deformations are modeled by the simpler two-dimensional theory. The manner in which this is achieved, on the basis of the three-dimensional parent theory, constitutes most of the contents of this book.

Few subjects exhibit such an intimate and beautiful interplay between Mechanics and Geometry as Shell Theory. However, precisely because of its connection to geometry, learning the theory is in many ways more challenging than learning standard continuum mechanics in the flat three-dimensional Euclidean space of our common experience. Accordingly, in this book, we begin with a fairly extensive exposition of the differential geometry of surfaces. This in turn requires a facility with tensor analysis in curvilinear coordinates, which is also covered in detail. As the book is concerned with the theory of elastic shells, further background on three-dimensional elasticity theory, both linear and nonlinear, is provided as a prelude to the discussion of the dimension reduction procedure.

With these prerequisites in hand, we go on to develop linear and nonlinear plate and shell models, including classical plate buckling theory, systematically. Our method is based on the expansion of the potential energy of the thin body in powers of its thickness. This is achieved by combining the Leibniz Rule, for differentiating an integral with respect to its integration limits, with a Taylor expansion. The expansion is terminated at the order of the cube of thickness to capture membrane and leading-order bending effects in a two-dimensional energy density defined over the shell midsurface. Though it is not strictly necessary to base the two-dimensional theory on the midsurface, it proves advantageous, from the viewpoint of the theoretical development, to do so.

Our procedure is reminiscent of the recent application, prevalent in the mathematical literature, of gamma convergence methods to thin bodies. In fact, our procedure delivers precisely the same two-dimensional models as those obtained by gamma convergence in circumstances where the latter may be applied. However, our procedure is far more accessible to those not trained in the sophisticated mathematics underpinning gamma convergence, and extends to circumstances where the latter—concerned exclusively with the limiting variational problem as thickness tends to zero—cannot be applied. That is, while gamma convergence has delivered rigorous limit models for membrane theory and pure bending theory separately, in contrast to our procedure it has not yielded models for combined bending and membrane behavior that are of principal interest to practitioners of shell theory.

Thus, we systematically derive existing classical linear and nonlinear theories of plates and shells in a conceptually clear manner. Importantly, we avoid the various ad hoc assumptions made in the historical development of the subject, most notably the classical Kirchhoff-Love hypothesis requiring that material lines initially normal to the shell surface remain so after deformation. Instead, such conditions, when appropriate, are here derived rather than postulated. We show how they emerge naturally in the course of our dimension-reduction procedure under certain conditions on the symmetries possessed by the underlying three-dimensional material. A further important example of a condition that was postulated historically but which we here derive is that of plane stress at the shell midsurface.

Though the book is primarily concerned with the development of theory, a number of simple explicit solutions are included with the aim of exhibiting the relative ease with which the theory can be used to make predictions. This contrasts sharply with the formidable obstacles one faces when using three-dimensional elasticity. Many problems are posed for the student. These are intended to promote engagement with the material and to reinforce understanding. However, as our intention is to use the book in our future teaching of shell theory we do not include solutions to these exercises.

Berkeley, CA, USA
September 2022

David J. Steigmann
Mircea Bîrsan
Milad Shirani

Acknowledgements The second author (M.B.) acknowledges support from the Deutsche Forschungsgemeinschaft (DFG, German Research Foundation)—Project no. 415894848.

Contents

Chapter 1
Tensor Analysis in Euclidean Space Using Curvilinear Coordinates

Abstract In this chapter, we review the main notations and results of tensor analysis in Euclidean space using curvilinear coordinates. These will be useful in the formulation of shell and plate models.

1.1 Cartesian and Curvilinear Coordinates

Let us consider three-dimensional Euclidean space, referred to the orthonormal basis $\{\boldsymbol{e}_i\}$, $i = 1, 2, 3$. Let x^1, x^2, x^3 be Cartesian coordinates. The position of a point P in space is determined by the position vector

$$\boldsymbol{x} = x^i \boldsymbol{e}_i \,.$$

Here we adopt the usual (Einstein) summation convention over repeated indices. Latin subscripts and superscripts $i, j, k, \ldots$ range over the set $\{1, 2, 3\}$, whereas Greek subscripts and superscripts $\alpha, \beta, \gamma, \ldots$ range over the set $\{1, 2\}$.

Let $(\theta^1, \theta^2, \theta^3)$ be a system of curvilinear coordinates in space. The position vector can be written as a function

$$\boldsymbol{x} = \boldsymbol{x}(\theta^j) = x^i(\theta^j)\boldsymbol{e}_i \,, \quad j = 1, 2, 3.$$

Locally, we suppose that the curvilinear coordinates θ^j are in a one-to-one correspondence with points P and that the x^i are functions of class C^1 of the parameters $\theta^j, i, j = 1, 2, 3$.

Then, the differential element can be expressed as

$$\mathrm{d}\boldsymbol{x} = \frac{\partial \boldsymbol{x}}{\partial x^i}\,\mathrm{d}x^i = \boldsymbol{e}_i\,\mathrm{d}x^i, \quad \text{or,} \quad \mathrm{d}\boldsymbol{x} = \frac{\partial \boldsymbol{x}}{\partial \theta^i}\,\mathrm{d}\theta^i = \boldsymbol{g}_i\,\mathrm{d}\theta^i,$$

where

$$\boldsymbol{g}_i = \frac{\partial \boldsymbol{x}}{\partial \theta^i} = \frac{\partial x^j}{\partial \theta^i}\,\boldsymbol{e}_j \,. \tag{1.1}$$

D. J. Steigmann et al., *Lecture Notes on the Theory of Plates and Shells*,
Solid Mechanics and Its Applications 274,
https://doi.org/10.1007/978-3-031-25674-5_1

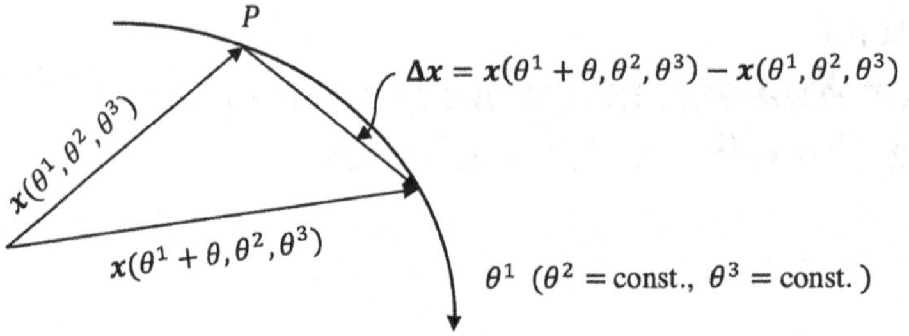

Fig. 1.1 The θ^1-curve through the current point $P(\theta^1, \theta^2, \theta^3)$

The vectors $\{g_1, g_2, g_3\}$ constitute the natural basis, induced by the curvilinear coordinates $\{\theta^i\}$. This basis varies with $\{\theta^i\}$, hence with P. For instance, the vector

$$g_1 = \left.\frac{\partial x}{\partial \theta^1}\right|_P = \lim_{\theta \to 0} \frac{\Delta x}{\theta}, \quad \text{with} \quad \Delta x = x(\theta^1 + \theta, \theta^2, \theta^3) - x(\theta^1, \theta^2, \theta^3),$$

is tangential to the θ^1-curve at the point $P(\theta^1, \theta^2, \theta^3)$ and directed along increasing θ^1, see Fig. 1.1.

One can show that the vectors g_i are linearly independent if and only if the following 3×3 determinant is non-vanishing:

$$\det\left(\frac{\partial x^i}{\partial \theta^j}\right) \neq 0. \tag{1.2}$$

Indeed, the three vectors are linearly independent if and only if the scalar triple product $[g_1, g_2, g_3]$ is not zero. Using (1.1) and $e_{ijk} = e_i \cdot (e_j \times e_k)$, we can write

$$\begin{aligned} [g_1, g_2, g_3] = g_1 \cdot (g_2 \times g_3) &= \frac{\partial x^i}{\partial \theta^1} e_i \cdot \left(\frac{\partial x^j}{\partial \theta^2} e_j \times \frac{\partial x^k}{\partial \theta^3} e_k\right) \\ &= e_{ijk} \frac{\partial x^i}{\partial \theta^1} \frac{\partial x^j}{\partial \theta^2} \frac{\partial x^k}{\partial \theta^3} = \det\left(\frac{\partial x^i}{\partial \theta^j}\right). \end{aligned} \tag{1.3}$$

In view of (1.2), the functions $x^i = \hat{x}^i(\theta^j)$ are locally invertible, by the inverse function theorem, if and only if $\{g_i\}$ at P is a basis. Then, we can define locally the functions

$$\theta^i = \hat{\theta}^i(x^j) = \hat{\theta}^i(e_j \cdot x) = \tilde{\theta}^i(x).$$

We choose the coordinates θ^i such that $g_1 \cdot (g_2 \times g_3) > 0$, so that (θ^i) is a "right-handed" coordinate system.

Note that $e^i = e_i$, because the vector basis $\{e_1, e_2, e_3\}$ is orthonormal. Therefore, we can write $x = x^i e_i = x_i e_i$ and we make no distinction between x^i and x_i.

Next, we define the vectors $\boldsymbol{g}^i = \nabla\theta^i(\boldsymbol{x})$. In general, the gradient of a scalar function $f(\boldsymbol{x}) = f(x^i)$ can be introduced as

$$\mathrm{d}f = (\nabla f)\cdot \mathrm{d}\boldsymbol{x}\,.$$

Taking into account the relations

$$\mathrm{d}f = \frac{\partial f}{\partial x^i}\,\mathrm{d}x^i = \frac{\partial f}{\partial x^i}\left(\boldsymbol{e}_i\cdot \mathrm{d}\boldsymbol{x}\right) = \left(\frac{\partial f}{\partial x^i}\,\boldsymbol{e}_i\right)\cdot \mathrm{d}\boldsymbol{x}\,,$$

we obtain that the gradient of a scalar function is given by

$$\nabla f = \frac{\partial f}{\partial x^i}\,\boldsymbol{e}_i\,.$$

Then, we have that for the vector $\boldsymbol{g}^i$,

$$\boldsymbol{g}^i = \nabla\theta^i = \frac{\partial\theta^i}{\partial x^j}\,\boldsymbol{e}_j\,. \tag{1.4}$$

We note that $\boldsymbol{g}^1$ is orthogonal to the surface $\theta^1 =$ const. and similarly for $\boldsymbol{g}^2$ and $\boldsymbol{g}^3$, see Fig. 1.2.

The set of vectors $\{\boldsymbol{g}^i\}$ is a basis if and only if $\{\boldsymbol{g}_i\}$ is a basis. Indeed, in view of (1.4) and $\dfrac{\partial\theta^i}{\partial x^j}\dfrac{\partial x^j}{\partial\theta^k} = \delta^i_k$ (the Kronecker symbol), we have

$$\begin{aligned}\boldsymbol{g}^1\cdot(\boldsymbol{g}^2\times\boldsymbol{g}^3) &= \frac{\partial\theta^1}{\partial x^i}\,\boldsymbol{e}_i\cdot\left(\frac{\partial\theta^2}{\partial x^j}\,\boldsymbol{e}_j\times\frac{\partial\theta^3}{\partial x^k}\,\boldsymbol{e}_k\right) = e_{ijk}\,\frac{\partial\theta^1}{\partial x^i}\,\frac{\partial\theta^2}{\partial x^j}\,\frac{\partial\theta^3}{\partial x^k}\\ &= \det\left(\frac{\partial\theta^i}{\partial x^j}\right) = \det\left(\frac{\partial x^i}{\partial\theta^j}\right)^{-1}.\end{aligned}$$

Thus, the basis $\{\boldsymbol{g}^i\}$ is "right-handed", i.e. $[\,\boldsymbol{g}^1, \boldsymbol{g}^2, \boldsymbol{g}^3] > 0$, according to our assumption on $\{\boldsymbol{g}_i\}$. The basis $\{\boldsymbol{g}^i\}$ is also called the *dual basis* associated with $\{\boldsymbol{g}_i\}$, since we have

$$\boldsymbol{g}^i\cdot\boldsymbol{g}_j = \frac{\partial\theta^i}{\partial x^k}\,\boldsymbol{e}_k\cdot\frac{\partial x^l}{\partial\theta^j}\,\boldsymbol{e}_l = \frac{\partial\theta^i}{\partial x^k}\,\frac{\partial x^k}{\partial\theta^j} = \delta^i_j\,.$$

Let us define the third order tensors ε_{rst} and ε^{rst} (see, e.g., [2])

$$\varepsilon_{rst} = \frac{\partial x^i}{\partial\theta^r}\,\frac{\partial x^j}{\partial\theta^s}\,\frac{\partial x^k}{\partial\theta^t}\,e_{ijk} = e_{rst}\,\det\left(\frac{\partial x^i}{\partial\theta^j}\right), \tag{1.5}$$

and

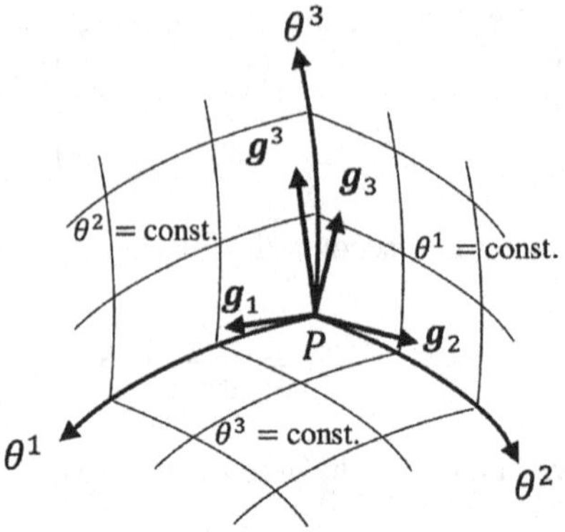

Fig. 1.2 The local basis $\{g_i\}$ and the vector g^3 of the local dual basis

$$\varepsilon^{rst} = \frac{\partial\theta^r}{\partial x^i}\frac{\partial\theta^s}{\partial x^j}\frac{\partial\theta^t}{\partial x^k}\, e^{ijk} = e^{rst}\,\det\left(\frac{\partial\theta^i}{\partial x^j}\right) = e^{rst}\,\det\left(\frac{\partial x^i}{\partial\theta^j}\right)^{-1}, \tag{1.6}$$

where $e^{ijk} = e_{ijk}$. With the help of these tensors, one can write the relations

$$\boldsymbol{g}_i \times \boldsymbol{g}_j = \varepsilon_{ijk}\,\boldsymbol{g}^k, \qquad \boldsymbol{g}^i \times \boldsymbol{g}^j = \varepsilon^{ijk}\,\boldsymbol{g}_k\,. \tag{1.7}$$

Indeed, to show the first equation in (1.7), we use $\boldsymbol{e}_k \times \boldsymbol{e}_l = e_{klm}\boldsymbol{e}_m$ and find

$$\boldsymbol{g}_i \times \boldsymbol{g}_j = \frac{\partial x^k}{\partial\theta^i}\frac{\partial x^l}{\partial\theta^j}\,\boldsymbol{e}_k \times \boldsymbol{e}_l = \frac{\partial x^k}{\partial\theta^i}\frac{\partial x^l}{\partial\theta^j}\, e_{klm}\boldsymbol{e}_m$$

and also

$$\boldsymbol{e}_m = \delta_i^m \boldsymbol{e}_i = \frac{\partial x^m}{\partial x^i}\,\boldsymbol{e}_i = \nabla x^m = \frac{\partial x^m}{\partial\theta^r}\,\nabla\theta^r = \frac{\partial x^m}{\partial\theta^r}\,\boldsymbol{g}^r. \tag{1.8}$$

From the last two relations and (1.5) we deduce $(1.7)_1$.

We introduce the *metric tensor components* [3]

$$g_{ij} = \boldsymbol{g}_i \cdot \boldsymbol{g}_j \qquad (g_{ij} = g_{ji}),$$

and the *dual metric tensor components*

$$g^{ij} = \boldsymbol{g}^i \cdot \boldsymbol{g}^j \qquad \left(g^{ij} = g^{ji}\right).$$

Let us show that

$$\boldsymbol{g}_i = g_{ij}\,\boldsymbol{g}^j \qquad \text{and} \qquad \boldsymbol{g}^i = g^{ij}\boldsymbol{g}_j\,. \tag{1.9}$$

Indeed, since $\{\boldsymbol{g}^i\}$ is a basis in the vector 3-space, there exist unique coefficients (a_{ij}) such that $\boldsymbol{g}_i = a_{ij}\,\boldsymbol{g}^j$. Then, we have

$$g_{ik} = \boldsymbol{g}_i \cdot \boldsymbol{g}_k = a_{ij}\, \boldsymbol{g}^j \cdot \boldsymbol{g}_k = a_{ij}\, \delta^j_k = a_{ik}$$

and the first relation (1.9) is proved. Similarly, we get the second relation (1.9).

Let us denote the determinant of the 3×3 matrix (g_{ij}) with

$$g = \det(g_{ij}). \tag{1.10}$$

In view of (1.2) and

$$g_{ij} = \boldsymbol{g}_i \cdot \boldsymbol{g}_j = \frac{\partial x^k}{\partial \theta^i} \boldsymbol{e}_k \cdot \frac{\partial x^l}{\partial \theta^j} \boldsymbol{e}_l = \sum_{k=1}^{3} \frac{\partial x^k}{\partial \theta^i} \frac{\partial x^k}{\partial \theta^j},$$

we see that

$$g = \det\left(\frac{\partial x^i}{\partial \theta^j}\right)^2 > 0, \tag{1.11}$$

so the matrix (g_{ij}) is invertible. The inverse of the matrix (g_{ij}) is

$$\left(g^{ij}\right) = (g_{ij})^{-1}. \tag{1.12}$$

Indeed, we can write

$$\delta^i_k\, \boldsymbol{g}^k = \boldsymbol{g}^i = g^{ij} \boldsymbol{g}_j = g^{ij} g_{jk}\, \boldsymbol{g}^k,$$

so that $\left(\delta^i_k - g^{ij} g_{jk}\right) \boldsymbol{g}^k = \boldsymbol{0}$, i.e. $g^{ij} g_{jk} = \delta^i_k$ and the relation (1.12) is proved. From Eq. (1.11) it follows that

$$\sqrt{g} = \det\left(\frac{\partial x^i}{\partial \theta^j}\right) \quad \text{and} \quad \frac{1}{\sqrt{g}} = \det\left(\frac{\partial \theta^i}{\partial x^j}\right). \tag{1.13}$$

With help of the metric tensor components we can express the squared length

$$0 < |\mathrm{d}\boldsymbol{x}|^2 = \mathrm{d}\boldsymbol{x} \cdot \mathrm{d}\boldsymbol{x} = \boldsymbol{g}_i \cdot \boldsymbol{g}_j\, \mathrm{d}\theta^i \mathrm{d}\theta^j = g_{ij}\, \mathrm{d}\theta^i \mathrm{d}\theta^j, \quad \text{for all } \mathrm{d}\theta^i \neq 0.$$

Therefore, the matrix (g_{ij}) is positive definite. Hence, the inverse matrix (g^{ij}) is also positive definite.

Finally, from (1.5), (1.6) and (1.13) we obtain the relations

$$\varepsilon_{ijk} = \sqrt{g}\, e_{ijk} \quad \text{and} \quad \varepsilon^{ijk} = \frac{1}{\sqrt{g}}\, e^{ijk}, \tag{1.14}$$

when the coordinate system (θ^i) is right-handed.

1.2 Representations of Vectors and Tensors

1.2.1 Vectors

Since $\{\boldsymbol{g}_i\}$ and $\{\boldsymbol{g}^i\}$ are bases in the vector 3-space, we can represent any vector $\boldsymbol{v}$ as a linear combination of the basis elements in the form

$$\boldsymbol{v} = v^i \boldsymbol{g}_i = v_i \boldsymbol{g}^i . \tag{1.15}$$

Thus, the vector $\boldsymbol{v}$ has the following components

$$\boldsymbol{v} \cdot \boldsymbol{g}^j = (v^i \boldsymbol{g}_i) \cdot \boldsymbol{g}^j = v^i \, \delta_i^j = v^j \qquad (\text{"contravariant components"}),$$
$$\boldsymbol{v} \cdot \boldsymbol{g}_j = (v_i \boldsymbol{g}^i) \cdot \boldsymbol{g}_j = v_i \, \delta_j^i = v_j \qquad (\text{"covariant components"})$$

and we can write

$$\boldsymbol{v} = (\boldsymbol{v} \cdot \boldsymbol{g}^i)\boldsymbol{g}_i = (\boldsymbol{v} \cdot \boldsymbol{g}_i)\boldsymbol{g}^i . \tag{1.16}$$

From (1.9) and (1.15) we get

$$v_j \boldsymbol{g}^j = \boldsymbol{v} = v^i \boldsymbol{g}_i = v^i g_{ij} \boldsymbol{g}^j ,$$

so

$$(v_j - v^i g_{ij})\boldsymbol{g}^j = \boldsymbol{0}$$

and we obtain

$$v_j = g_{ij} \, v^i . \tag{1.17}$$

Likewise, we can show that

$$v^j = g^{ij} \, v_i . \tag{1.18}$$

The relation (1.17) is also referred to as "lowering the index i", while (1.18) as "raising the index i".

Then, the *scalar product* between vectors can be expressed as

$$\boldsymbol{u} \cdot \boldsymbol{v} = (u^i \boldsymbol{g}_i) \cdot (v^j \boldsymbol{g}_j) = u^i \, v^j g_{ij} = u^i v_i = u_j \, v_i \, g^{ij} = u_j \, v^j .$$

1.2.2 Representation in Cartesian Coordinates

If the curvilinear coordinate system (θ^i) coincides with the Cartesian system, i.e. $\theta^i = x^i$, then we have

$$\boldsymbol{g}_i = \boldsymbol{g}^i = \boldsymbol{e}_i , \qquad g_{ij} = g^{ij} = \delta_{ij} .$$

So, for Cartesian coordinates, there is no distinction between covariant and contravariant components, since

$$v_i = g_{ij}\, v^j = \delta_{ij}\, v^j = v^i .$$

Also, if $(\bar{\theta}^k)$ is a different curvilinear coordinate system with natural basis $\{\bar{\boldsymbol{g}}_k\}$ and dual basis $\{\bar{\boldsymbol{g}}^k\}$, then we can write

$$\begin{aligned} \boldsymbol{e}_i &= \frac{\partial \boldsymbol{x}}{\partial x^i} = \frac{\partial \boldsymbol{x}}{\partial \bar{\theta}^k}\,\frac{\partial \bar{\theta}^k}{\partial x^i} = \bar{\boldsymbol{g}}_k\,\frac{\partial \bar{\theta}^k}{\partial x^i} \\ \text{and} \quad \boldsymbol{e}_i &= \nabla x^i = \frac{\partial x^i}{\partial \bar{\theta}^k}\,\nabla \bar{\theta}^k = \bar{\boldsymbol{g}}^k\,\frac{\partial x^i}{\partial \bar{\theta}^k} . \end{aligned} \tag{1.19}$$

Let $\boldsymbol{v} = \bar{v}_k \bar{\boldsymbol{g}}^k = \bar{v}^k \bar{\boldsymbol{g}}_k$ be a vector field. If we denote for the moment by v_i^c the components of $\boldsymbol{v}$ in the Cartesian system, we deduce from (1.19) that

$$\begin{aligned} v_i^c &= \boldsymbol{v}\cdot\boldsymbol{e}_i = \boldsymbol{v}\cdot\left(\bar{\boldsymbol{g}}_k\,\frac{\partial \bar{\theta}^k}{\partial x^i}\right) = \frac{\partial \bar{\theta}^k}{\partial x^i}\,\bar{v}_k \\ \text{and} \quad v_i^c &= \boldsymbol{v}\cdot\boldsymbol{e}_i = \boldsymbol{v}\cdot\left(\bar{\boldsymbol{g}}^k\,\frac{\partial x^i}{\partial \bar{\theta}^k}\right) = \frac{\partial x^i}{\partial \bar{\theta}^k}\,\bar{v}^k . \end{aligned}$$

It follows that

$$\bar{v}_i = \frac{\partial x^j}{\partial \bar{\theta}^i}\, v_j^c \quad \text{and} \quad \bar{v}^i = \frac{\partial \bar{\theta}^i}{\partial x^j}\, v_j^c . \tag{1.20}$$

These are examples of "transformation laws" for vector components induced by coordinate changes, which will be presented in general in Sect. 1.2.4.

1.2.3 Tensors

Let $\boldsymbol{A}$ be a second order tensor with Cartesian components A_{ij}^c. In view of (1.19) we have

$$\boldsymbol{A} = A_{ij}^c \boldsymbol{e}_i \otimes \boldsymbol{e}_j = A_{ij}^c\,\frac{\partial \theta^k}{\partial x^i}\,\frac{\partial \theta^l}{\partial x^j}\,\boldsymbol{g}_k \otimes \boldsymbol{g}_l = A^{kl}\boldsymbol{g}_k \otimes \boldsymbol{g}_l , \tag{1.21}$$

where

$$A^{kl} = \frac{\partial \theta^k}{\partial x^i}\,\frac{\partial \theta^l}{\partial x^j}\,A_{ij}^c \tag{1.22}$$

are the so-called "contravariant components" of $\boldsymbol{A}$ in the curvilinear coordinate system (θ^i). Note that

$$\boldsymbol{A}\boldsymbol{g}^j = (A^{kl}\boldsymbol{g}_k \otimes \boldsymbol{g}_l)\boldsymbol{g}^j = A^{kl}\delta_l^j\,\boldsymbol{g}_k = A^{kj}\boldsymbol{g}_k ,$$

so that

$$\boldsymbol{g}^i \cdot \boldsymbol{A}\boldsymbol{g}^j = A^{kj}\boldsymbol{g}_k \cdot \boldsymbol{g}^i = A^{ij},$$

i.e.

$$\boldsymbol{A} = (\boldsymbol{g}^i \cdot \boldsymbol{A}\boldsymbol{g}^j)\, \boldsymbol{g}_i \otimes \boldsymbol{g}_j .$$

Alternatively, we can write

$$\boldsymbol{A} = A_{ij}\, \boldsymbol{g}^i \otimes \boldsymbol{g}^j = A^{i}_{\cdot j}\, \boldsymbol{g}_i \otimes \boldsymbol{g}^j = A_{i\cdot}^{\;j}\, \boldsymbol{g}^i \otimes \boldsymbol{g}_j , \tag{1.23}$$

where

$$\begin{aligned}
A_{ij} &= \boldsymbol{g}_i \cdot \boldsymbol{A}\boldsymbol{g}_j = \frac{\partial x^k}{\partial \theta^i}\frac{\partial x^l}{\partial \theta^j} A^c_{kl} && \text{(“covariant components”)},\\
A^{i}_{\cdot j} &= \boldsymbol{g}^i \cdot \boldsymbol{A}\boldsymbol{g}_j = \frac{\partial \theta^i}{\partial x^k}\frac{\partial x^l}{\partial \theta^j} A^c_{kl} && \text{(“left-mixed components”)},\\
A_{i}^{\cdot j} &= \boldsymbol{g}_i \cdot \boldsymbol{A}\boldsymbol{g}^j = \frac{\partial x^k}{\partial \theta^i}\frac{\partial \theta^j}{\partial x^l} A^c_{kl} && \text{(“right-mixed components”)},
\end{aligned} \tag{1.24}$$

Further, applying the tensor $A_{kl}\, \boldsymbol{g}^k \otimes \boldsymbol{g}^l = A^{ij}\, \boldsymbol{g}_i \otimes \boldsymbol{g}_j$ to the vector $\boldsymbol{g}^m$, we find

$$A_{kl}\, \boldsymbol{g}^k(\boldsymbol{g}^l \cdot \boldsymbol{g}^m) = A^{ij}\, \boldsymbol{g}_i(\boldsymbol{g}_j \cdot \boldsymbol{g}^m), \qquad \text{or} \qquad g^{lm} A_{kl}\, \boldsymbol{g}^k = A^{im}\, \boldsymbol{g}_i .$$

Taking the scalar product of the last relation with $\boldsymbol{g}^n$, we get

$$A^{nm} = g^{kn} g^{lm} A_{kl} . \tag{1.25}$$

Similarly, we can show that

$$A^{ij} = g^{ik} A_{k}^{\cdot j} = g^{jk} A^{i}_{\cdot k} , \qquad A_{ij} = g_{ik}\, g_{jl}\, A^{kl} = g_{ik} A^{k}_{\cdot j} = g_{jl} A_{i}^{\cdot l} , \quad \text{etc.} \tag{1.26}$$

The relations of the type (1.25) and (1.26) express the “raising” and “lowering” of indices for the components of second order tensors.

Let $\boldsymbol{I}$ be the *unit tensor*, which satisfies $\boldsymbol{I}\boldsymbol{v} = \boldsymbol{v}$, for any vector $\boldsymbol{v}$. By virtue of (1.23) and (1.9), we have

$$\boldsymbol{I} = (\boldsymbol{g}_i \cdot \boldsymbol{I}\boldsymbol{g}_j)\, \boldsymbol{g}^i \otimes \boldsymbol{g}^j = g_{ij}\, \boldsymbol{g}^i \otimes \boldsymbol{g}^j = \boldsymbol{g}_i \otimes \boldsymbol{g}^i = g^{ij}\, \boldsymbol{g}_i \otimes \boldsymbol{g}_j = \boldsymbol{g}^i \otimes \boldsymbol{g}_i . \tag{1.27}$$

Note that

$$\begin{aligned}
\boldsymbol{I} &= \boldsymbol{g}_j \otimes \boldsymbol{g}^j = \delta^i_j\, \boldsymbol{g}_i \otimes \boldsymbol{g}^j\\
&= \boldsymbol{g}^j \otimes \boldsymbol{g}_j = \delta^j_i\, \boldsymbol{g}^i \otimes \boldsymbol{g}_j .
\end{aligned}$$

So the mixed components of the unit tensor are the same in all coordinate systems.

1.2.4 Coordinate Transformations

In the relations (1.20), (1.22) and (1.24) we have already seen some examples of transformation laws for components, when we change from a Cartesian to a curvilinear coordinate system.

Let us consider now two different curvilinear coordinate systems (θ^i) and $(\bar{\theta}^i)$. These induce two natural bases $\{\boldsymbol{g}_i\}$ and $\{\bar{\boldsymbol{g}}_i\}$, respectively, at any point P.

Thus, we have

$$\bar{\boldsymbol{g}}_i = \frac{\partial \boldsymbol{x}}{\partial \bar{\theta}^i} = \frac{\partial \boldsymbol{x}}{\partial \theta^j}\,\frac{\partial \theta^j}{\partial \bar{\theta}^i} = \frac{\partial \theta^j}{\partial \bar{\theta}^i}\,\boldsymbol{g}_j\,. \tag{1.28}$$

Also, for the induced dual bases $\{\boldsymbol{g}^i\}$ and $\{\bar{\boldsymbol{g}}^i\}$ we have at P

$$\bar{\boldsymbol{g}}^i = \nabla\bar{\theta}^i = \frac{\partial \bar{\theta}^i}{\partial \theta^j}\,\nabla\theta^j = \frac{\partial \bar{\theta}^i}{\partial \theta^j}\,\boldsymbol{g}^j\,. \tag{1.29}$$

For any vector $\boldsymbol{v}$, we denote by v^i its components in the basis $\{\boldsymbol{g}_i\}$ and by $\bar{v}^i$ its components in the basis $\{\bar{\boldsymbol{g}}_i\}$. By virtue of (1.28), we can write

$$\bar{v}^j\,\bar{\boldsymbol{g}}_j = \boldsymbol{v} = v^i\,\boldsymbol{g}_i = v^i\,\frac{\partial \bar{\theta}^j}{\partial \theta^i}\,\bar{\boldsymbol{g}}_j\,,$$

so we obtain the transformation law for contravariant components

$$\bar{v}^i = \frac{\partial \bar{\theta}^i}{\partial \theta^j}\,v^j\,.$$

Likewise, we get the transformation law for covariant components

$$\bar{v}_i = \frac{\partial \theta^j}{\partial \bar{\theta}^i}\,v_j\,.$$

Clearly, these relations can be inverted, i.e.

$$v^i = \frac{\partial \theta^i}{\partial \bar{\theta}^j}\,\bar{v}^j \qquad \text{and} \qquad v_i = \frac{\partial \bar{\theta}^j}{\partial \theta^i}\,\bar{v}_j\,. \tag{1.30}$$

We can proceed analogously for tensors. Let $\boldsymbol{A}$ be a second order tensor having the covariant components A_{ij} in the coordinate system (θ^i) and the covariant components $\bar{A}_{ij}$ in the coordinate system $(\bar{\theta}^i)$, i.e.

$$\bar{A}_{ij}\,\bar{\boldsymbol{g}}^i \otimes \bar{\boldsymbol{g}}^j = \boldsymbol{A} = A_{kl}\,\boldsymbol{g}^k \otimes \boldsymbol{g}^l\,.$$

Using here the relation (1.29), we deduce the transformation law

$$\bar{A}_{ij} = A_{kl} \frac{\partial \theta^k}{\partial \bar{\theta}^i} \frac{\partial \theta^l}{\partial \bar{\theta}^j} . \tag{1.31}$$

Similarly, we obtain the following transformation laws for contravariant and mixed components of $\boldsymbol{A}$:

$$\bar{A}^{ij} = A^{kl} \frac{\partial \bar{\theta}^i}{\partial \theta^k} \frac{\partial \bar{\theta}^j}{\partial \theta^l} , \qquad \bar{A}^{i}_{\cdot j} = A^{k}_{\cdot l} \frac{\partial \bar{\theta}^i}{\partial \theta^k} \frac{\partial \theta^l}{\partial \bar{\theta}^j} , \qquad \bar{A}_{i}^{\cdot j} = A_{k}^{\cdot l} \frac{\partial \theta^k}{\partial \bar{\theta}^i} \frac{\partial \bar{\theta}^j}{\partial \theta^l} , \tag{1.32}$$

where $\bar{A}^{ij} = \bar{\boldsymbol{g}}^i \cdot \boldsymbol{A} \bar{\boldsymbol{g}}^j$, $\bar{A}^{i}_{\cdot j} = \bar{\boldsymbol{g}}^i \cdot \boldsymbol{A} \bar{\boldsymbol{g}}_j$ and $\bar{A}_{i}^{\cdot j} = \bar{\boldsymbol{g}}_i \cdot \boldsymbol{A} \bar{\boldsymbol{g}}^j$.

1.3 Differential Operators in Curvilinear Coordinates

Let us denote the partial derivative with respect to the curvilinear coordinates θ^i with a subscript comma preceding the index i , i.e. $f_{,i} = \dfrac{\partial f}{\partial \theta^i}$.

1.3.1 Gradients

The gradient of a scalar function $\phi(\boldsymbol{x})$ is designated by $\nabla\phi$ and is defined as

$$\mathrm{d}\phi = \nabla\phi \cdot \mathrm{d}\boldsymbol{x} .$$

In order to find the gradient of the function ϕ we start by differentiating ϕ as

$$\mathrm{d}\phi = \frac{\partial \phi}{\partial x^i} \mathrm{d}x^i .$$

On the other hand, we have $\boldsymbol{x} = x^i \boldsymbol{e}_i$, so we can write $\mathrm{d}x^i = \boldsymbol{e}^i \cdot \mathrm{d}\boldsymbol{x}$. As a result we get

$$\mathrm{d}\phi = \frac{\partial \phi}{\partial x^i} \left(\boldsymbol{e}^i \cdot \mathrm{d}\boldsymbol{x}\right) = \left(\frac{\partial \phi}{\partial x^i} \boldsymbol{e}^i\right) \cdot \mathrm{d}\boldsymbol{x} .$$

By comparing this relation with $\mathrm{d}\phi = \nabla\phi \cdot \mathrm{d}\boldsymbol{x}$, we find the gradient of the scalar function ϕ as

$$\nabla\phi = \frac{\partial \phi}{\partial x^i} \boldsymbol{e}^i = \frac{\partial \phi}{\partial x^i} \boldsymbol{e}_i .$$

Moreover, we have

$$\frac{\partial \phi}{\partial x^i}\, \boldsymbol{e}_i = \frac{\partial \phi}{\partial \theta^k}\,\frac{\partial \theta^k}{\partial x^i}\, \boldsymbol{e}_i = \frac{\partial \phi}{\partial \theta^k}\, \boldsymbol{g}^k .$$

Therefore, we conclude

$$\nabla \phi = \frac{\partial \phi}{\partial x^i}\, \boldsymbol{e}_i = \frac{\partial \phi}{\partial \theta^k}\, \boldsymbol{g}^k . \tag{1.33}$$

The last relation can also be written in the form

$$\nabla \phi = \frac{\partial \phi}{\partial \theta^i}\, \nabla \theta^i = \phi_{,i}\, \boldsymbol{g}^i = g^{ij} \phi_{,i}\, \boldsymbol{g}_j . \tag{1.34}$$

The gradient of a vector valued function $\boldsymbol{v}(\boldsymbol{x})$ is designated by $\nabla \boldsymbol{v}$ and is defined as

$$\mathrm{d}\boldsymbol{v} = (\nabla \boldsymbol{v})\, \mathrm{d}\boldsymbol{x}$$

On the other hand, by differentiating the vector $\boldsymbol{v}$ we find

$$\mathrm{d}\boldsymbol{v} = \frac{\partial \boldsymbol{v}}{\partial x^i}\, \mathrm{d}x^i$$

and since $\mathrm{d}x^i = \boldsymbol{e}^i \cdot \mathrm{d}\boldsymbol{x}$, we find

$$\mathrm{d}\boldsymbol{v} = \frac{\partial \boldsymbol{v}}{\partial x^i}\, \mathrm{d}x^i = \frac{\partial \boldsymbol{v}}{\partial x^i}\left(\boldsymbol{e}^i \cdot \mathrm{d}\boldsymbol{x}\right) = \left(\frac{\partial \boldsymbol{v}}{\partial x^i} \otimes \boldsymbol{e}^i\right) \mathrm{d}\boldsymbol{x} .$$

By comparing this relation with $\mathrm{d}\boldsymbol{v} = (\nabla \boldsymbol{v})\, \mathrm{d}\boldsymbol{x}$ we conclude

$$\nabla \boldsymbol{v} = \frac{\partial \boldsymbol{v}}{\partial x^i} \otimes \boldsymbol{e}^i = \frac{\partial \boldsymbol{v}}{\partial x^i} \otimes \boldsymbol{e}_i .$$

Let us denote the components of the vector field $\boldsymbol{v}$ by $\boldsymbol{v} = v_i^c \boldsymbol{e}_i = v^j \boldsymbol{g}_j = v_j \boldsymbol{g}^j$. Since basis vectors $\boldsymbol{e}_i = \boldsymbol{e}^i$ are constant, we have

$$\frac{\partial \boldsymbol{v}}{\partial x^i} = \frac{\partial v_j^c}{\partial x^i}\, \boldsymbol{e}_j .$$

Hence, we find the gradient of the vector $\boldsymbol{v}$ as

$$\nabla \boldsymbol{v} = \frac{\partial v_j^c}{\partial x^i}\, \boldsymbol{e}_j \otimes \boldsymbol{e}_i . \tag{1.35}$$

Also, we can use chain rule to rewrite the gradient of a vector field $\boldsymbol{v}$ as

$$\nabla \boldsymbol{v} = \frac{\partial \boldsymbol{v}}{\partial x^i} \otimes \boldsymbol{e}_i = \frac{\partial \boldsymbol{v}}{\partial \theta^k} \otimes \frac{\partial \theta^k}{\partial x^i}\, \boldsymbol{e}_i = \frac{\partial \boldsymbol{v}}{\partial \theta^k} \otimes \boldsymbol{g}^k , \tag{1.36}$$

i.e., we have

$$\nabla \boldsymbol{v} = \boldsymbol{v}_{,i} \otimes \boldsymbol{g}^i \,. \tag{1.37}$$

Here, the derivative $\boldsymbol{v}_{,i}$ can be written

$$\boldsymbol{v}_{,i} = (v^j \, \boldsymbol{g}_j)_{,i} = v^j{}_{,i} \, \boldsymbol{g}_j + v^j \, \boldsymbol{g}_{j,i} \,. \tag{1.38}$$

Since $\{\boldsymbol{g}_k\}$ is a basis, we can decompose the vector $\boldsymbol{g}_{j,i}$ as

$$\boldsymbol{g}_{j,i} = (\boldsymbol{g}^k \cdot \boldsymbol{g}_{j,i})\boldsymbol{g}_k = \Gamma^k_{ji} \, \boldsymbol{g}_k \,, \tag{1.39}$$

where we have denoted by Γ^k_{ji} the so-called *Christoffel symbols* given by

$$\Gamma^k_{ji} = \boldsymbol{g}^k \cdot \boldsymbol{x}_{,ji} = \boldsymbol{g}^k \cdot \boldsymbol{x}_{,ij} = \Gamma^k_{ij} \,. \tag{1.40}$$

Here, we assume that the function $\boldsymbol{x}(\theta^i)$ is of class C^2. In view of (1.38) and (1.39) we obtain

$$\boldsymbol{v}_{,i} = v^k{}_{|i} \, \boldsymbol{g}_k \,, \qquad \text{where} \qquad v^k{}_{|i} = v^k{}_{,i} + v^j \, \Gamma^k_{ji} \tag{1.41}$$

is the *covariant derivative* for the contravariant components v^k.

Similarly, we have

$$\boldsymbol{v}_{,i} = (v_j \, \boldsymbol{g}^j)_{,i} = v_{j,i} \, \boldsymbol{g}^j + v_j \, \boldsymbol{g}^j{}_{,i} = v_{j,i} \, \boldsymbol{g}^j + v_j \, (\boldsymbol{g}_k \cdot \boldsymbol{g}^j{}_{,i})\boldsymbol{g}^k. \tag{1.42}$$

If we differentiate the relation $\boldsymbol{g}_k \cdot \boldsymbol{g}^j = \delta^j_k$ with respect to θ^i, we get

$$\boldsymbol{g}_k \cdot \boldsymbol{g}^j{}_{,i} = -\Gamma^j_{ki} \tag{1.43}$$

and substituting this in (1.42) yields

$$\boldsymbol{v}_{,i} = v_{k|i} \, \boldsymbol{g}^k \,, \qquad \text{where} \qquad v_{k|i} = v_{k,i} - v_j \, \Gamma^j_{ki} \tag{1.44}$$

is the *covariant derivative* for the covariant components v_k.

By virtue of (1.37), (1.41) and (1.44), the gradient of $\boldsymbol{v}$ has the form

$$\nabla \boldsymbol{v} = \boldsymbol{v}_{,i} \otimes \boldsymbol{g}^i = v_{k|i} \, \boldsymbol{g}^k \otimes \boldsymbol{g}^i = v^k{}_{|i} \, \boldsymbol{g}_k \otimes \boldsymbol{g}^i \,. \tag{1.45}$$

1.3.2 Divergence

The divergence of $\boldsymbol{v}$ is defined as the trace of the gradient. Thus, in view of (1.36) and (1.45), we obtain

$$\operatorname{div} \boldsymbol{v} = \operatorname{tr}(\nabla \boldsymbol{v}) = \boldsymbol{v}_{,i} \cdot \boldsymbol{g}^i \,, \qquad \text{or} \qquad \operatorname{div} \boldsymbol{v} = \frac{\partial v_i^c}{\partial x^i} = g^{ki} v_{k|i} = v^i{}_{|i} \,. \tag{1.46}$$

Also, for any scalar field ϕ of class C^2, the Laplace operator is

$$\begin{aligned} \Delta\phi &= \operatorname{div}(\nabla\phi) = \operatorname{tr}\nabla(\nabla\phi) = \operatorname{tr}[(\nabla\phi)_{,i} \otimes \boldsymbol{g}^i] = \boldsymbol{g}^i \cdot (\nabla\phi)_{,i} = \boldsymbol{g}^i \cdot (\phi_{,j}\, \boldsymbol{g}^j)_{,i} \\ &= \boldsymbol{g}^i \cdot (\phi_{,ij}\, \boldsymbol{g}^j + \phi_{,j}\, \boldsymbol{g}^j{}_{,i}) = \boldsymbol{g}^i \cdot (\phi_{,ik} - \phi_{,j}\, \Gamma^j_{ik})\, \boldsymbol{g}^k = g^{ik} \phi_{|ik} \,, \end{aligned} \tag{1.47}$$

where

$$\phi_{|ik} = \phi_{,ik} - \phi_{,j}\, \Gamma^j_{ik} \,. \tag{1.48}$$

Let $\boldsymbol{A}$ be a second order tensor field with Cartesian components A^c_{ij} , see (1.21). The divergence is then given by

$$\operatorname{div} \boldsymbol{A} = \frac{\partial A^c_{ij}}{\partial x^j}\, \boldsymbol{e}_i = \frac{\partial (A^c_{ij}\, \boldsymbol{e}_i)}{\partial x^j} = \frac{\partial (A^c_{ik}\, \boldsymbol{e}_i \otimes \boldsymbol{e}_k) \boldsymbol{e}_j}{\partial x^j} = \left(\frac{\partial \boldsymbol{A}}{\partial x^j} \right) \boldsymbol{e}_j \,.$$

Using here the relation (1.8), we obtain

$$\operatorname{div} \boldsymbol{A} = \left(\frac{\partial \boldsymbol{A}}{\partial \theta^i} \right) \frac{\partial \theta^i}{\partial x^j}\, \boldsymbol{e}_j = \left(\frac{\partial \boldsymbol{A}}{\partial \theta^i} \right) \boldsymbol{g}^i = (\boldsymbol{A}_{,i})\, \boldsymbol{g}^i \,. \tag{1.49}$$

For the derivative $\boldsymbol{A}_{,i}$ we deduce with the help of (1.39)

$$\begin{aligned} \boldsymbol{A}_{,j} &= (A^{ik} \boldsymbol{g}_i \otimes \boldsymbol{g}_k)_{,j} = A^{ik}{}_{,j}\, \boldsymbol{g}_i \otimes \boldsymbol{g}_k + A^{ik} \boldsymbol{g}_{i,j} \otimes \boldsymbol{g}_k + A^{ik} \boldsymbol{g}_i \otimes \boldsymbol{g}_{k,j} \\ &= A^{ik}{}_{,j}\, \boldsymbol{g}_i \otimes \boldsymbol{g}_k + A^{ik} \Gamma^l_{ij}\, \boldsymbol{g}_l \otimes \boldsymbol{g}_k + A^{ik} \Gamma^l_{kj}\, \boldsymbol{g}_i \otimes \boldsymbol{g}_l \,, \end{aligned}$$

and therefore

$$\boldsymbol{A}_{,j} = A^{ik}{}_{|j}\, \boldsymbol{g}_i \otimes \boldsymbol{g}_k \,, \qquad \text{where} \qquad A^{ik}{}_{|j} = A^{ik}{}_{,j} + A^{il} \Gamma^k_{lj} + A^{lk} \Gamma^i_{lj} \tag{1.50}$$

is the covariant derivative for the contravariant components A^{ik}. Thus, from Eqs. (1.49) and (1.50) we obtain for the divergence the expression

$$\operatorname{div} \boldsymbol{A} = A^{ij}{}_{|j}\, \boldsymbol{g}_i \,. \tag{1.51}$$

Alternatively, we can use the covariant components of $\boldsymbol{A}$ and (1.43) to write

$$\begin{aligned} \boldsymbol{A}_{,k} &= (A_{ij}\, \boldsymbol{g}^i \otimes \boldsymbol{g}^j)_{,k} = A_{ij,k}\, \boldsymbol{g}^i \otimes \boldsymbol{g}^j + A_{ij}\, \boldsymbol{g}^i{}_{,k} \otimes \boldsymbol{g}^j + A_{ij}\, \boldsymbol{g}^i \otimes \boldsymbol{g}^j{}_{,k} \\ &= A_{ij,k}\, \boldsymbol{g}^i \otimes \boldsymbol{g}^j - A_{ij}\, \Gamma^i_{kl}\, \boldsymbol{g}^l \otimes \boldsymbol{g}^j - A_{ij}\, \Gamma^j_{kl}\, \boldsymbol{g}^i \otimes \boldsymbol{g}^l \,, \end{aligned}$$

i.e.

$$\boldsymbol{A}_{,k} = A_{ij|k}\, \boldsymbol{g}^i \otimes \boldsymbol{g}^j \,, \qquad \text{where} \qquad A_{ij|k} = A_{ij,k} - A_{lj} \Gamma^l_{ik} - A_{il} \Gamma^l_{kj} \tag{1.52}$$

is the covariant derivative for the covariant components A_{ij}. Then, from (1.49) and (1.52) we obtain the divergence in the form

$$\operatorname{div} \boldsymbol{A} = A_{ij|k}\, g^{jk}\, \boldsymbol{g}^i \, . \tag{1.53}$$

We can prove the remarkable relation

$$g_{ij|k} = 0. \tag{1.54}$$

Indeed, taking into account (1.39), we have

$$g_{ij,k} = (\boldsymbol{g}_i \cdot \boldsymbol{g}_j)_{,k} = \boldsymbol{g}_i \cdot (\Gamma^l_{jk}\, \boldsymbol{g}_l) + (\Gamma^l_{ik}\, \boldsymbol{g}_l) \cdot \boldsymbol{g}_j = g_{il}\, \Gamma^l_{jk} + g_{lj}\, \Gamma^l_{ik} \, , \tag{1.55}$$

which is just Eq. (1.54).

If we introduce the symbols Γ_{ijk} given by

$$\Gamma_{ijk} = \Gamma^l_{ij}\, g_{lk} = \Gamma_{jik} \, , \tag{1.56}$$

then the relation (1.55) can be written as

$$g_{ij,k} = \Gamma_{ikj} + \Gamma_{jki} \, . \tag{1.57}$$

From (1.56) and (1.57) it follows that

$$g_{jk,i} + g_{ki,j} - g_{ij,k} = 2\Gamma_{ijk} = 2 g_{kl}\, \Gamma^l_{ij}$$

and then

$$\Gamma^m_{ij} = \frac{1}{2}\, g^{mk}(g_{jk,i} + g_{ki,j} - g_{ij,k}). \tag{1.58}$$

Thus, we have

$$\begin{aligned} 2\Gamma^i_{ij} &= g^{ik}(g_{ik,j} + g_{jk,i} - g_{ij,k}) = g^{ik} g_{ik,j} + g^{ik} g_{jk,i} - g^{ki} g_{ij,k} \\ &= g^{ik} g_{ik,j} + g^{ik} g_{jk,i} - g^{ik} g_{kj,i} = g^{ik} g_{ik,j} \, , \end{aligned}$$

i.e.

$$\Gamma^i_{ij} = \frac{1}{2}\, g^{ik} g_{ik,j} \, . \tag{1.59}$$

Since $g = \det(g_{ij})$ and (g^{ij}) is the inverse of the matrix (g_{ij}), we deduce that

$$\frac{\partial g}{\partial g_{ik}} = g\, g^{ik} \, , \tag{1.60}$$

and the relation (1.59) implies that

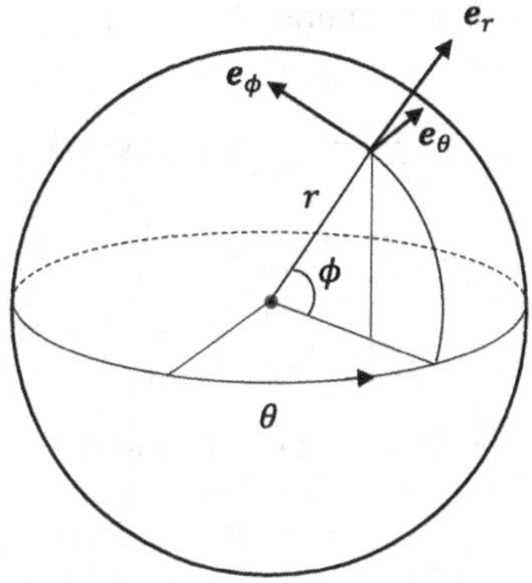

Fig. 1.3 Spherical coordinates

$$\Gamma^i_{ij} = \frac{1}{2g}\frac{\partial g}{\partial g_{ik}}\, g_{ik,j} = \frac{1}{2g}\, g_{,j} = \frac{1}{\sqrt{g}}\,(\sqrt{g}\,)_{,j}\,, \tag{1.61}$$

yielding

$$\Gamma^i_{ij} = \frac{\partial}{\partial \theta^j}\left(\ln\sqrt{g}\right). \tag{1.62}$$

The relations (1.46) and (1.61) yield the useful results

$$\operatorname{div}\boldsymbol{v} = v^i{}_{|i} = v^i{}_{,i} + v^j\,\Gamma^i_{ij} = v^i{}_{,i} + \frac{1}{\sqrt{g}}\,v^j\,(\sqrt{g}\,)_{,j} = \frac{1}{\sqrt{g}}\left[\sqrt{g}\,v^i{}_{,i} + v^i\,(\sqrt{g}\,)_{,i}\right],$$

or

$$\operatorname{div}\boldsymbol{v} = v^i{}_{|i} = \frac{1}{\sqrt{g}}\left(\sqrt{g}\,v^i\right)_{,i}. \tag{1.63}$$

If the vector field $\boldsymbol{v}$ is a gradient field, i.e. there exists a scalar field f with $\boldsymbol{v} = \nabla f$, then from (1.34) we see that $v^i = g^{ij} f_{,j}$ and from (1.63) we get

$$\Delta f = \operatorname{div}(\nabla f) = \frac{1}{\sqrt{g}}\left(\sqrt{g}\,g^{ij}\,f_{,j}\right)_{,i}. \tag{1.64}$$

Example: Spherical coordinates

Let us consider spherical coordinates $(\theta^1, \theta^2, \theta^3) = (r, \theta, \phi)$ as an example of curvilinear coordinates, see Fig. 1.3.

The Cartesian coordinates (x^i) are expressed as functions of spherical coordinates by

$$\begin{aligned} x^1 &= r\cos\phi\cos\theta = \theta^1\cos\theta^3\cos\theta^2\,,\\ x^2 &= r\cos\phi\sin\theta = \theta^1\cos\theta^3\sin\theta^2\,,\\ x^3 &= r\sin\phi = \theta^1\sin\theta^3\,. \end{aligned} \tag{1.65}$$

Then, the natural basis $\{\boldsymbol{g}_i\}$ is given by

$$\begin{aligned}
\boldsymbol{g}_1 &= \frac{\partial x^j}{\partial \theta^1}\,\boldsymbol{e}_j &&= \cos\phi(\cos\theta\,\boldsymbol{e}_1 + \sin\theta\,\boldsymbol{e}_2) + \sin\phi\,\boldsymbol{e}_3\,,\\
\boldsymbol{g}_2 &= \frac{\partial x^j}{\partial \theta^2}\,\boldsymbol{e}_j &&= r\cos\phi(-\sin\theta\,\boldsymbol{e}_1 + \cos\theta\,\boldsymbol{e}_2),\\
\boldsymbol{g}_3 &= \frac{\partial x^j}{\partial \theta^3}\,\boldsymbol{e}_j &&= r\big[-\sin\phi(\cos\theta\,\boldsymbol{e}_1 + \sin\theta\,\boldsymbol{e}_2) + \cos\phi\,\boldsymbol{e}_3\big].
\end{aligned} \tag{1.66}$$

If we introduce the notations

$$\begin{aligned}
\boldsymbol{e}_\rho &= \cos\theta\,\boldsymbol{e}_1 + \sin\theta\,\boldsymbol{e}_2\,, \qquad & \boldsymbol{e}_r &= \cos\phi\boldsymbol{e}_\rho + \sin\phi\,\boldsymbol{e}_3\,,\\
\boldsymbol{e}_\theta &= -\sin\theta\,\boldsymbol{e}_1 + \cos\theta\,\boldsymbol{e}_2\,, \qquad & \boldsymbol{e}_\phi &= -\sin\phi\boldsymbol{e}_\rho + \cos\phi\,\boldsymbol{e}_3\,,
\end{aligned} \tag{1.67}$$

we notice that $\boldsymbol{e}_r$, $\boldsymbol{e}_\theta$, $\boldsymbol{e}_\phi$ are unit vectors and the relations (1.66) can be written

$$\boldsymbol{g}_1 = \boldsymbol{e}_r\,, \qquad \boldsymbol{g}_2 = r\cos\phi\,\boldsymbol{e}_\theta\,, \qquad \boldsymbol{g}_3 = r\,\boldsymbol{e}_\phi\,. \tag{1.68}$$

The metric tensor components are then given by

$$(g_{ij}) = \begin{pmatrix} 1 & 0 & 0\\ 0 & r^2\cos^2\phi & 0\\ 0 & 0 & r^2 \end{pmatrix} \quad \text{and} \quad \left(g^{ij}\right) = \begin{pmatrix} 1 & 0 & 0\\ 0 & r^{-2}\cos^{-2}\phi & 0\\ 0 & 0 & r^{-2} \end{pmatrix} \tag{1.69}$$

Since the matrix is diagonal, we see that the spherical coordinates are orthogonal. The general orthogonal coordinates will be discussed in Sect. 1.4.

In view of (1.10), we find $g = r^4\cos^2\phi$. If we assume $\phi \in \left(-\frac{\pi}{2}\,,\,\frac{\pi}{2}\right)$, then

$$\sqrt{g} \;=\; r^2\cos\phi \;>\; 0. \tag{1.70}$$

The dual basis $\{\boldsymbol{g}^i\}$ can be obtained from the relations $\boldsymbol{g}^i = g^{ij}\boldsymbol{g}_j$ and (1.68), (1.69):

$$\boldsymbol{g}^1 = \boldsymbol{g}_1 = \boldsymbol{e}_r\,, \qquad \boldsymbol{g}^2 = \frac{1}{r\cos\phi}\,\boldsymbol{e}_\theta\,, \qquad \boldsymbol{g}^3 = \frac{1}{r}\,\boldsymbol{e}_\phi\,. \tag{1.71}$$

Taking into account (1.3), (1.13) and (1.70), the elemental volume can be expressed as

$$\begin{aligned}
\mathrm{d}V &= \mathrm{d}\boldsymbol{x}_1\cdot(\mathrm{d}\boldsymbol{x}_2\times\mathrm{d}\boldsymbol{x}_3) = \boldsymbol{g}_1\mathrm{d}\theta^1\cdot(\boldsymbol{g}_2\mathrm{d}\theta^2\times\boldsymbol{g}_3\mathrm{d}\theta^3)\\
&= \det\left(\frac{\partial x^i}{\partial\theta^j}\right)\mathrm{d}\theta^1\mathrm{d}\theta^2\mathrm{d}\theta^3 = \sqrt{g}\,\mathrm{d}\theta^1\mathrm{d}\theta^2\mathrm{d}\theta^3 = r^2\cos\phi\,\mathrm{d}\theta^1\mathrm{d}\theta^2\mathrm{d}\theta^3.
\end{aligned} \tag{1.72}$$

The gradient of a scalar function f can be expressed as

$$\nabla f \;=\; f_{,i}\,\boldsymbol{g}^i \;=\; \frac{\partial f}{\partial r}\,\boldsymbol{e}_r + \frac{1}{r\cos\phi}\,\frac{\partial f}{\partial\theta}\,\boldsymbol{e}_\theta + \frac{1}{r}\,\frac{\partial f}{\partial\phi}\,\boldsymbol{e}_\phi \tag{1.73}$$

and the Laplacian of f can be calculated from (cf. (1.64))

$$\Delta f = \frac{1}{\sqrt{g}} \left(\sqrt{g}\, g^{ij} f_{,j} \right)_{,i} \,. \tag{1.74}$$

In view of (1.69) and (1.70), we have

$$\sqrt{g}\, g^{1j} f_{,j} = r^2 \cos\phi \, f_{,r} \,, \qquad \sqrt{g}\, g^{2j} f_{,j} = \frac{1}{\cos\phi} f_{,\theta} \,, \qquad \sqrt{g}\, g^{3j} f_{,j} = \cos\phi \, f_{,\phi}$$

and from (1.74) we find

$$\begin{aligned} \Delta f &= \frac{1}{r^2 \cos\phi} \left(\left(r^2 \cos\phi f_{,r} \right)_{,r} + \frac{1}{\cos\phi} f_{,\theta\theta} + \left(\cos\phi f_{,\phi} \right)_{,\phi} \right) \\ &= \frac{1}{r^2} \left(r^2 f_{,r} \right)_{,r} + \frac{1}{r^2 \cos^2\phi} f_{,\theta\theta} + \frac{1}{r^2 \cos\phi} \left(\cos\phi f_{,\phi} \right)_{,\phi} \,. \end{aligned} \tag{1.75}$$

1.3.3 Product Rule for Covariant Derivatives

The well-known product rule for differentiation holds also for the covariant derivative. We will show this with a simple example.

Let $\boldsymbol{u}$, $\boldsymbol{v}$ be vector fields and $\boldsymbol{A}$ a tensor field such that $\boldsymbol{v} = \boldsymbol{A}\boldsymbol{u}$. Then, we have

$$\boldsymbol{v} = \boldsymbol{A}\boldsymbol{u} = (A^{ij} \boldsymbol{g}_i \otimes \boldsymbol{g}_j)(u_k \boldsymbol{g}^k) = A^{ij} u_j \boldsymbol{g}_i \,, \quad \text{so} \quad v^i = A^{ij} u_j \tag{1.76}$$

and using (1.41)

$$\begin{aligned} v^i{}_{|k} &= v^i{}_{,k} + v^j \Gamma^i_{jk} = A^{ij}{}_{,k} u_j + A^{ij} u_{j,k} + A^{jl} u_l \Gamma^i_{jk} \\ &= A^{ij}{}_{,k} u_j + A^{ij} u_{j,k} + A^{lj} \Gamma^i_{lk} u_j + A^{il} \Gamma^j_{lk} u_j - A^{ij} \Gamma^l_{jk} u_l \\ &= \left(A^{ij}{}_{,k} + A^{il} \Gamma^j_{lk} + A^{lj} \Gamma^i_{lk} \right) u_j + A^{ij} \left(u_{j,k} - u_l \Gamma^l_{jk} \right). \end{aligned} \tag{1.77}$$

Thus, in view of (1.50), (1.76) and (1.77), we have proved the following product rule

$$\left(A^{ij} u_j \right)_{|k} = A^{ij}{}_{|k} \, u_j + A^{ij} u_{j|k} \,. \tag{1.78}$$

Similarly, we can show further formulas of this type for covariant derivatives, such as for instance

$$\left(A^{ij} B_{kl} \right)_{|m} = A^{ij}{}_{|m} \, B_{kl} + A^{ij} B_{kl|m}$$

and

$$(\phi v^i)_{|k} = \phi_{|k} v^i + \phi v^i{}_{|k} \,,$$

where for a scalar field we have $\phi_{|k} = \phi_{,k}$.

1.4 Orthogonal Coordinates

Let us consider the case of orthogonal coordinates, i.e. curvilinear coordinates (θ^i) such that

$$\boldsymbol{g}_i \cdot \boldsymbol{g}_j = 0 \quad \text{for} \quad i \neq j.$$

We denote by $\alpha = |\boldsymbol{g}_1|$, $\beta = |\boldsymbol{g}_2|$, $\gamma = |\boldsymbol{g}_3|$ the lengths of the natural basis vectors. Then, the components of the metric tensor and dual metric tensor are

$$(g_{ij}) = \begin{pmatrix} \alpha^2 & 0 & 0 \\ 0 & \beta^2 & 0 \\ 0 & 0 & \gamma^2 \end{pmatrix} \quad \text{and} \quad \left(g^{ij}\right) = \begin{pmatrix} \alpha^{-2} & 0 & 0 \\ 0 & \beta^{-2} & 0 \\ 0 & 0 & \gamma^{-2} \end{pmatrix}. \tag{1.79}$$

Let us rename the coordinates $\theta^1 = \xi$, $\theta^2 = \eta$, $\theta^3 = \zeta$ and introduce the unit vectors

$$\boldsymbol{e}_\xi = \frac{1}{\alpha}\boldsymbol{g}_1, \qquad \boldsymbol{e}_\eta = \frac{1}{\beta}\boldsymbol{g}_2, \qquad \boldsymbol{e}_\zeta = \frac{1}{\gamma}\boldsymbol{g}_3. \tag{1.80}$$

Then, any vector $\boldsymbol{v}$ can be decomposed as

$$\boldsymbol{v} = v^i\boldsymbol{g}_i = v^1\alpha\,\boldsymbol{e}_\xi + v^2\beta\,\boldsymbol{e}_\eta + v^3\gamma\,\boldsymbol{e}_\zeta = v_\xi\,\boldsymbol{e}_\xi + v_\eta\,\boldsymbol{e}_\eta + v_\zeta\,\boldsymbol{e}_\zeta, \tag{1.81}$$

where we have denoted

$$v_\xi = \alpha\,v^1, \qquad v_\eta = \beta\,v^2, \qquad v_\zeta = \gamma\,v^3. \tag{1.82}$$

These are the so-called "physical components" of $\boldsymbol{v}$, i.e. they have the same dimensions as $\boldsymbol{v}$, whereas the dimensions of the components v^i are generally different.

Alternatively, we can use the dual basis $\{\boldsymbol{g}^i\}$ and write $\boldsymbol{v} = v_i\,\boldsymbol{g}^i$ with

$$\boldsymbol{g}^1 = g^{11}\boldsymbol{g}_1 = \alpha^{-1}\boldsymbol{e}_\xi, \qquad \boldsymbol{g}^2 = \beta^{-1}\boldsymbol{e}_\eta, \qquad \boldsymbol{g}^3 = \gamma^{-1}\boldsymbol{e}_\zeta. \tag{1.83}$$

Hence, we obtain from (1.81) the alternative expressions of the physical components

$$v_\xi = \alpha^{-1}v_1, \qquad v_\eta = \beta^{-1}v_2, \qquad v_\zeta = \gamma^{-1}v_3. \tag{1.84}$$

Similarly, any second order tensor $\boldsymbol{A} = A^{ij}\,\boldsymbol{g}_i \otimes \boldsymbol{g}_j = A^i_{\cdot j}\,\boldsymbol{g}_i \otimes \boldsymbol{g}^j = A_i^{\cdot j}\,\boldsymbol{g}^i \otimes \boldsymbol{g}_j = A_{ij}\,\boldsymbol{g}^i \otimes \boldsymbol{g}^j$ admits the following representation in terms of physical components

$$\begin{aligned} \boldsymbol{A} = \; & A_{\xi\xi}\,\boldsymbol{e}_\xi \otimes \boldsymbol{e}_\xi + A_{\xi\eta}\,\boldsymbol{e}_\xi \otimes \boldsymbol{e}_\eta + A_{\xi\zeta}\,\boldsymbol{e}_\xi \otimes \boldsymbol{e}_\zeta \\ & + A_{\eta\xi}\,\boldsymbol{e}_\eta \otimes \boldsymbol{e}_\xi + A_{\eta\eta}\,\boldsymbol{e}_\eta \otimes \boldsymbol{e}_\eta + A_{\eta\zeta}\,\boldsymbol{e}_\eta \otimes \boldsymbol{e}_\zeta \\ & + A_{\zeta\xi}\,\boldsymbol{e}_\zeta \otimes \boldsymbol{e}_\xi + A_{\zeta\eta}\,\boldsymbol{e}_\zeta \otimes \boldsymbol{e}_\eta + A_{\zeta\zeta}\,\boldsymbol{e}_\zeta \otimes \boldsymbol{e}_\zeta, \end{aligned} \tag{1.85}$$

In view of (1.80) and (1.83), the physical components are expressed by

$$\begin{aligned}
A_{\xi\xi} &= \alpha^2 A^{11} = \alpha^{-2} A_{11} = A^{1}_{\cdot 1} = A_{1}{}^{1}_{\cdot}, \\
A_{\xi\eta} &= \alpha\beta A^{12} = (\alpha\beta)^{-1} A_{12} = \frac{\alpha}{\beta} A^{1}_{\cdot 2} = \frac{\beta}{\alpha} A_{1}{}^{2}_{\cdot}, \\
A_{\xi\zeta} &= \alpha\gamma A^{13} = (\alpha\gamma)^{-1} A_{13} = \frac{\alpha}{\gamma} A^{1}_{\cdot 3} = \frac{\gamma}{\alpha} A_{1}{}^{3}_{\cdot}, \qquad \text{etc.}
\end{aligned} \tag{1.86}$$

We mention that physical components can be defined also for non-orthogonal coordinates, but they are not unique. Therefore, this idea is not so useful in the general case of curvilinear coordinates.

1.5 Exercises

1.1 Recall the tensor representation formulas $\mathbf{A} = A^{i}_{\cdot j}\mathbf{g}_i \otimes \mathbf{g}^j = A_{i}^{\cdot j}\mathbf{g}^i \otimes \mathbf{g}_j$. Show that the mixed components of $\mathbf{A}^T$ are $\left(\mathbf{A}^T\right)^{i}_{\cdot j} = A_{j}^{\cdot i}$ and $\left(\mathbf{A}^T\right)_{i}^{\cdot j} = A^{j}_{\cdot i}$.

1.2 Find $\{\mathbf{g}_i\}$, $\{\mathbf{g}^i\}$, g_{ij}, g^{ij} and Γ^{i}_{jk} in terms of $\{\theta^i\} = \{\xi, \eta, \phi\}$ for the following coordinate systems:

   a. Parabolic coordinates: $x^1 = \xi\eta\cos\phi$, $x^2 = \xi\eta\sin\phi$ and $x^3 = \frac{1}{2}\left(\xi^2 - \eta^2\right)$. Write out the Laplacian $\Delta f\,(\xi, \eta, \phi)$ explicitly in terms of f and its partial derivatives with respect to ξ, η, ϕ.
   b. Elliptic cylindrical coordinates: $x^1 = \cosh\xi\,\cos\eta$, $x^2 = \sinh\xi\,\sin\eta$, $x^3 = \phi$.

1.3 Consider the complex-valued analytic function $g\,(z)$ of the complex variable $z = x + iy$, where $x\left(= x^1\right)$ and $y\left(= x^2\right)$ are Cartesian coordinates in the plane. Let $\{\theta^i\} = \{\phi, \psi, x^3\}$, where $\phi = Re\,(g)$ and $\psi = Im\,(g)$. Consider the example $g\,(z) = z^2$ and compute $\{\mathbf{g}_i\}$, $\{\mathbf{g}^i\}$, g_{ij}, g^{ij} in terms of the x^i. Sketch the coordinate surfaces (the surfaces on which each coordinate is constant in succession).

1.4 Suppose the complex variables $z = x^1 + ix^2$ and $\varsigma = u + iv$ are related by $z = F\,(z)$, where F is an analytic (i.e., differentiable) function. Show that the Laplacian may be written as

$$\Delta f = h^{-2}\left(\frac{\partial^2 f}{\partial u^2} + \frac{\partial^2 f}{\partial v^2}\right) + \frac{\partial^2 f}{\partial w^2},$$

where $w = x^3$ and $h = \left|F'\,(\varsigma)\right|$.

1.5 Consider an oblique coordinate system $\{\theta^i\}$ for which $\mathbf{g}_1 = \mathbf{e}_1$, $\mathbf{g}_2 = \cos\phi\mathbf{e}_2 + \sin\phi\mathbf{e}_1$, $\mathbf{g}_3 = \mathbf{e}_3$ with ϕ a constant angle. Find the x^i as functions of the θ^i (Hint: construct the position vector and suppose the θ^i vanish at the origin.). Invert these relations and also obtain g_{ij}, g^{ij} and find $\{\mathbf{g}^i\}$ in terms of $\{\mathbf{e}_i\}$. Finally, if v_i^c are the Cartesian components of a vector, what are the co- and contra-variant components of the vector in the $\{\theta^i\}$ system?

1.6 In cylindrical polar coordinates $\{r, \theta, z\}$ the position function is given by $\mathbf{x} = r\mathbf{e}_r(\theta) + z\mathbf{e}_3$, where r is the radius from the z- axis, z measures distance along this axis, θ is the azimuthal angle, and $\mathbf{e}_r(\theta) = \cos\theta\mathbf{e}_1 + \sin\theta\mathbf{e}_2$. In some problems involving cone-shaped surfaces, it may be ore convenient to use a 'conical-polar' coordinate system (our terminology!). For this purpose we introduce coordinates $\{\bar{r}, \theta, \bar{z}\}$, where $r = \bar{r}\cos\phi$, $\bar{z} = z - \bar{r}\sin\phi$ and ϕ is the *constant* cone angle. The position function may then be written $\mathbf{x} = \bar{r}\bar{\mathbf{e}}_r(\theta) + \bar{z}\mathbf{e}_3$, where $\bar{\mathbf{e}}_r(\theta) = \cos\phi\mathbf{e}_r(\theta) + \sin\phi\mathbf{e}_3$ (draw a figure).
Let $\{\theta^i\} = \{\bar{r}, \theta, \bar{z}\}$ and write out the divergence $A^{ij}_{\ |j}$ explicitly in terms of A^{ij} (contravariant components of a tensor) and their partial derivatives. You may assume the A^{ij} to be independent of θ for simplicity.

1.7 Position in a helical coordinate system $\{r, \theta, s\}$ (see Fig. 1.4) is described by

$$\mathbf{x}(r, \theta, s) = \mathbf{R}(s) + r\cos\theta\mathbf{N}(s) + r\sin\theta\mathbf{B}(s),$$

where $\mathbf{R}(s)$ is the equation of the centerline, s is arclength along the centerline, $\mathbf{N}(s)$ is the *principal normal* to the centerline and $\mathbf{B}(s)$ is the *binormal.* The curvature of the centerline is $\kappa(s) = |\mathbf{T}'(s)|$, where $\mathbf{T}(s) = \mathbf{R}'(s)$ is the unit tangent to the centerline. Wherever κ is non-zero, we stipulate that $\mathbf{N} = \kappa^{-1}\mathbf{T}'(s)$ and $\mathbf{B} = \mathbf{T} \times \mathbf{N}$ so that $\{\mathbf{T}, \mathbf{N}, \mathbf{B}\}$ is an orthonormal basis at every point of the curve.
The Serret-Frenet equations for smooth space curves are (see any elementary geometry text)

$$\mathbf{N}'(s) = \tau\mathbf{B} - \kappa\mathbf{T} \qquad \text{and} \qquad \mathbf{B}'(s) = -\tau\mathbf{N},$$

where $\tau(s)$ is the *torsion* of the centerline.
Let $\{\theta^i\} = \{r, \theta, s\}$ and obtain $\{\mathbf{g}_i\}$, $\{\mathbf{g}^i\}$, g_{ij}, g^{ij} in terms of $\mathbf{T}$, $\mathbf{N}$, $\mathbf{B}$, κ, τ and the coordinates. Show that this coordinate system is *non-orthogonal.*
Helices of uniform radius and pitch are distinguished by property that the curvature and torsion are constants, independent of arclength (see Fig. 1.5). Obtain an explicit expression for he Laplacian $\Delta f(r, \theta)$ in this special case, for a function f that is independent of arclength. This problem is of obvious importance in diverse applications including, for example, heat conduction in coiled wires. A number of faulty papers have been written on the latter subject based on the incorrect assumption that the coordinates are orthogonal.

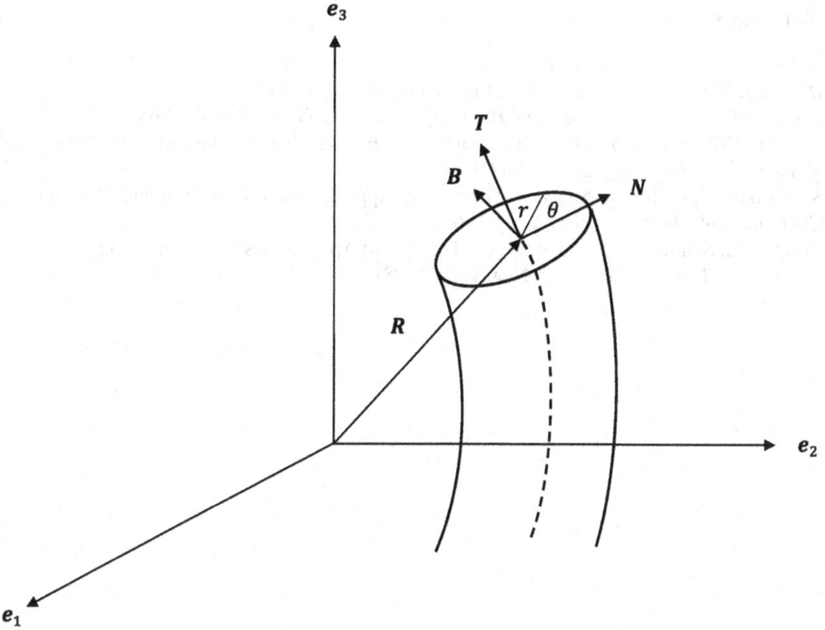

Fig. 1.4 Helical coordinates

Fig. 1.5 Uniform helix

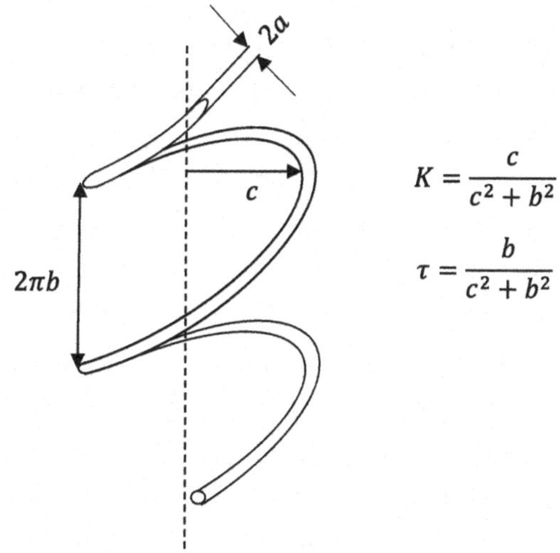

References

1. Dickmen, M.: Theory of Thin Elastic Shells. Pitman, Boston (1982)
2. Green, A.E., Zerna, W.: Theoretical Elasticity. Clarendon Press, Oxford (1968)
3. Naghdi, P.M.: The theory of shells and plates. In: Flügge, W. (ed.) Handbuch der Physik, vol. VIa/2, pp. 425–640. Springer, Berlin (1972)
4. Sokolnikoff, I.S.: Tensor Analysis: Theory and Applications to Geometry and Mechanics of Continua, 2nd edn. Wiley, New York (1964)
5. Synge, J.L., Schild, A.: Tensor Analysis: Theory and Applications to Geometry and Mechanics of Continua. Dover Publications, New York (1978)

Chapter 2
Local Geometry of Deformation

Abstract In this chapter, we describe the deformation of continua and define the strain and stress tensors. Then, we review the main results of the differential geometry of surfaces in the Euclidean space.

2.1 Description of the Deformation

Let us consider a deformable body $\mathscr{B}$, which occupies in its reference configuration the domain $\mathscr{K}$ of Euclidean space. The reference configuration $\mathscr{K}$ is referred to a system of curvilinear coordinates (θ^i) and every point P will be identified by the position vector $\boldsymbol{x} = \boldsymbol{x}(\theta^i)$, see Fig. 2.1.

The natural basis in the reference configuration will be denoted by $\{\boldsymbol{G}_i\}$, with

$$\boldsymbol{G}_i = \boldsymbol{x}_{,i}\,, \qquad G_{ij} = \boldsymbol{G}_i \cdot \boldsymbol{G}_j\,, \qquad G = \det(G_{ij}), \tag{2.1}$$

whereas $\{\boldsymbol{G}^i\}$ is the dual basis and $G^{ij} = \boldsymbol{G}^i \cdot \boldsymbol{G}^j$.

We consider "convected" (embedded) coordinates (θ^i). This means that a material point P is identified by fixed values of the coordinates (θ^i) in all configurations. The θ^i-curves stretch and rotate with the material, i.e. they are "material" curves.

Let us denote the deformed configuration of the body at time t by $\mathscr{R}$, i.e. $\mathscr{R} = \chi(\mathscr{K}, t)$, where χ is the deformation function. Then, the position vector of any material point $P(\theta^i)$ at time t is given by

$$\boldsymbol{y} = \boldsymbol{y}(\theta^i, t) = \chi\big(\boldsymbol{x}(\theta^i), t\big).$$

Regarding θ^i and t as independent variables is tantamount to a *Lagrangian* formulation. Thus, for the natural basis in the deformed configuration we have

$$\boldsymbol{g}_i = \boldsymbol{y}_{,i}\,, \qquad g_{ij} = \boldsymbol{g}_i \cdot \boldsymbol{g}_j\,, \qquad g = \det(g_{ij}) \tag{2.2}$$

D. J. Steigmann et al., *Lecture Notes on the Theory of Plates and Shells*,
Solid Mechanics and Its Applications 274,
https://doi.org/10.1007/978-3-031-25674-5_2

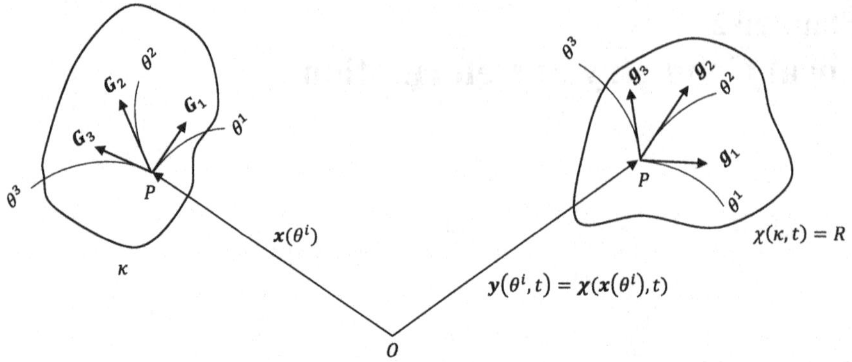

Fig. 2.1 Reference configuration and deformed configuration of the body

and for the dual basis

$$\boldsymbol{g}^i \cdot \boldsymbol{g}_j = \delta^i_j\,, \qquad g^{ij} = \boldsymbol{g}^i \cdot \boldsymbol{g}^j\,.$$

2.1.1 Deformation Tensor

In view of (2.1) and (2.2), the differential elements in the reference and deformed configurations, respectively, are

$$\mathrm{d}\boldsymbol{x} = \boldsymbol{x}_{,i}\,\mathrm{d}\theta^i = \boldsymbol{G}_i\,\mathrm{d}\theta^i \qquad \text{and} \qquad \mathrm{d}\boldsymbol{y} = \boldsymbol{y}_{,i}\,\mathrm{d}\theta^i = \boldsymbol{g}_i\,\mathrm{d}\theta^i\,. \tag{2.3}$$

Then, it follows that

$$\boldsymbol{G}^j \cdot \mathrm{d}\boldsymbol{x} = \boldsymbol{G}^j \cdot \boldsymbol{G}_i\,\mathrm{d}\theta^i = \delta^j_i\,\mathrm{d}\theta^i = \mathrm{d}\theta^j\,,$$

i.e.

$$\mathrm{d}\theta^i = \boldsymbol{G}^i \cdot \mathrm{d}\boldsymbol{x}\,. \tag{2.4}$$

From (2.3) and (2.4) we obtain $\mathrm{d}\boldsymbol{y} = \boldsymbol{g}_i(\boldsymbol{G}^i \cdot \mathrm{d}\boldsymbol{x}) = (\boldsymbol{g}_i \otimes \boldsymbol{G}^i)\mathrm{d}\boldsymbol{x}$, that is

$$\mathrm{d}\boldsymbol{y} = \boldsymbol{F}\,\mathrm{d}\boldsymbol{x}\,, \quad \text{where} \quad \boldsymbol{F} = \boldsymbol{g}_i \otimes \boldsymbol{G}^i \tag{2.5}$$

is the *deformation gradient*. In view of (1.1) and (1.4), the deformation gradient can be written in Cartesian representation as

$$\boldsymbol{F} = \frac{\partial y^j}{\partial \theta^i}\boldsymbol{e}_j \otimes \frac{\partial \theta^i}{\partial x^k}\boldsymbol{e}_k = \frac{\partial y^j}{\partial x^k}\boldsymbol{e}_j \otimes \boldsymbol{e}_k\,, \tag{2.6}$$

where $\boldsymbol{y} = y^j \boldsymbol{e}_j$ and $\boldsymbol{x} = x^k \boldsymbol{e}_k$. Let us denote its determinant by

$$J = \det \boldsymbol{F} = \det \left(\frac{\partial y^i}{\partial x^j} \right). \tag{2.7}$$

The *Cauchy-Green deformation tensor* $\boldsymbol{C}$ is defined by

$$\boldsymbol{C} = \boldsymbol{F}^T \boldsymbol{F}. \tag{2.8}$$

By virtue of (2.5), we have $\boldsymbol{F}^T \boldsymbol{g}_j = g_{ij} \, \boldsymbol{G}^i$, so the relation (2.8) becomes

$$\boldsymbol{C} = (\boldsymbol{F}^T \boldsymbol{g}_j) \otimes \boldsymbol{G}^j, \qquad \text{i.e.} \qquad \boldsymbol{C} = g_{ij} \, \boldsymbol{G}^i \otimes \boldsymbol{G}^j, \tag{2.9}$$

where g_{ij} is the "deformed metric", see (2.2).

Notice that the components of $\boldsymbol{C}$ in the tensor basis $\{\boldsymbol{G}^i \otimes \boldsymbol{G}^j\}$ coincide with g_{ij}, that is

$$C_{ij} = g_{ij}, \tag{2.10}$$

but $C^{ij} \neq g^{ij}$ in general, rather $C^{ij} = \boldsymbol{G}^i \cdot \boldsymbol{C} \boldsymbol{G}^j = G^{ik} G^{jl} g_{kl}$.

In view of (2.5) and $(\boldsymbol{G}_i \otimes \boldsymbol{g}^i)(\boldsymbol{g}_j \otimes \boldsymbol{G}^j) = \boldsymbol{G}_i \otimes \boldsymbol{G}^i = \boldsymbol{I}$, we see that

$$\boldsymbol{F}^{-1} = \boldsymbol{G}_i \otimes \boldsymbol{g}^i. \tag{2.11}$$

Thus, it follows that

$$\boldsymbol{C}^{-1} = (\boldsymbol{F}^T \boldsymbol{F})^{-1} = \boldsymbol{F}^{-1} \boldsymbol{F}^{-T} = (\boldsymbol{G}_i \otimes \boldsymbol{g}^i)(\boldsymbol{g}^j \otimes \boldsymbol{G}_j) = g^{ij} \boldsymbol{G}_i \otimes \boldsymbol{G}_j, \tag{2.12}$$

so we have $g^{ij} = \boldsymbol{G}_i \cdot \boldsymbol{C}^{-1} \boldsymbol{G}_j = \left(\boldsymbol{C}^{-1}\right)^{ij}$, i.e. the contravariant components of $\boldsymbol{C}^{-1}$.

2.1.2 Volumes and Areas

Taking into account (1.3) and (1.13), the volume elements in the reference configuration and deformed configuration are

$$\begin{aligned} \mathrm{d}V &= \left[\boldsymbol{G}_1 \, \mathrm{d}\theta^1, \; \boldsymbol{G}_2 \, \mathrm{d}\theta^2, \; \boldsymbol{G}_3 \, \mathrm{d}\theta^3 \right] = \sqrt{G} \, \mathrm{d}\theta^1 \mathrm{d}\theta^2 \mathrm{d}\theta^3 \qquad \text{and} \\ \mathrm{d}v &= \left[\boldsymbol{g}_1 \, \mathrm{d}\theta^1, \; \boldsymbol{g}_2 \, \mathrm{d}\theta^2, \; \boldsymbol{g}_3 \, \mathrm{d}\theta^3 \right] = \sqrt{g} \, \mathrm{d}\theta^1 \mathrm{d}\theta^2 \mathrm{d}\theta^3, \end{aligned}$$

respectively. Further, since the relation (2.7) implies that $\mathrm{d}v = J \, \mathrm{d}V$, we get

$$J = \sqrt{\frac{g}{G}}. \tag{2.13}$$

Fig. 2.2 Oriented area element in the reference configuration

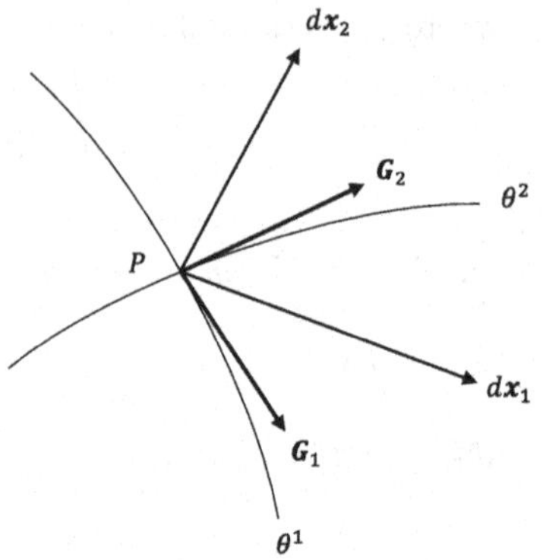

Further, $\rho_K \mathrm{d}V = \rho \mathrm{d}v$, where ρ_K and ρ are the mass densities in the reference and deformed configurations, respectively. Then, in view of (2.13), we obtain

$$\rho_K = \rho\, J\,, \qquad \text{or} \qquad \rho_K \sqrt{G} = \rho \sqrt{g}\,. \tag{2.14}$$

Consider an oriented area element $\boldsymbol{N}\,\mathrm{d}A = \mathrm{d}\boldsymbol{x}_1 \times \mathrm{d}\boldsymbol{x}_2$ in the reference configuration, where $\mathrm{d}\boldsymbol{x}_1 = \boldsymbol{G}_i\,\mathrm{d}\overset{(1)}{\theta}{}^i$ and $\mathrm{d}\boldsymbol{x}_2 = \boldsymbol{G}_i\,\mathrm{d}\overset{(2)}{\theta}{}^i$ are two differential elements at P, see Fig. 2.2.

Let us define the third order tensor

$$\mu_{ijk} = \sqrt{G}\, e_{ijk}\,, \qquad \mu^{ijk} = \frac{1}{\sqrt{G}}\, e^{ijk}, \tag{2.15}$$

where $e_{ijk} = e^{ijk} = \boldsymbol{e}_i \cdot (\boldsymbol{e}_j \times \boldsymbol{e}_k)$. Then, we have

$$\boldsymbol{G}_i \times \boldsymbol{G}_j = \mu_{ijk}\boldsymbol{G}^k \qquad \text{and} \qquad \boldsymbol{G}^i \times \boldsymbol{G}^j = \mu^{ijk}\boldsymbol{G}_k\,.$$

Hence, it follows that

$$\boldsymbol{N}\,\mathrm{d}A = \mathrm{d}\boldsymbol{x}_1 \times \mathrm{d}\boldsymbol{x}_2 = \boldsymbol{G}_i \times \boldsymbol{G}_j\,\mathrm{d}\overset{(1)}{\theta}{}^i\mathrm{d}\overset{(2)}{\theta}{}^j = \mu_{ijk}\,\boldsymbol{G}^k\,\mathrm{d}\overset{(1)}{\theta}{}^i\mathrm{d}\overset{(2)}{\theta}{}^j\,.$$

Thus, using (2.15) and the notation $\boldsymbol{N} = N_k\boldsymbol{G}^k$, we get

$$N_k\mathrm{d}A = \sqrt{G}\, e_{ijk}\,\mathrm{d}\overset{(1)}{\theta}{}^i\mathrm{d}\overset{(2)}{\theta}{}^j\,. \tag{2.16}$$

Consider now the corresponding area element in the deformed configuration (at P), denoted by

$$\boldsymbol{n}\,\mathrm{d}a = \mathrm{d}\boldsymbol{y}_1 \times \mathrm{d}\boldsymbol{y}_2\,, \qquad \text{where} \qquad \mathrm{d}\boldsymbol{y}_1 = \boldsymbol{F}\,\mathrm{d}\boldsymbol{x}_1\,, \quad \mathrm{d}\boldsymbol{y}_2 = \boldsymbol{F}\,\mathrm{d}\boldsymbol{x}_2\,.$$

In view of (2.5), we find

$$\mathrm{d}\boldsymbol{y}_1 = (\boldsymbol{g}_j \otimes \boldsymbol{G}^j)(\boldsymbol{G}_i \,\mathrm{d}\overset{(1)}{\theta}{}^i) = \boldsymbol{g}_i \,\mathrm{d}\overset{(1)}{\theta}{}^i \qquad \text{and} \qquad \mathrm{d}\boldsymbol{y}_2 = \boldsymbol{g}_i \,\mathrm{d}\overset{(2)}{\theta}{}^i .$$

With the help of ε_{ijk} given by (1.14) and $\boldsymbol{n} = n_k \boldsymbol{g}^k$, we can write similarly

$$\boldsymbol{n}\,da = \mathrm{d}\boldsymbol{y}_1 \times \mathrm{d}\boldsymbol{y}_2 = \boldsymbol{g}_i \times \boldsymbol{g}_j \,\mathrm{d}\overset{(1)}{\theta}{}^i \mathrm{d}\overset{(2)}{\theta}{}^j = \varepsilon_{ijk}\, \boldsymbol{g}^k \,\mathrm{d}\overset{(1)}{\theta}{}^i \mathrm{d}\overset{(2)}{\theta}{}^j ,$$

i.e.

$$n_k da = \sqrt{g}\, e_{ijk} \,\mathrm{d}\overset{(1)}{\theta}{}^i \mathrm{d}\overset{(2)}{\theta}{}^j . \tag{2.17}$$

From (2.16) and (2.17) we obtain

$$n_k da = \sqrt{\frac{g}{G}}\, N_k \mathrm{d}A = J\, N_k \mathrm{d}A . \tag{2.18}$$

The relation (2.18) holds for *covariant components only*, in the sense that $\boldsymbol{n}\,da \neq J\,\boldsymbol{N}\,\mathrm{d}A$. Indeed, in view of (2.11) and (2.18), we have

$$\boldsymbol{n}\,da = \boldsymbol{g}^k n_k da = J\, \boldsymbol{g}^k N_k \mathrm{d}A = J(\boldsymbol{g}^i \otimes \boldsymbol{G}_i)\boldsymbol{G}^k N_k \mathrm{d}A = J\,\boldsymbol{F}^{-T}\boldsymbol{N}\,\mathrm{d}A ,$$

so we obtain the *Piola-Nanson formula* for the area elements

$$\boldsymbol{n}\,da = \boldsymbol{F}^* \boldsymbol{N}\,\mathrm{d}A , \qquad \text{with} \qquad \boldsymbol{F}^* = J\,\boldsymbol{F}^{-T} . \tag{2.19}$$

The notation $\boldsymbol{F}^*$ stands for the cofactor of $\boldsymbol{F}$.

2.2 Displacement, Strain, and Stress

In order to write the strain-displacement relations, let us mention precisely the notations that we use for the two different configurations $\mathscr{K}$ and $\mathscr{R}$.

For the reference configuration $\mathscr{K}$, we denote by $\boldsymbol{G}_i$ the natural basis, G_{ij} the metric tensor components, G^{ij} the dual metric tensor components, and $\bar{\Gamma}^k_{ij} = \boldsymbol{G}^k \cdot \boldsymbol{G}_{i,j}$ the Christoffel symbols. The covariant derivative (based on the Christoffel symbols $\bar{\Gamma}^k_{ij}$) will be denoted by a subscript vertical bar, e.g. $v_{i|j}$, see (1.44). We also employ the notations μ_{ijk} and μ^{ijk} given by (2.15).

For the deformed configuration $\mathscr{R} = \chi(\mathscr{K}, t)$, we have the natural basis $\boldsymbol{g}_i$, the metric tensor components g_{ij} and g^{ij}, and the Christoffel symbols $\Gamma^k_{ij} = \boldsymbol{g}^k \cdot \boldsymbol{g}_{i,j}$. The covariant derivative in the deformed configuration (based on Γ^k_{ij}) will be denoted by a subscript semicolon, e.g. $v^i{}_{;j}$, see (1.41). We also use $\varepsilon_{ijk} = \sqrt{g}\, e_{ijk}$ and $\varepsilon^{ijk} = (\sqrt{g})^{-1}\, e^{ijk}$, according to (1.14).

The *displacement vector* is

$$\boldsymbol{u}(\theta^i, t) = \boldsymbol{y}(\theta^i, t) - \boldsymbol{x}(\theta^i).$$

Differentiating with respect to θ^i, we get

$$\boldsymbol{g}_i = \boldsymbol{G}_i + \boldsymbol{u}_{,i}\,, \qquad \text{with} \qquad \boldsymbol{u}_{,i} = u_{k|i}\, \boldsymbol{G}^k = u^k{}_{|i}\, \boldsymbol{G}_k\,. \tag{2.20}$$

Then, we obtain

$$g_{ij} = \boldsymbol{g}_i \cdot \boldsymbol{g}_j = (\boldsymbol{G}_i + \boldsymbol{u}_{,i}) \cdot (\boldsymbol{G}_j + \boldsymbol{u}_{,j}) = \boldsymbol{G}_i \cdot \boldsymbol{G}_j + \boldsymbol{G}_i \cdot \boldsymbol{u}_{,j} + \boldsymbol{u}_{,i} \cdot \boldsymbol{G}_j + \boldsymbol{u}_{,i} \cdot \boldsymbol{u}_{,j}\,,$$

so

$$g_{ij} = G_{ij} + u_{i|j} + u_{j|i} + u_{k|i}\, u^k{}_{|j}\,. \tag{2.21}$$

The *strain tensor* $\boldsymbol{E}$ is defined by

$$\boldsymbol{E} = \frac{1}{2}\,(\boldsymbol{C} - \boldsymbol{I}). \tag{2.22}$$

In view of (1.27) and (2.9), it follows that

$$\boldsymbol{E} = \frac{1}{2}\,(g_{ij} - G_{ij}) \boldsymbol{G}^i \otimes \boldsymbol{G}^j = E_{ij}\, \boldsymbol{G}^i \otimes \boldsymbol{G}^j, \tag{2.23}$$

where, by virtue of (2.21),

$$E_{ij} = \frac{1}{2}\,(u_{i|j} + u_{j|i} + u_{k|i}\, u^k{}_{|j}). \tag{2.24}$$

These are the so-called *strain-displacement relations*.

Let $\boldsymbol{t}$ be the stress vector acting on a material surface in $\mathscr{R}$ with unit normal $\boldsymbol{n}$ and let $\boldsymbol{T}$ be the *Cauchy stress tensor*. Then,

$$\boldsymbol{t} = \boldsymbol{T}\boldsymbol{n}\,. \tag{2.25}$$

We refer $\boldsymbol{T}$ to the tensor basis $\{\boldsymbol{g}_i \otimes \boldsymbol{g}_j\}$ and write, in view of (1.51),

$$\boldsymbol{T} = T^{ij} \boldsymbol{g}_i \otimes \boldsymbol{g}_j \qquad \text{and} \qquad \operatorname{div} \boldsymbol{T} = T^{ij}{}_{;j}\, \boldsymbol{g}_i\,. \tag{2.26}$$

From the principle of linear momentum we deduce the local equations of motion

$$\operatorname{div} \boldsymbol{T} + \rho \boldsymbol{b} = \rho \boldsymbol{a}\,, \qquad \text{or} \qquad T^{ij}{}_{;j} + \rho b^i = \rho a^i\,, \tag{2.27}$$

where $\boldsymbol{b} = b^i \boldsymbol{g}_i$ denotes the body force per unit mass and $\boldsymbol{a} = a^i \boldsymbol{g}_i$ the acceleration vector. If we designate the time derivative with a superposed dot, then we have $\boldsymbol{a} = \dot{\boldsymbol{v}}$, where $\boldsymbol{v} = v^i \boldsymbol{g}_i = \dot{\boldsymbol{y}}$ is the velocity vector. Thus, it follows that

$$a^i = \dot{\boldsymbol{v}} \cdot \boldsymbol{g}^i = (\dot{v}^j \boldsymbol{g}_j + v^j \dot{\boldsymbol{g}}_j) \cdot \boldsymbol{g}^i = \dot{v}^i + v^i{}_{;j}\, v^j\,, \tag{2.28}$$

since $\dot{\boldsymbol{g}}_j = (\boldsymbol{y}_{,j})^{\cdot} = \dot{\boldsymbol{y}}_{,j} = \boldsymbol{v}_{,j} = v^k{}_{;j}\, \boldsymbol{g}_k$.

From the moment of momentum principle we derive that the Cauchy stress tensor is symmetric, i.e.

$$\boldsymbol{T} = \boldsymbol{T}^T\,, \qquad \text{or} \qquad T^{ij} = T^{ji}\,, \tag{2.29}$$

which can be written in the equivalent form

$$T^{ij}\, \boldsymbol{g}_i \times \boldsymbol{g}_j = \boldsymbol{0}. \tag{2.30}$$

Let $\boldsymbol{p}$ be the stress vector acting on a material surface in the deformed configuration $\mathscr{R}$ (with unit normal $\boldsymbol{n}$), but measured per unit area of the corresponding material surface in the reference configuration $\mathscr{K}$ with unit normal $\boldsymbol{N}$. Let $\boldsymbol{P}$ be the *Piola stress tensor* such that

$$\boldsymbol{p} = \boldsymbol{P}\,\boldsymbol{N}\,. \tag{2.31}$$

The Piola stress tensor $\boldsymbol{P}$ is given by

$$\boldsymbol{P} = \boldsymbol{T}\,\boldsymbol{F}^* = J\,\boldsymbol{T}\,\boldsymbol{F}^{-T}\,. \tag{2.32}$$

Indeed, from (2.19), (2.25) and (2.31) we have

$$\boldsymbol{t}\,\mathrm{d}a = \boldsymbol{T}\boldsymbol{n}\,\mathrm{d}a = J\,\boldsymbol{T}\,\boldsymbol{F}^{-T}\boldsymbol{N}\,\mathrm{d}A = \boldsymbol{P}\boldsymbol{N}\,\mathrm{d}A = \boldsymbol{p}\,\mathrm{d}A\,.$$

With the help of the components T^{ij} and relation (2.11), the Piola stress tensor (2.32) can be written in the form

$$\boldsymbol{P} = J(T^{ik}\boldsymbol{g}_i \otimes \boldsymbol{g}_k)(\boldsymbol{g}^j \otimes \boldsymbol{G}_j) = J\,T^{ij}\boldsymbol{g}_i \otimes \boldsymbol{G}_j\,, \tag{2.33}$$

i.e.

$$\boldsymbol{P} = \boldsymbol{P}^i \otimes \boldsymbol{G}_i\,, \qquad \text{where} \qquad \boldsymbol{P}^i = J\,T^{ij}\boldsymbol{g}_j \tag{2.34}$$

are the so-called *stress vectors*. Note that the symmetry relation (2.30) can be written in terms of the stress vectors as

$$\boldsymbol{g}_i \times \boldsymbol{P}^i = \boldsymbol{0}, \tag{2.35}$$

while the traction $\boldsymbol{p}$ is given by

$$\boldsymbol{p} = \boldsymbol{P}\,\boldsymbol{N} = (\boldsymbol{P}^i \otimes \boldsymbol{G}_i)\boldsymbol{N} = \boldsymbol{P}^i(\boldsymbol{G}_i \cdot \boldsymbol{N}) = \boldsymbol{P}^i N_i\,. \tag{2.36}$$

Also, we remark that

$$\boldsymbol{P}^i = \boldsymbol{P}^j\delta^i_j = \boldsymbol{P}^j(\boldsymbol{G}_j \cdot \boldsymbol{G}^i) = (\boldsymbol{P}^j \otimes \boldsymbol{G}_j)\boldsymbol{G}^i = \boldsymbol{P}\,\boldsymbol{G}^i = \boldsymbol{P}(\nabla_{\mathrm{x}}\theta^i), \tag{2.37}$$

so the stress vector $\boldsymbol{P}^i$ is proportional to the traction $\boldsymbol{p}$ on the surface $\theta^i = \text{const.}$ Indeed, let us show this in detail for the surface $\theta^1 = \text{const}$: the unit normal vector is $\boldsymbol{N} = \frac{\nabla_{\mathrm{x}}\theta^1}{|\nabla_{\mathrm{x}}\theta^1|}$ and the traction is

$$\boldsymbol{p} = \boldsymbol{P}\,\boldsymbol{N} = \frac{1}{|\nabla_{\mathrm{x}}\theta^1|}\,\boldsymbol{P}(\nabla_{\mathrm{x}}\theta^1) = \frac{1}{|\boldsymbol{G}^1|}\,\boldsymbol{P}^1 = \frac{1}{\sqrt{G^{11}}}\,\boldsymbol{P}^1\,,$$

in view of (2.37) and $|\boldsymbol{G}^1| = \sqrt{\boldsymbol{G}^1 \cdot \boldsymbol{G}^1} = \sqrt{G^{11}}\,.$

The local equations of motion (2.27) can be written in terms of the Piola stress as

$$\mathrm{Div}\,\boldsymbol{P} + \rho_K\boldsymbol{b} = \rho_K\dot{\boldsymbol{v}}\,, \qquad \text{with} \qquad \mathrm{Div}\,\boldsymbol{P} = P^{ij}{}_{|j}\,\boldsymbol{G}_i\,, \tag{2.38}$$

where $\boldsymbol{P} = P^{ij}\boldsymbol{G}_i \otimes \boldsymbol{G}_j$. In view of (1.49) and (1.39), the divergence of $\boldsymbol{P}$ can be written equivalently

$$\mathrm{Div}\,\boldsymbol{P} = (\boldsymbol{P}_{,j})\boldsymbol{G}^j\,, \tag{2.39}$$

where

$$\boldsymbol{P}_{,j} = (\boldsymbol{P}^i \otimes \boldsymbol{G}_i)_{,j} = \boldsymbol{P}^i{}_{,j} \otimes \boldsymbol{G}_i + \boldsymbol{P}^i \otimes \boldsymbol{G}_{i,j} = \boldsymbol{P}^i{}_{,j} \otimes \boldsymbol{G}_i + \bar{\Gamma}^k_{ij}\,\boldsymbol{P}^i \otimes \boldsymbol{G}_k\,.$$

Thus, from (2.39) we obtain the more convenient form

$$\mathrm{Div}\,\boldsymbol{P} = (\boldsymbol{P}^i{}_{,j} \otimes \boldsymbol{G}_i + \bar{\Gamma}^k_{ij}\,\boldsymbol{P}^i \otimes \boldsymbol{G}_k)\boldsymbol{G}^j = \boldsymbol{P}^i{}_{,i} + \bar{\Gamma}^k_{ik}\,\boldsymbol{P}^i = \boldsymbol{P}^i{}_{|i}\,, \tag{2.40}$$

where we have introduced the notation $\boldsymbol{P}^i{}_{|i} = \boldsymbol{P}^i{}_{,i} + \bar{\Gamma}^k_{ik}\,\boldsymbol{P}^i$, in analogy to $(1.41)_2$. Using (1.61), we can write

$$\bar{\Gamma}^k_{ik} = \frac{1}{\sqrt{G}}\left(\sqrt{G}\right)_{,i} = \left(\ln\sqrt{G}\right)_{,i}$$

and the equations of motion (2.38) become

$$\boldsymbol{P}^i{}_{,i} + \frac{1}{\sqrt{G}}\left(\sqrt{G}\right)_{,i}\boldsymbol{P}^i + \rho_K\boldsymbol{b} = \rho_K\dot{\boldsymbol{v}}\,,$$

i.e.

$$\frac{1}{\sqrt{G}}\left(\sqrt{G}\,\boldsymbol{P}^i\right)_{,i} + \rho_K \boldsymbol{b} = \rho_K \dot{\boldsymbol{v}}\,. \tag{2.41}$$

Note that the differential operator $G^{-1/2}\left(G^{1/2}\boldsymbol{P}^i\right)_{,i}$ is invariant under coordinate transformations, see Exercise 2.2. Further, relation (2.34) yields $\sqrt{G}\,\boldsymbol{P}^i = \sqrt{G}\,J\,T^{ij}\boldsymbol{g}_j = \sqrt{g}\,T^{ij}\boldsymbol{g}_j$ and from (2.41) and (2.14) we get

$$\frac{1}{\sqrt{g}}\left(\sqrt{g}\,T^{ij}\boldsymbol{g}_j\right)_{,i} + \rho\boldsymbol{b} = \rho\dot{\boldsymbol{v}}\,, \tag{2.42}$$

which is equivalent to (2.27). Thus, convected coordinates facilitate the trivial transformation between equations of motion based on Piola or Cauchy stresses. This transformation (based on $\boldsymbol{P} = \boldsymbol{T}\,\boldsymbol{F}^*$) accounts automatically for the relation

$$\operatorname{Div}\boldsymbol{F}^* = 0. \tag{2.43}$$

Indeed, to prove (2.43), we use (2.11) to write $\boldsymbol{F}^* = J\,\boldsymbol{F}^{-T} = (J\,\boldsymbol{g}^i)\otimes\boldsymbol{G}_i$ and we can show, as in (2.41), that

$$\operatorname{Div}\boldsymbol{F}^* = \frac{1}{\sqrt{G}}\left(\sqrt{G}\,J\,\boldsymbol{g}^i\right)_{,i}\,.$$

Then, from $J\sqrt{G} = \sqrt{g}$ and (1.61) it follows that

$$\operatorname{Div}\boldsymbol{F}^* = \frac{J}{\sqrt{g}}\left(\sqrt{g}\,\boldsymbol{g}^i\right)_{,i} = J\Big[\frac{(\sqrt{g})_{,i}}{\sqrt{g}}\,\boldsymbol{g}^i + \boldsymbol{g}^i{}_{,i}\Big] = J\Big[\frac{(\sqrt{g})_{,i}}{\sqrt{g}}\,\boldsymbol{g}^i - \Gamma^i_{ik}\,\boldsymbol{g}^k\Big] = 0.$$

The *second Piola-Kirchhoff stress tensor* $\boldsymbol{S}$ is defined by

$$\boldsymbol{S} = \boldsymbol{F}^{-1}\boldsymbol{P}\,, \tag{2.44}$$

In view of (2.11) and $(2.34)_1$, we obtain

$$\boldsymbol{S} = (\boldsymbol{G}_k\otimes\boldsymbol{g}^k)(\boldsymbol{P}^i\otimes\boldsymbol{G}_i) = (\boldsymbol{g}^k\cdot\boldsymbol{P}^i)\boldsymbol{G}_k\otimes\boldsymbol{G}_i\,,$$

i.e.

$$\boldsymbol{S} = S^{ij}\boldsymbol{G}_i\otimes\boldsymbol{G}_j\,,\qquad\text{with}\qquad S^{ij} = \boldsymbol{g}^i\cdot\boldsymbol{P}^j = \boldsymbol{g}^i\cdot(J\,T^{kj}\boldsymbol{g}_k) = J\,T^{ij}\,. \tag{2.45}$$

From the last relation we see that the symmetry of the Cauchy stress tensor (2.29) is equivalent to the symmetry of the second Piola-Kirchhoff stress tensor $\boldsymbol{S}$, i.e.

$$\boldsymbol{S} = \boldsymbol{S}^T\,. \tag{2.46}$$

2.3 Local Geometry of Surfaces

Let us consider a two-dimensional surface $\mathscr{S}$ having the local parametrization $\boldsymbol{r} = \boldsymbol{r}(\theta^1, \theta^2)$, where $\boldsymbol{r}$ is the position vector of a generic point P on $\mathscr{S}$ and θ^1, θ^2 are the two surface parameters, see Fig. 2.3.

We regard the surface as being immersed in Euclidean 3-space and write the decomposition

$$\boldsymbol{r} = r^i(\theta^\alpha)\boldsymbol{e}_i\,, \qquad i = 1, 2, 3, \quad \alpha = 1, 2. \tag{2.47}$$

Denote the tangent plane at P with T_P, which is a two-dimensional Euclidean space. The *natural basis vectors* for the surface are defined by

$$\boldsymbol{a}_\alpha = \frac{\partial \boldsymbol{r}}{\partial \theta^\alpha} = \boldsymbol{r}_{,\alpha} = r^i_{,\alpha}\,\boldsymbol{e}_i\,, \qquad \alpha = 1, 2. \tag{2.48}$$

The vectors $\boldsymbol{a}_\alpha$ span the tangent plane T_P provided that $\boldsymbol{a}_1 \times \boldsymbol{a}_2 \neq \boldsymbol{0}$.

We assume that to each pair (θ^1, θ^2) there corresponds a unique single-valued $\boldsymbol{r}(\theta^1, \theta^2)$, but we do *not* assume a one-to-one correspondence. Thus, the function $\boldsymbol{r}(\theta^\alpha)$ is not necessarily bijective, so $\boldsymbol{r}(\overset{(1)}{\theta}{}^\alpha) = \boldsymbol{r}(\overset{(2)}{\theta}{}^\alpha)$ is possible, see Fig. 2.4.

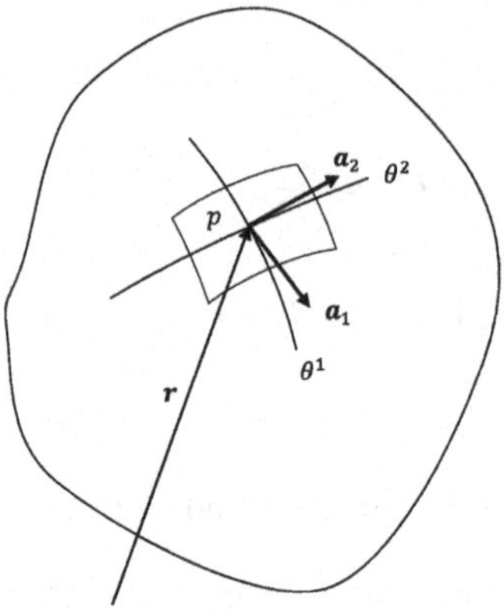

Fig. 2.3 Local parametrization of the surface

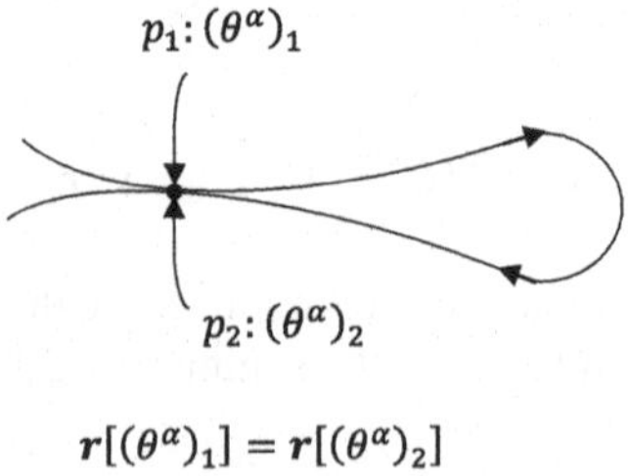

Fig. 2.4 The parametrization is not necessarily a bijective function

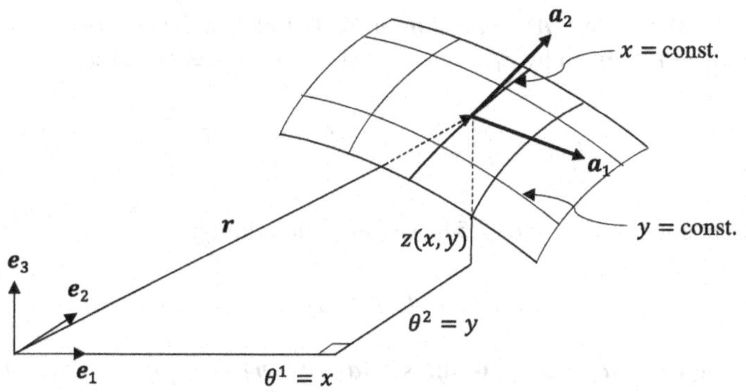

Fig. 2.5 Parametrization in terms of Cartesian coordinates

For example, using the Cartesian coordinates (x, y, z) in the Euclidean 3-space, we can choose the surface parameters $\theta^1 = x$, $\theta^2 = y$ (see Fig. 2.5) and write the local parametrization in the form

$$\boldsymbol{r} = x\,\boldsymbol{e}_1 + y\,\boldsymbol{e}_2 + z(x, y)\boldsymbol{e}_3 = \theta^1\boldsymbol{e}_1 + \theta^2\boldsymbol{e}_2 + z(\theta^1, \theta^2)\boldsymbol{e}_3\,. \tag{2.49}$$

Here we restrict the parameters θ^1, θ^2 such that the function $z(\theta^1, \theta^2)$ is single-valued.

In view of (2.47) and (2.48), we have in general

$$\mathrm{d}\boldsymbol{r} = \boldsymbol{a}_\alpha\,\mathrm{d}\theta^\alpha \quad \text{and} \quad |\mathrm{d}\boldsymbol{r}|^2 = \mathrm{d}\boldsymbol{r}\cdot\mathrm{d}\boldsymbol{r} = a_{\alpha\beta}\,\mathrm{d}\theta^\alpha\mathrm{d}\theta^\beta\,, \tag{2.50}$$

where

$$a_{\alpha\beta} = \boldsymbol{a}_\alpha\cdot\boldsymbol{a}_\beta = a_{\beta\alpha} \tag{2.51}$$

are the components of the *surface metric tensor*. Let us denote by a the 2×2 determinant $a = \det(a_{\alpha\beta})$. In view of the Lagrange identity for vector products

$$(\boldsymbol{a}\times\boldsymbol{b})\cdot(\boldsymbol{c}\times\boldsymbol{d}) = (\boldsymbol{a}\cdot\boldsymbol{c})(\boldsymbol{b}\cdot\boldsymbol{d}) - (\boldsymbol{b}\cdot\boldsymbol{c})(\boldsymbol{a}\cdot\boldsymbol{d}),$$

which holds for any vectors $\boldsymbol{a}, \boldsymbol{b}, \boldsymbol{c}, \boldsymbol{d}$, we obtain by putting $\boldsymbol{a} = \boldsymbol{c} = \boldsymbol{a}_1$ and $\boldsymbol{b} = \boldsymbol{d} = \boldsymbol{a}_2$ that

$$|\boldsymbol{a}_1\times\boldsymbol{a}_2|^2 = a_{11}a_{22} - (a_{12})^2 = \det(a_{\alpha\beta}) = a \geq 0, \quad \text{so} \quad |\boldsymbol{a}_1\times\boldsymbol{a}_2| = \sqrt{a}\,. \tag{2.52}$$

According to the assumption $\boldsymbol{a}_1\times\boldsymbol{a}_2 \neq \boldsymbol{0}$, we see that the surface metric tensor $a_{\alpha\beta}$ is positive definite.

We consider the local orientation of the surface given by the unit normal vector

$$\boldsymbol{n} = \frac{1}{\sqrt{a}}\,\boldsymbol{a}_1\times\boldsymbol{a}_2\,. \tag{2.53}$$

In view of (2.52), we have $|\boldsymbol{n}| = 1$ and $\boldsymbol{n} \cdot \boldsymbol{a}_\alpha = 0$. Let us define the two-dimensional alternator $e_{\alpha\beta} = e^{\alpha\beta}$ given by $e_{12} = -e_{21} = 1$, $e_{11} = e_{22} = 0$ and set

$$\varepsilon_{\alpha\beta} = \sqrt{a}\, e_{\alpha\beta}\,, \qquad \varepsilon^{\alpha\beta} = \frac{1}{\sqrt{a}}\, e^{\alpha\beta}\,. \tag{2.54}$$

Taking into account (2.53) and (2.54), we have the relation

$$\boldsymbol{a}_\alpha \times \boldsymbol{a}_\beta = \varepsilon_{\alpha\beta}\, \boldsymbol{n}\,. \tag{2.55}$$

Then, provided that $\boldsymbol{a}_1 \times \boldsymbol{a}_2 \neq \boldsymbol{0}$, the set $\{\boldsymbol{a}_1, \boldsymbol{a}_2, \boldsymbol{n}\}$ is a basis for vectors in the Euclidean 3-space, i.e. we can decompose any vector $\boldsymbol{v}$ as

$$\boldsymbol{v} = v^\alpha \boldsymbol{a}_\alpha + v\, \boldsymbol{n}\,. \tag{2.56}$$

Let us introduce the dual components $a^{\alpha\beta}$ of the surface metric tensor through the relations

$$a^{\alpha\gamma} a_{\gamma\beta} = \delta^\alpha_\beta\,, \qquad \text{i.e.} \qquad \left(a^{\alpha\beta}\right) = (a_{\alpha\beta})^{-1} = \frac{1}{a}\begin{pmatrix} a_{22} & -a_{12} \\ -a_{12} & a_{11} \end{pmatrix}. \tag{2.57}$$

The *dual basis vectors* $\boldsymbol{a}^\alpha$ in the tangent plane are then given by

$$\boldsymbol{a}^\alpha = a^{\alpha\beta} \boldsymbol{a}_\beta \tag{2.58}$$

and we note that

$$\begin{aligned} &\boldsymbol{a}^\alpha \cdot \boldsymbol{a}_\beta = a^{\alpha\gamma} \boldsymbol{a}_\gamma \cdot \boldsymbol{a}_\beta = a^{\alpha\gamma} a_{\gamma\beta} = \delta^\alpha_\beta \qquad \text{and} \\ &\boldsymbol{a}_\alpha = \delta^\beta_\alpha\, \boldsymbol{a}_\beta = a^{\beta\gamma} a_{\gamma\alpha}\, \boldsymbol{a}_\beta = a_{\alpha\gamma}\, \boldsymbol{a}^\gamma\,. \end{aligned}$$

Also, from (2.55) and (2.58) we get

$$\boldsymbol{a}^\alpha \times \boldsymbol{a}^\beta = a^{\alpha\gamma} a^{\beta\delta} \boldsymbol{a}_\gamma \times \boldsymbol{a}_\delta = a^{\alpha\gamma} a^{\beta\delta} \varepsilon_{\gamma\delta}\, \boldsymbol{n}\,. \tag{2.59}$$

By a direct calculation, we can show that (see Exercise 2.8)

$$a^{\alpha\gamma} a^{\beta\delta} \varepsilon_{\gamma\delta} = \varepsilon^{\alpha\beta}\,,$$

and the relation (2.59) becomes

$$\boldsymbol{a}^\alpha \times \boldsymbol{a}^\beta = \varepsilon^{\alpha\beta} \boldsymbol{n}\,. \tag{2.60}$$

Further, since $\boldsymbol{n} \perp T_P$ we have $\boldsymbol{n} \times \boldsymbol{a}_\alpha = u_{\alpha\beta}\, \boldsymbol{a}^\beta$ for some coefficients $u_{\alpha\beta}$. Then, from (2.55) we have $u_{\alpha\beta} = \boldsymbol{a}_\beta \cdot (\boldsymbol{n} \times \boldsymbol{a}_\alpha) = \boldsymbol{n} \cdot (\boldsymbol{a}_\alpha \times \boldsymbol{a}_\beta) = \varepsilon_{\alpha\beta}$, so that

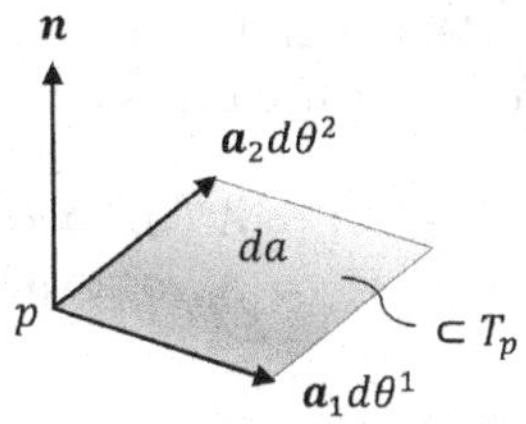

Fig. 2.6 The oriented elemental area on the surface

$$\boldsymbol{n} \times \boldsymbol{a}_\alpha = \varepsilon_{\alpha\beta}\, \boldsymbol{a}^\beta \tag{2.61}$$

and, similarly, from (2.60) we find

$$\boldsymbol{n} \times \boldsymbol{a}^\alpha = \varepsilon^{\alpha\beta}\, \boldsymbol{a}_\beta\,. \tag{2.62}$$

For the oriented *elemental area* we consider $\mathrm{d}\boldsymbol{r}_1 = \boldsymbol{a}_1 \mathrm{d}\theta^1$ and $\mathrm{d}\boldsymbol{r}_2 = \boldsymbol{a}_2 \mathrm{d}\theta^2$, see Fig. 2.6, and obtain

$$\boldsymbol{n}\, \mathrm{d}a = \mathrm{d}\boldsymbol{r}_1 \times \mathrm{d}\boldsymbol{r}_2 = \boldsymbol{a}_1 \times \boldsymbol{a}_2\, \mathrm{d}\theta^1 \mathrm{d}\theta^2 = \boldsymbol{n}\, \varepsilon_{12}\, \mathrm{d}\theta^1 \mathrm{d}\theta^2 = \boldsymbol{n}\sqrt{a}\, \mathrm{d}\theta^1 \mathrm{d}\theta^2 \tag{2.63}$$

where $a = \det(a_{\alpha\beta})$ and we have used (2.55). Thus, for the area element we have

$$\mathrm{d}a = \sqrt{a}\, \mathrm{d}\theta^1 \mathrm{d}\theta^2\,, \tag{2.64}$$

which generates coordinate-invariant area and can be employed to write the area of a patch $\mathscr{T}$:

$$A(\mathscr{T}) = \int_{\mathscr{T}} \mathrm{d}a = \int_T \sqrt{a}\, \mathrm{d}\theta^1 \mathrm{d}\theta^2\,, \qquad \text{with} \qquad (\theta^1, \theta^2) \in T.$$

The relation (2.64) is useful in writing surface integrals as iterated integrals.

By virtue of (2.54), (2.55) and (2.60), if a is positive and finite, both sets $\{\boldsymbol{a}_1\,, \boldsymbol{a}_2\}$ and $\{\boldsymbol{a}^1\,, \boldsymbol{a}^2\}$ are bases that span the tangent plane T_P and for any vector $\boldsymbol{v}$ we have

$$\boldsymbol{v} - v\,\boldsymbol{n} = v^\alpha \boldsymbol{a}_\alpha = v_\alpha \boldsymbol{a}^\alpha\,, \qquad \text{i.e.} \qquad v^\alpha = \boldsymbol{v} \cdot \boldsymbol{a}^\alpha\,, \quad v_\alpha = \boldsymbol{v} \cdot \boldsymbol{a}_\alpha\,.$$

All the usual rules relating to covariant and contravariant components apply on T_P, e.g. the raising and lowering of indices (analogous to (1.17), (1.18))

$$v^\alpha = a^{\alpha\beta} v_\beta\,, \qquad v_\alpha = a_{\alpha\beta} v^\beta\,,$$

and the transformation rules for coordinates changes (similar to (1.30))

$$\bar{v}^\alpha = \frac{\partial \bar{\theta}^\alpha}{\partial \theta^\beta}\, v^\beta \qquad \text{and} \qquad \bar{v}_\alpha = \frac{\partial \theta^\beta}{\partial \bar{\theta}^\alpha}\, v_\beta\,, \quad \text{etc.}$$

Moreover, for any second order tensor $\boldsymbol{A}$ in the Euclidean 3-space, we have the following decompositions

$$\begin{aligned}\boldsymbol{A} &= A^{\alpha\beta}\boldsymbol{a}_\alpha \otimes \boldsymbol{a}_\beta + A^{\alpha 3}\boldsymbol{a}_\alpha \otimes \boldsymbol{n} + A^{3\alpha}\boldsymbol{n} \otimes \boldsymbol{a}_\alpha + A^{33}\boldsymbol{n} \otimes \boldsymbol{n} \\ &= A_{\alpha\beta}\boldsymbol{a}^\alpha \otimes \boldsymbol{a}^\beta + A_{\alpha 3}\boldsymbol{a}^\alpha \otimes \boldsymbol{n} + A_{3\alpha}\boldsymbol{n} \otimes \boldsymbol{a}^\alpha + A_{33}\boldsymbol{n} \otimes \boldsymbol{n}\,,\end{aligned} \tag{2.65}$$

where

$$A^{\alpha\beta} = a^{\alpha\gamma}a^{\beta\delta}A_{\gamma\delta}\,, \qquad A^{\alpha 3} = a^{\alpha\beta}A_{\beta 3}\,, \qquad A^{3\alpha} = a^{\alpha\beta}A_{3\beta}\,, \qquad A^{33} = A_{33}$$

and the following transformation laws for coordinate changes hold (see (1.32))

$$\bar{A}^{\alpha\beta} = \frac{\partial\bar{\theta}^\alpha}{\partial\theta^\gamma}\frac{\partial\bar{\theta}^\beta}{\partial\theta^\delta}A^{\gamma\delta}\,, \qquad \bar{A}_{\alpha 3} = \frac{\partial\theta^\beta}{\partial\bar{\theta}^\alpha}A_{\beta 3}\,, \qquad \bar{A}_{33} = A_{33}\,, \qquad \text{etc.}$$

The second order tensor $\boldsymbol{A}$ is called a *surface tensor*, if $\boldsymbol{A}$ admits a representation of the type

$$\boldsymbol{A} = A^{\alpha\beta}\boldsymbol{a}_\alpha \otimes \boldsymbol{a}_\beta = A_{\alpha\beta}\boldsymbol{a}^\alpha \otimes \boldsymbol{a}^\beta\,,$$

i.e. the components $A^{\alpha 3}$, $A^{3\alpha}$, A^{33} and $A_{\alpha 3}\,,\ A_{3\alpha}\,,\ A_{33}$ vanish. Then, the tensor $\boldsymbol{A}$ can also be written with help of the mixed components in the form

$$\boldsymbol{A} = A^{\alpha}_{\cdot\beta}\,\boldsymbol{a}_\alpha \otimes \boldsymbol{a}^\beta = A_{\alpha}^{\cdot\beta}\boldsymbol{a}^\alpha \otimes \boldsymbol{a}_\beta\,.$$

For instance, the *unit tensor* in the tangent plane is a surface tensor given by

$$\mathbf{1} = a_{\alpha\beta}\boldsymbol{a}^\alpha \otimes \boldsymbol{a}^\beta = a^{\alpha\beta}\boldsymbol{a}_\alpha \otimes \boldsymbol{a}_\beta = \delta^\alpha_\beta\boldsymbol{a}_\alpha \otimes \boldsymbol{a}^\beta = \delta^\beta_\alpha\boldsymbol{a}^\alpha \otimes \boldsymbol{a}_\beta = \boldsymbol{a}^\alpha \otimes \boldsymbol{a}_\alpha = \boldsymbol{a}_\alpha \otimes \boldsymbol{a}^\alpha, \tag{2.66}$$

see the relation (1.27) for comparison with $\boldsymbol{I}$. In view of (2.66), we have $\mathbf{1}\boldsymbol{u} = \boldsymbol{u}$, for any vector $\boldsymbol{u} \in T_P$. Let $\boldsymbol{c}, \boldsymbol{d} \in T_P$ be two orthonormal vectors in the tangent plane. Then, we have

$$\begin{aligned}\boldsymbol{a}_\alpha &= (\boldsymbol{a}_\alpha \cdot \boldsymbol{c})\boldsymbol{c} + (\boldsymbol{a}_\alpha \cdot \boldsymbol{d})\boldsymbol{d} = c_\alpha\,\boldsymbol{c} + d_\alpha\,\boldsymbol{d} \qquad \text{and} \\ \boldsymbol{a}^\alpha &= (\boldsymbol{a}^\alpha \cdot \boldsymbol{c})\boldsymbol{c} + (\boldsymbol{a}^\alpha \cdot \boldsymbol{d})\boldsymbol{d} = c^\alpha\,\boldsymbol{c} + d^\alpha\,\boldsymbol{d}\,,\end{aligned}$$

so that

$$a_{\alpha\beta} = \boldsymbol{a}_\alpha \cdot \boldsymbol{a}_\beta = c_\alpha\,c_\beta + d_\alpha\,d_\beta\,, \qquad a^{\alpha\beta} = \boldsymbol{a}^\alpha \cdot \boldsymbol{a}^\beta = c^\alpha\,c^\beta + d^\alpha\,d^\beta$$

and

$$\mathbf{1} = a_{\alpha\beta}\boldsymbol{a}^\alpha \otimes \boldsymbol{a}^\beta = (c_\alpha\,c_\beta + d_\alpha\,d_\beta)\boldsymbol{a}^\alpha \otimes \boldsymbol{a}^\beta = \boldsymbol{c} \otimes \boldsymbol{c} + \boldsymbol{d} \otimes \boldsymbol{d}\,. \tag{2.67}$$

Thus, in view of (1.27) we have

$$\boldsymbol{I} = \boldsymbol{c} \otimes \boldsymbol{c} + \boldsymbol{d} \otimes \boldsymbol{d} + \boldsymbol{n} \otimes \boldsymbol{n} \qquad \text{and} \qquad \mathbf{1} = \boldsymbol{I} - \boldsymbol{n} \otimes \boldsymbol{n}\,. \tag{2.68}$$

Examples:

(i) Let us consider the example mentioned above in (2.49), also called the *Monge parametrization*, in which the two surface parameters $\theta^1 = x\,, \theta^2 = y$ coincide with the Cartesian coordinates, i.e.

$$\boldsymbol{r} = \theta^1 \boldsymbol{e}_1 + \theta^2 \boldsymbol{e}_2 + z(\theta^1, \theta^2)\boldsymbol{e}_3\,.$$

Then, from (2.48) and (2.51) we deduce in this case

$$\boldsymbol{a}_1 = \boldsymbol{e}_1 + z_{,1}\boldsymbol{e}_3\,, \quad \boldsymbol{a}_2 = \boldsymbol{e}_2 + z_{,2}\boldsymbol{e}_3\,, \quad (a_{\alpha\beta}) = \begin{pmatrix} 1 + (z_{,1})^2 & z_{,1}\, z_{,2} \\ z_{,1}\, z_{,2} & 1 + (z_{,2})^2 \end{pmatrix}.$$

We remark that the surface coordinate system is non-orthogonal in general, since $\boldsymbol{a}_1 \cdot \boldsymbol{a}_2 = a_{12} = z_{,1}\, z_{,2} \neq 0$. Further, we obtain the relations

$$a = \det(a_{\alpha\beta}) = 1 + (z_{,1})^2 + (z_{,2})^2 \qquad \text{and}$$
$$\boldsymbol{n} = \frac{1}{\sqrt{a}}\, \boldsymbol{a}_1 \times \boldsymbol{a}_2 = \frac{1}{\sqrt{1 + (z_{,1})^2 + (z_{,2})^2}}\,(\boldsymbol{e}_3 - z_{,1}\boldsymbol{e}_1 - z_{,2}\boldsymbol{e}_2).$$

(ii) Let us consider a circular cylindrical surface of radius R. The surface parameters are $\theta^1 = \theta\,, \theta^2 = z$, where the angle θ and the distance z to the x_1x_2-plane are the usual cylindrical coordinates, see Fig. 2.7. Then, the parametrization of the surface has the form

$$\boldsymbol{r} = R(\cos\theta^1 \boldsymbol{e}_1 + \sin\theta^1 \boldsymbol{e}_2) + \theta^2 \boldsymbol{e}_3\,.$$

Hence, we obtain

$$\boldsymbol{a}_1 = \boldsymbol{r}_{,1} = R(-\sin\theta^1 \boldsymbol{e}_1 + \cos\theta^2 \boldsymbol{e}_2) = R\,\boldsymbol{e}_\theta\,, \qquad \boldsymbol{a}_2 = \boldsymbol{r}_{,2} = \boldsymbol{e}_3\,,$$

where $\boldsymbol{e}_\theta = -\sin\theta^1 \boldsymbol{e}_1 + \cos\theta^2 \boldsymbol{e}_2$. Also, we find

$$(a_{\alpha\beta}) = \begin{pmatrix} R^2 & 0 \\ 0 & 1 \end{pmatrix}, \qquad a = R^2\,, \qquad \boldsymbol{n} = \frac{1}{\sqrt{a}}\, \boldsymbol{a}_1 \times \boldsymbol{a}_2 = \cos\theta^1 \boldsymbol{e}_1 + \sin\theta^1 \boldsymbol{e}_2 = \boldsymbol{e}_\rho\,,$$

where we have designated by $\boldsymbol{e}_\rho = \cos\theta^1 \boldsymbol{e}_1 + \sin\theta^1 \boldsymbol{e}_2$. Notice that the surface coordinates are orthogonal, since the matrix $(a_{\alpha\beta})$ has a diagonal form.

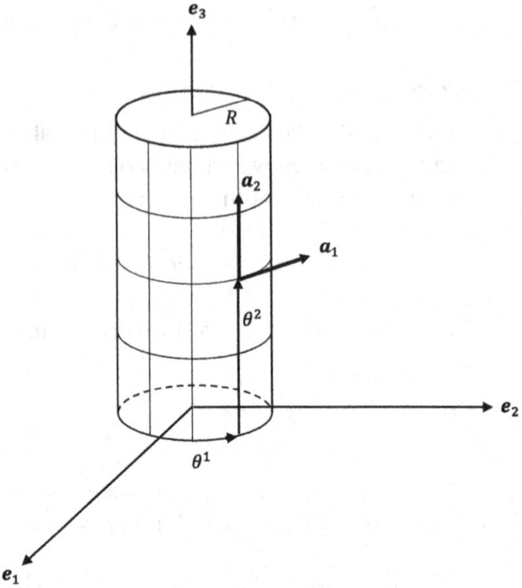

Fig. 2.7 Circular cylindrical surface of radius R

2.4 Differentiation on Surfaces. Curvature

By differentiating the relation $\boldsymbol{n} \cdot \boldsymbol{n} = 1$ with respect to θ^α we deduce that $\boldsymbol{n} \cdot \boldsymbol{n}_{,\alpha} = 0$, so $\boldsymbol{n}_{,\alpha}$ lies in the tangent plane T_P . Thus, we can decompose it in the basis $\{\boldsymbol{a}^1, \boldsymbol{a}^2\}$ as follows

$$\boldsymbol{n}_{,\alpha} = (\boldsymbol{a}_\beta \cdot \boldsymbol{n}_{,\alpha})\boldsymbol{a}^\beta ,$$

or

$$\boldsymbol{n}_{,\alpha} = -b_{\alpha\beta}\, \boldsymbol{a}^\beta , \qquad \text{where} \qquad b_{\alpha\beta} = -\boldsymbol{a}_\beta \cdot \boldsymbol{n}_{,\alpha} . \tag{2.69}$$

These are called the *Weingarten equations.*

If we differentiate the relation $\boldsymbol{n} \cdot \boldsymbol{a}_\beta = 0$ with respect to θ^α, we get $\boldsymbol{n}_{,\alpha} \cdot \boldsymbol{a}_\beta = -\boldsymbol{n} \cdot \boldsymbol{a}_{\beta,\alpha}$ and from (2.69) we obtain

$$b_{\alpha\beta} = -\boldsymbol{n}_{,\alpha} \cdot \boldsymbol{a}_\beta = \boldsymbol{n} \cdot \boldsymbol{a}_{\beta,\alpha} = \boldsymbol{n} \cdot \boldsymbol{r}_{,\beta\alpha} = b_{\beta\alpha} , \tag{2.70}$$

if $\boldsymbol{r}(\theta^\alpha)$ is a function of class C^2. Further, we can show that the transformation laws for coordinate changes hold for the components $b_{\alpha\beta}$, i.e.

$$\bar{b}_{\alpha\beta} = \frac{\partial\theta^\gamma}{\partial\bar{\theta}^\alpha}\,\frac{\partial\theta^\delta}{\partial\bar{\theta}^\beta}\, b_{\gamma\delta} ,$$

and so we can define the surface tensor $\boldsymbol{b}$, also called the *curvature tensor*, by

$$\boldsymbol{b} = b_{\alpha\beta}\, \boldsymbol{a}^{\alpha} \otimes \boldsymbol{a}^{\beta} = b^{\alpha}_{\cdot\beta}\, \boldsymbol{a}_{\alpha} \otimes \boldsymbol{a}^{\beta} = b_{\alpha}^{\cdot\beta} \boldsymbol{a}^{\alpha} \otimes \boldsymbol{a}_{\beta} = b^{\alpha\beta} \boldsymbol{a}_{\alpha} \otimes \boldsymbol{a}_{\beta}\,, \tag{2.71}$$

where

$$b^{\alpha}_{\cdot\beta} = a^{\alpha\gamma} b_{\gamma\beta}\,, \qquad b_{\alpha}^{\cdot\beta} = a^{\beta\gamma} b_{\alpha\gamma}\,, \qquad b^{\alpha\beta} = a^{\alpha\gamma} a^{\beta\delta} b_{\gamma\delta}\,. \tag{2.72}$$

From (2.70) we see that the tensor $\boldsymbol{b}$ is symmetric, i.e.

$$\boldsymbol{b} = \boldsymbol{b}^T\,, \qquad \text{so} \qquad b^{\alpha\beta} = b^{\beta\alpha} \qquad \text{and} \qquad b^{\alpha}_{\cdot\beta} = b_{\beta}^{\cdot\alpha} =: b^{\alpha}_{\beta}\,. \tag{2.73}$$

Next, using the general decomposition (2.56) we can write for the derivative $\boldsymbol{a}_{\alpha,\beta}$ the relations

$$\boldsymbol{a}_{\alpha,\beta} = (\boldsymbol{a}^{\gamma} \cdot \boldsymbol{a}_{\alpha,\beta})\boldsymbol{a}_{\gamma} + (\boldsymbol{n} \cdot \boldsymbol{a}_{\alpha,\beta})\boldsymbol{n}\,. \tag{2.74}$$

In analogy to (1.40), we define the surface Christoffel symbols

$$\Gamma^{\gamma}_{\alpha\beta} = \boldsymbol{a}^{\gamma} \cdot \boldsymbol{a}_{\alpha,\beta} = \boldsymbol{a}^{\gamma} \cdot \boldsymbol{r}_{,\alpha\beta} = \Gamma^{\gamma}_{\beta\alpha}\,. \tag{2.75}$$

Substituting (2.70) and (2.75) into (2.74), we obtain the *Gauss relations*

$$\boldsymbol{a}_{\alpha,\beta} = \Gamma^{\gamma}_{\alpha\beta}\, \boldsymbol{a}_{\gamma} + b_{\alpha\beta}\, \boldsymbol{n}\,. \tag{2.76}$$

Also, for the derivative $\boldsymbol{a}^{\alpha}{}_{,\beta}$ we write

$$\boldsymbol{a}^{\alpha}{}_{,\beta} = (\boldsymbol{a}_{\gamma} \cdot \boldsymbol{a}^{\alpha}{}_{,\beta})\boldsymbol{a}^{\gamma} + (\boldsymbol{n} \cdot \boldsymbol{a}^{\alpha}{}_{,\beta})\boldsymbol{n}\,. \tag{2.77}$$

By differentiating the relations $\boldsymbol{a}^{\alpha} \cdot \boldsymbol{a}_{\gamma} = \delta^{\alpha}_{\gamma}$ and $\boldsymbol{a}^{\alpha} \cdot \boldsymbol{n} = 0$ with respect to θ^{β} we find, respectively,

$$\begin{aligned} \boldsymbol{a}^{\alpha}{}_{,\beta} \cdot \boldsymbol{a}_{\gamma} &= -\boldsymbol{a}^{\alpha} \cdot \boldsymbol{a}_{\gamma,\beta} = -\Gamma^{\alpha}_{\gamma\beta} \qquad \text{and} \\ \boldsymbol{a}^{\alpha}{}_{,\beta} \cdot \boldsymbol{n} &= -\boldsymbol{a}^{\alpha} \cdot \boldsymbol{n}_{,\beta} = -a^{\alpha\gamma} \boldsymbol{a}_{\gamma} \cdot \boldsymbol{n}_{,\beta} = a^{\alpha\gamma} b_{\beta\gamma} = b^{\alpha}_{\beta}\,. \end{aligned} \tag{2.78}$$

Then, inserting (2.78) into (2.77) we get another set of Gauss relations

$$\boldsymbol{a}^{\alpha}{}_{,\beta} = -\Gamma^{\alpha}_{\gamma\beta}\, \boldsymbol{a}^{\gamma} + b^{\alpha}_{\beta}\, \boldsymbol{n}\,. \tag{2.79}$$

Let $\boldsymbol{v}(\theta^{\alpha})$ be a 3-vector field on the surface with components

$$\boldsymbol{v} = v^{\alpha} \boldsymbol{a}_{\alpha} + v\, \boldsymbol{n} = v_{\alpha} \boldsymbol{a}^{\alpha} + v\, \boldsymbol{n}\,.$$

Then, the derivative with respect to θ^{β} can be written as

$$\boldsymbol{v}_{,\beta} = v^{\alpha}{}_{,\beta} \boldsymbol{a}_{\alpha} + v^{\alpha} \boldsymbol{a}_{\alpha,\beta} + v_{,\beta}\, \boldsymbol{n} + v\, \boldsymbol{n}_{,\beta}$$

and using (2.69) and (2.76) we find

$$\boldsymbol{v}_{,\beta} = \big(v^{\alpha}{}_{|\beta} - v\, b^{\alpha}_{\beta}\big)\boldsymbol{a}_{\alpha} + \big(v_{,\beta} + v^{\alpha} b_{\alpha\beta}\big)\,\boldsymbol{n}\,, \tag{2.80}$$

where we have denoted by $v^{\alpha}{}_{|\beta}$ the surface *covariant derivative*

$$v^{\alpha}{}_{|\beta} = v^{\alpha}{}_{,\beta} + v^{\gamma}\,\Gamma^{\alpha}_{\gamma\beta}\,, \tag{2.81}$$

in analogy with (1.41). Also, we can write

$$\boldsymbol{v}_{,\beta} = v_{\alpha,\beta}\boldsymbol{a}^{\alpha} + v_{\alpha}\boldsymbol{a}^{\alpha}{}_{,\beta} + v_{,\beta}\,\boldsymbol{n} + v\,\boldsymbol{n}_{,\beta}$$

and using (2.79) we get

$$\boldsymbol{v}_{,\beta} = \big(v_{\alpha|\beta} - v\, b_{\alpha\beta}\big)\boldsymbol{a}^{\alpha} + \big(v_{,\beta} + v_{\alpha} b^{\alpha}_{\beta}\big)\,\boldsymbol{n}\,, \tag{2.82}$$

where

$$v_{\alpha|\beta} = v_{\alpha,\beta} - v_{\gamma}\,\Gamma^{\gamma}_{\alpha\beta} \tag{2.83}$$

is the surface covariant derivative.

In a similar way, we can calculate the derivatives of a surface tensor field $\boldsymbol{A}(\theta^{\gamma}) = A_{\alpha\beta}\,\boldsymbol{a}^{\alpha} \otimes \boldsymbol{a}^{\beta}$. Making use of (2.79) we obtain the relation

$$\boldsymbol{A}_{,\gamma} = A_{\alpha\beta|\gamma}\,\boldsymbol{a}^{\alpha} \otimes \boldsymbol{a}^{\beta} + A_{\alpha\beta}\big(b^{\alpha}_{\gamma}\,\boldsymbol{n} \otimes \boldsymbol{a}^{\beta} + b^{\beta}_{\gamma}\,\boldsymbol{a}^{\alpha} \otimes \boldsymbol{n}\big), \tag{2.84}$$

where we denote by

$$A_{\alpha\beta|\gamma} = A_{\alpha\beta,\gamma} - A_{\alpha\lambda}\,\Gamma^{\lambda}_{\beta\gamma} - A_{\lambda\beta}\,\Gamma^{\lambda}_{\alpha\gamma} \tag{2.85}$$

the *covariant derivative* for the covariant components $A_{\alpha\beta}$, which corresponds to (1.52). For the contravariant components $A^{\alpha\beta}$ and the mixed components $A^{\alpha}{}_{\cdot\beta}$ we can deduce analogous formulas. Thus, if we differentiate $\boldsymbol{A} = A^{\alpha\beta}\,\boldsymbol{a}_{\alpha} \otimes \boldsymbol{a}_{\beta}$ with respect to θ^{γ} and use (2.76), we deduce that

$$\boldsymbol{A}_{,\gamma} = A^{\alpha\beta}{}_{|\gamma}\,\boldsymbol{a}_{\alpha} \otimes \boldsymbol{a}_{\beta} + A^{\alpha\beta}\big(b_{\alpha\gamma}\,\boldsymbol{n} \otimes \boldsymbol{a}_{\beta} + b_{\beta\gamma}\,\boldsymbol{a}_{\alpha} \otimes \boldsymbol{n}\big), \tag{2.86}$$

where we denote by

$$A^{\alpha\beta}{}_{|\gamma} = A^{\alpha\beta}{}_{,\gamma} + A^{\alpha\lambda}\,\Gamma^{\beta}_{\lambda\gamma} + A^{\lambda\beta}\,\Gamma^{\alpha}_{\lambda\gamma} \tag{2.87}$$

the covariant derivative which corresponds to (1.50). Similar to (1.54), on the basis of (2.76) and (2.79) we can prove that

$$a_{\alpha\beta|\gamma} = 0 \quad \text{and} \quad a^{\alpha\beta}{}_{|\gamma} = 0. \tag{2.88}$$

Hence, for the mixed tensor components of $\boldsymbol{A} = A^{\alpha}{}_{\cdot\beta}\, \boldsymbol{a}_\alpha \otimes \boldsymbol{a}^\beta = A_{\alpha}{}^{\cdot\beta} \boldsymbol{a}^\alpha \otimes \boldsymbol{a}_\beta$ we have

$$\begin{aligned} A^{\alpha}{}_{\cdot\beta|\gamma} &= \left(a_{\lambda\beta}\, A^{\alpha\lambda}\right)_{|\gamma} = a_{\lambda\beta}\, A^{\alpha\lambda}{}_{|\gamma} = A^{\alpha}{}_{\cdot\beta,\gamma} + A^{\lambda}{}_{\cdot\beta}\, \Gamma^{\alpha}_{\lambda\gamma} - A^{\alpha}{}_{\cdot\lambda}\, \Gamma^{\lambda}_{\beta\gamma} \quad \text{and} \\ A_{\alpha}{}^{\cdot\beta}{}_{|\gamma} &= \left(a^{\lambda\beta}\, A_{\alpha\lambda}\right)_{|\gamma} = a^{\lambda\beta}\, A_{\alpha\lambda|\gamma} = A_{\alpha}{}^{\cdot\beta}{}_{,\gamma} - A_{\lambda}{}^{\cdot\beta}\, \Gamma^{\lambda}_{\alpha\gamma} + A_{\alpha}{}^{\cdot\lambda}\, \Gamma^{\beta}_{\lambda\gamma}\,, \quad \text{etc.} \end{aligned}$$

For the surface Christoffel symbols we can show in the same manner as (1.58) and (1.61) that the following relations hold

$$\Gamma^{\gamma}_{\alpha\beta} = \frac{1}{2}\, a^{\gamma\lambda}\left(a_{\beta\lambda,\alpha} + a_{\lambda\alpha,\beta} - a_{\alpha\beta,\lambda}\right) \quad \text{and} \quad \Gamma^{\alpha}_{\alpha\beta} = \frac{1}{\sqrt{a}}\left(\sqrt{a}\right)_{,\beta} = \left(\ln\sqrt{a}\right)_{,\beta} \tag{2.89}$$

and in analogy with (1.63) we have

$$v^{\alpha}{}_{|\alpha} = \frac{1}{\sqrt{a}}\left(\sqrt{a}\, v^{\alpha}\right)_{,\alpha}\,. \tag{2.90}$$

Indeed, from (2.81) and (2.89) we get

$$v^{\alpha}{}_{|\alpha} = v^{\alpha}{}_{,\alpha} + v^{\gamma}\, \Gamma^{\alpha}_{\gamma\alpha} = v^{\alpha}{}_{,\alpha} + \frac{1}{\sqrt{a}}\left(\sqrt{a}\right)_{,\gamma} v^{\gamma} = \frac{1}{\sqrt{a}}\left(\sqrt{a}\, v^{\alpha}\right)_{,\alpha}\,.$$

2.4.1 Curvature of a Curve on the Surface

Consider a curve $\mathscr{C}$ on the surface $\mathscr{S}$ with unit tangent vector $\boldsymbol{\lambda}$ at P, see Fig. 2.8. Let the parameter u along the curve measure the arclength on $\mathscr{C}$. Then, the curve is parametrized by $\boldsymbol{r} = \boldsymbol{r}(\theta^1(u), \theta^2(u))$ and we have

$$\boldsymbol{\lambda} = \frac{\mathrm{d}}{\mathrm{d}u}\boldsymbol{r} = \lambda^{\alpha}\boldsymbol{a}_\alpha = \lambda_\alpha\, \boldsymbol{a}^\alpha\,, \qquad \boldsymbol{\lambda} \in T_P\,, \qquad |\boldsymbol{\lambda}| = 1. \tag{2.91}$$

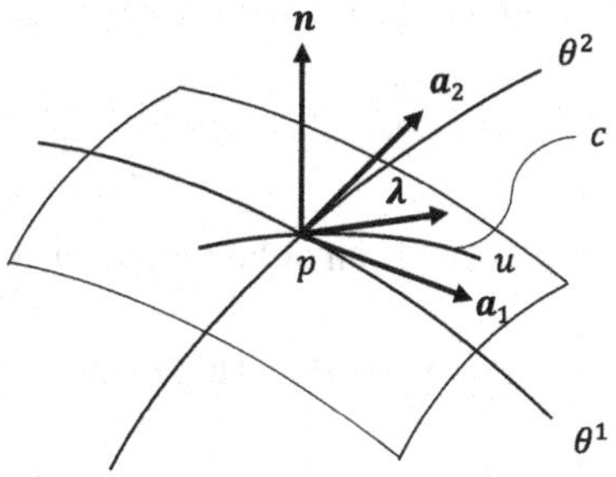

Fig. 2.8 The curve $\mathscr{C}$ on the surface with unit tangent $\boldsymbol{\lambda}$

The *curvature vector* for the curve $\mathscr{C}$ is given by

$$\boldsymbol{\kappa} = \frac{\mathrm{d}}{\mathrm{d}u}\boldsymbol{\lambda}\,. \tag{2.92}$$

Since $\boldsymbol{\lambda}\cdot\boldsymbol{\lambda} = 1$, we have

$$0 = \boldsymbol{\lambda}\cdot\frac{\mathrm{d}\boldsymbol{\lambda}}{\mathrm{d}u} = \boldsymbol{\lambda}\cdot\boldsymbol{\kappa}\,, \qquad \text{so} \qquad \boldsymbol{\lambda}\perp\boldsymbol{\kappa}\,.$$

If we define the vector $\boldsymbol{\mu}$ by

$$\boldsymbol{\mu} = \boldsymbol{n}\times\boldsymbol{\lambda}\,, \tag{2.93}$$

then the set $\{\boldsymbol{\lambda}, \boldsymbol{\mu}, \boldsymbol{n}\}$ is an orthonormal basis and the curvature vector $\boldsymbol{\kappa}$ belongs to the plane spanned by $\{\boldsymbol{\mu}, \boldsymbol{n}\}$. Thus, we have the decomposition

$$\boldsymbol{\kappa} = \kappa_g\,\boldsymbol{\mu} + \kappa_n\,\boldsymbol{n}\,, \qquad \text{with} \qquad \kappa_g = \boldsymbol{\mu}\cdot\boldsymbol{\kappa}\,, \quad \kappa_n = \boldsymbol{n}\cdot\boldsymbol{\kappa}\,, \tag{2.94}$$

where κ_g is called the *geodesic curvature* (which generalizes the notion of curvature of a *plane* curve) and κ_n is the *normal curvature*.

One can show that if the curve $\mathscr{C}$ is a *geodesic* (i.e., a C^2 path on $\mathscr{S}$ of minimum arc length connecting two points of the surface), then the geodesic curvature vanishes, i.e. $\kappa_g = 0$. Thus, the geodesic curves of a surface generalize the straight lines on a plane.

We have

$$\begin{aligned}\boldsymbol{\kappa} &= \frac{\mathrm{d}}{\mathrm{d}u}\boldsymbol{\lambda} = \frac{\mathrm{d}}{\mathrm{d}u}(\lambda^\alpha \boldsymbol{a}_\alpha) = \frac{\mathrm{d}\lambda^\alpha}{\mathrm{d}u}\boldsymbol{a}_\alpha + \lambda^\alpha\frac{\mathrm{d}}{\mathrm{d}u}\boldsymbol{a}_\alpha \\ &= \frac{\mathrm{d}\lambda^\alpha}{\mathrm{d}u}\boldsymbol{a}_\alpha + \lambda^\alpha\,\boldsymbol{a}_{\alpha,\beta}\frac{\mathrm{d}\theta^\beta}{\mathrm{d}u}\,.\end{aligned} \tag{2.95}$$

From (2.91) and $\boldsymbol{\lambda} = \frac{\mathrm{d}}{\mathrm{d}u}\boldsymbol{r} = \boldsymbol{r}_{,\alpha}\frac{\mathrm{d}\theta^\alpha}{\mathrm{d}u} = \boldsymbol{a}_\alpha\frac{\mathrm{d}\theta^\alpha}{\mathrm{d}u}$ we deduce that

$$\lambda^\alpha = \frac{\mathrm{d}\theta^\alpha}{\mathrm{d}u}\,. \tag{2.96}$$

Substituting (2.76) and (2.96) into (2.95) we get

$$\boldsymbol{\kappa} = \Big(\frac{\mathrm{d}\lambda^\gamma}{\mathrm{d}u} + \lambda^\alpha\lambda^\beta\Gamma^\gamma_{\alpha\beta}\Big)\boldsymbol{a}_\gamma + b_{\alpha\beta}\,\lambda^\alpha\lambda^\beta\,\boldsymbol{n}\,. \tag{2.97}$$

Since $\{\boldsymbol{\lambda}, \boldsymbol{\mu}\}$ is an orthonormal basis in the tangent plane T_P, we can write

$$\boldsymbol{a}_\gamma = (\boldsymbol{\lambda}\cdot\boldsymbol{a}_\gamma)\boldsymbol{\lambda} + (\boldsymbol{\mu}\cdot\boldsymbol{a}_\gamma)\boldsymbol{\mu} = \lambda_\gamma\,\boldsymbol{\lambda} + \mu_\gamma\,\boldsymbol{\mu} \qquad \text{and} \qquad \boldsymbol{a}^\gamma = \lambda^\gamma\,\boldsymbol{\lambda} + \mu^\gamma\,\boldsymbol{\mu}\,. \tag{2.98}$$

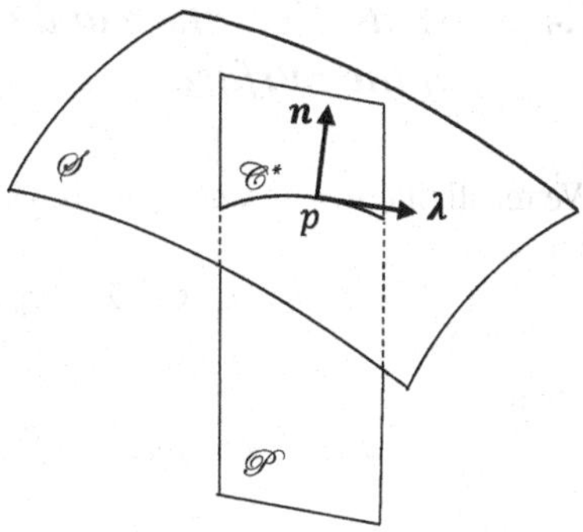

Fig. 2.9 The normal plane $\mathscr{P}$ and the plane curve $\mathscr{C}^* \subset \mathscr{S} \cap \mathscr{P}$

Inserting $(2.98)_1$ into (2.97) we obtain the decomposition of the curvature vector $\boldsymbol{\kappa}$ in the vector basis $\{\boldsymbol{\lambda}, \boldsymbol{\mu}, \boldsymbol{n}\}$. Thus, from $\boldsymbol{\lambda} \perp \boldsymbol{\kappa} = 0$ we deduce the equation

$$\left(\frac{\mathrm{d}\lambda^\gamma}{\mathrm{d}u} + \lambda^\alpha \lambda^\beta \Gamma^\gamma_{\alpha\beta}\right)\lambda_\gamma = 0, \tag{2.99}$$

together with

$$\kappa_g = \boldsymbol{\mu} \cdot \boldsymbol{\kappa} = \left(\frac{\mathrm{d}\lambda^\gamma}{\mathrm{d}u} + \lambda^\alpha \lambda^\beta \Gamma^\gamma_{\alpha\beta}\right)\mu_\gamma \tag{2.100}$$

and

$$\begin{aligned}\kappa_n &= \boldsymbol{n} \cdot \boldsymbol{\kappa} = b_{\alpha\beta}\, \lambda^\alpha \lambda^\beta \\ &= b_{\alpha\beta}\, (\boldsymbol{\lambda} \cdot \boldsymbol{a}^\alpha)(\boldsymbol{\lambda} \cdot \boldsymbol{a}^\beta) = \boldsymbol{\lambda} \cdot \boldsymbol{b}\, \boldsymbol{\lambda}\,.\end{aligned} \tag{2.101}$$

From Eq. (2.100) we remark that the geodesic curvature κ_g depends only on the curve $\mathscr{C}$ (through the components $\lambda^\alpha = \frac{\mathrm{d}\theta^\alpha}{\mathrm{d}u}$) and the surface metric $a_{\alpha\beta}$ (through the Christoffel symbols $\Gamma^\gamma_{\alpha\beta}$, see (2.89)). Further, the components μ_γ depend on the same quantities, since

$$\boldsymbol{\mu} = \boldsymbol{n} \times \boldsymbol{\lambda} = \boldsymbol{n} \times \boldsymbol{a}_\alpha\, \lambda^\alpha = \varepsilon_{\alpha\beta}\, \lambda^\alpha \boldsymbol{a}^\beta\,,$$

so

$$\mu_\beta = \varepsilon_{\alpha\beta}\, \lambda^\alpha = \sqrt{a}\, e_{\alpha\beta}\, \frac{\mathrm{d}\theta^\alpha}{\mathrm{d}u}\,. \tag{2.102}$$

Further equations for the geodesic curvature κ_g are given in Sect. 2.7 using surface differential operators, see (2.163).

Concerning the normal curvature κ_n , we mention the following interpretation: let $\mathscr{P}$ be the plane which passes through the point P and contains the vectors $\{\boldsymbol{n}, \boldsymbol{\lambda}\}$, see Fig. 2.9. Further, let $\mathscr{C}^*$ be the intersection curve between the surface $\mathscr{S}$ and the normal plane $\mathscr{P}$, i.e. $\mathscr{C}^* \subset \mathscr{S} \cap \mathscr{P}$. Then, the normal curvature κ_n is equal to the curvature of the plane curve $\mathscr{C}^*$ at point P on plane $\mathscr{P}$.

2.4.2 Mean Curvature and Gaussian Curvature of the Surface

We recall that $\boldsymbol{b} = b_{\alpha\beta}\, \boldsymbol{a}^\alpha \otimes \boldsymbol{a}^\beta$ is a symmetric surface tensor. Using here $(2.98)_2$ we get

$$\boldsymbol{b} = \kappa_\lambda\, \boldsymbol{\lambda} \otimes \boldsymbol{\lambda} + \kappa_\mu\, \boldsymbol{\mu} \otimes \boldsymbol{\mu} + \tau(\boldsymbol{\lambda} \otimes \boldsymbol{\mu} + \boldsymbol{\mu} \otimes \boldsymbol{\lambda}), \tag{2.103}$$

where

$$\kappa_\lambda = b_{\alpha\beta}\, \lambda^\alpha \lambda^\beta \qquad \text{and} \qquad \kappa_\mu = b_{\alpha\beta}\, \mu^\alpha \mu^\beta \tag{2.104}$$

are the *normal curvatures*, whereas

$$\tau = b_{\alpha\beta}\, \lambda^\alpha \mu^\beta = b_{\alpha\beta}\, \mu^\alpha \lambda^\beta \tag{2.105}$$

is the *twist*. Due to its symmetry, the tensor $\boldsymbol{b}$ admits the spectral decomposition

$$\boldsymbol{b} = \kappa_1\, \boldsymbol{u}_1 \otimes \boldsymbol{u}_1 + \kappa_2\, \boldsymbol{u}_2 \otimes \boldsymbol{u}_2 \qquad (\text{with } \kappa_1\, , \kappa_2 \in \mathbb{R}), \tag{2.106}$$

where $\kappa_1\, , \kappa_2$ are the *principal curvatures* and $\{\boldsymbol{u}_1\, , \boldsymbol{u}_2\}$ are the orthonormal *principal vectors*. Then, the principal curvatures $\kappa_1\, , \kappa_2$ are the roots of the quadratic equation

$$\kappa^2 - (\operatorname{tr} \boldsymbol{b})\kappa + \det \boldsymbol{b} = 0. \tag{2.107}$$

In view of (2.103) and (2.106), we have

$$\begin{aligned} \operatorname{tr} \boldsymbol{b} &= \kappa_\lambda + \kappa_\mu = \kappa_1 + \kappa_2 = 2H\, , \\ \det \boldsymbol{b} &= \kappa_\lambda\, \kappa_\mu - \tau^2 = \kappa_1\, \kappa_2 = K\, , \end{aligned} \tag{2.108}$$

where we have denoted by $H = \frac{1}{2} \operatorname{tr} \boldsymbol{b}$ the *mean curvature* and by $K = \det \boldsymbol{b}$ the *Gaussian curvature* of the surface.

Thus, in view of the Cayley-Hamilton theorem, the tensor $\boldsymbol{b}$ satisfies the equation

$$\boldsymbol{b}^2 - 2H\, \boldsymbol{b} + K\, \mathbf{1} = \mathbf{0}\, . \tag{2.109}$$

We can assume without loss of generality that $\kappa_1 \leq \kappa_2$. Then, we can prove that

$$\kappa_1 \leq \kappa_\lambda \leq \kappa_2 \qquad \text{and} \qquad \kappa_1 \leq \kappa_\mu \leq \kappa_2\, . \tag{2.110}$$

Indeed, if we denote by ϕ the angle between the vectors $\boldsymbol{\lambda}$ and $\boldsymbol{u}_1$ in the tangent plane, we have

$$\boldsymbol{\lambda} = \cos\phi\, \boldsymbol{u}_1 + \sin\phi\, \boldsymbol{u}_2\, , \qquad \boldsymbol{\mu} = -\sin\phi\, \boldsymbol{u}_1 + \cos\phi\, \boldsymbol{u}_2\, .$$

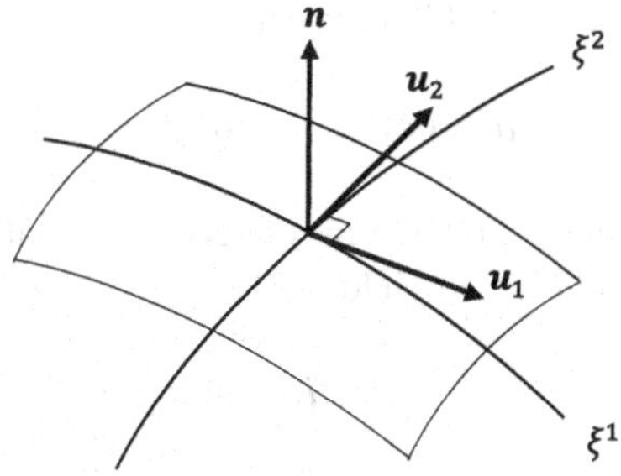

Fig. 2.10 Lines-of-curvature coordinates

Hence, we find for the normal curvatures the Euler relations

$$\begin{aligned}\kappa_\lambda &= \boldsymbol{\lambda}\cdot\boldsymbol{b}\boldsymbol{\lambda} = \boldsymbol{\lambda}\cdot(\kappa_1\,\boldsymbol{u}_1\otimes\boldsymbol{u}_1 + \kappa_2\,\boldsymbol{u}_2\otimes\boldsymbol{u}_2)(\cos\phi\,\boldsymbol{u}_1 + \sin\phi\,\boldsymbol{u}_2)\\ &= (\cos\phi\,\boldsymbol{u}_1 + \sin\phi\,\boldsymbol{u}_2)\cdot(\kappa_1\,\cos\phi\,\boldsymbol{u}_1 + \kappa_2\,\sin\phi\,\boldsymbol{u}_2) = \kappa_1\,\cos^2\phi + \kappa_2\,\sin^2\phi\,,\\ \kappa_\mu &= \boldsymbol{\mu}\cdot\boldsymbol{b}\boldsymbol{\mu} = \kappa_1\,\sin^2\phi + \kappa_2\,\cos^2\phi\,,\end{aligned} \tag{2.111}$$

Thus, since $\kappa_1 \le \kappa_2$ we deduce

$$\begin{aligned}\kappa_1 &= \kappa_1(\cos^2\phi + \sin^2\phi) \le \kappa_1\,\cos^2\phi + \kappa_2\,\sin^2\phi = \kappa_\lambda\,,\\ \kappa_2 &= \kappa_2(\cos^2\phi + \sin^2\phi) \ge \kappa_1\,\cos^2\phi + \kappa_2\,\sin^2\phi = \kappa_\lambda\,,\end{aligned}$$

and analogously $\kappa_1 \le \kappa_\mu \le \kappa_2$, i.e. the inequalities (2.110) are proved.

Moreover, we see from (2.111) that

$$\kappa_1 = \min_\phi \kappa_\lambda = \min_\phi \kappa_\mu \quad \text{and} \quad \kappa_2 = \max_\phi \kappa_\lambda = \max_\phi \kappa_\mu\,, \tag{2.112}$$

i.e. the principal curvatures κ_1, κ_2 are the extreme normal curvatures.

If $\kappa_1 = \kappa_2$ at point P, then $\kappa_\lambda = \kappa_\mu = \kappa_1$ for any angle ϕ and the point P is called an *umbilic point*.

If $\kappa_\lambda = \boldsymbol{\lambda}\cdot\boldsymbol{b}\boldsymbol{\lambda} = 0$ along the curve, then the $\boldsymbol{\lambda}$-curve is called an *asymptotic line*.

Consider now a parametrization $\boldsymbol{r}(\xi^1, \xi^2)$ of the surface, where (ξ^1, ξ^2) are orthogonal coordinates along the $\boldsymbol{u}_1$-curves and $\boldsymbol{u}_2$-curves, respectively, where $\boldsymbol{u}_1$, $\boldsymbol{u}_2$ are the principal vectors of $\boldsymbol{b}$. Such coordinates (ξ^1, ξ^2) are also called *lines-of-curvature* coordinates, see Fig. 2.10. There exists a local unique lines-of-curvature system through any point P, provided $\kappa_1 \ne \kappa_2$, cf. [5].

For the surface coordinates $\theta^\alpha = \xi^\alpha$ we have the natural basis vectors

$$\boldsymbol{a}_1 = \alpha\,\boldsymbol{u}_1\,,\quad \boldsymbol{a}_2 = \beta\,\boldsymbol{u}_2\,,\qquad \text{where}\qquad \alpha = |\boldsymbol{a}_1|\,,\quad \beta = |\boldsymbol{a}_2|\,.$$

Hence, in view of $\boldsymbol{u}_1\cdot\boldsymbol{u}_2 = 0$, $\boldsymbol{b}\,\boldsymbol{u}_1 = \kappa_1\,\boldsymbol{u}_1$ and $\boldsymbol{b}\,\boldsymbol{u}_2 = \kappa_2\,\boldsymbol{u}_2$, we obtain

$$(a_{\gamma\delta}) = (\boldsymbol{a}_\gamma\cdot\boldsymbol{a}_\delta) = \begin{pmatrix}\alpha^2 & 0\\ 0 & \beta^2\end{pmatrix},\qquad (b_{\gamma\delta}) = (\boldsymbol{a}_\gamma\cdot\boldsymbol{b}\,\boldsymbol{a}_\delta) = \begin{pmatrix}\alpha^2\kappa_1 & 0\\ 0 & \beta^2\kappa_2\end{pmatrix}.$$

Using $(2.108)_2$ we get

$$a = \det(a_{\gamma\delta}) = \alpha^2\beta^2\,, \qquad b = \det(b_{\gamma\delta}) = \alpha^2\beta^2\kappa_1\,\kappa_2 = a\,K\,, \tag{2.113}$$

where we have introduced the notation $b = \det(b_{\alpha\beta})$. Then, the relations $(2.108)_1$ and (2.113) yield

$$\begin{aligned} 2H &= \operatorname{tr}\boldsymbol{b} = \operatorname{tr}\,(b_{\alpha\beta}\boldsymbol{a}^\alpha \otimes \boldsymbol{a}^\beta) = b_{\alpha\beta}\,a^{\alpha\beta} = b^\alpha_\alpha\,,\\ K &= \frac{b}{a} = \frac{\det(b_{\alpha\beta})}{\det(a_{\alpha\beta})} = \det(a^{\alpha\gamma}\,b_{\gamma\beta}) = \det(b^\alpha_\beta)\,. \end{aligned} \tag{2.114}$$

Note that the relations (2.114) hold for general surface coordinates θ^α (not only for lines-of-curvature coordinates).

2.5 Compatibility Conditions for the Fundamental Tensors

We can write the Gauss equations (2.76) and the Weingarten relations (2.69), respectively, in the forms

$$\begin{aligned} \boldsymbol{r}_{,\alpha\beta} &= \Gamma^\gamma_{\alpha\beta}\,\boldsymbol{r}_{,\gamma} + b_{\alpha\beta}\,\boldsymbol{n} \qquad \text{and}\\ \boldsymbol{n}_{,\alpha} &= -b^\beta_\alpha\,\boldsymbol{r}_{,\beta}\,. \end{aligned} \tag{2.115}$$

These relations represent in total 15 scalar equations for 6 unknown variables (the components of $\boldsymbol{r}$ and $\boldsymbol{n}$). Thus, (2.115) is an overdetermined system of equations, with coefficients depending only on the metric tensor $a_{\alpha\beta}$ and the curvature tensor $b_{\alpha\beta}$, cf. (2.89) and (2.72). Therefore, arbitrary components $a_{\alpha\beta}$ and $b_{\alpha\beta}$ will not be the metric and curvature for a surface unless they satisfy certain *compatibility conditions*.

For example, from the Weingarten relations (2.69) we deduce that

$$\boldsymbol{n}_{,\alpha\beta} = -(b_{\alpha\gamma}\,\boldsymbol{a}^\gamma)_{,\beta} = -(b_{\alpha\gamma,\beta} - b_{\alpha\lambda}\Gamma^\lambda_{\gamma\beta})\boldsymbol{a}^\gamma - b_{\alpha\gamma}\,b^\gamma_\beta\,\boldsymbol{n} \tag{2.116}$$

and simultaneously

$$\boldsymbol{n}_{,\beta\alpha} = -(b_{\beta\gamma}\,\boldsymbol{a}^\gamma)_{,\alpha} = -(b_{\beta\gamma,\alpha} - b_{\beta\lambda}\Gamma^\lambda_{\gamma\alpha})\boldsymbol{a}^\gamma - b_{\beta\gamma}\,b^\gamma_\alpha\,\boldsymbol{n}\,. \tag{2.117}$$

If we compare the relations (2.116) and (2.117) and equate the coefficients of $\boldsymbol{n}$ we obtain the identity

$$b_{\alpha\gamma}\,b^\gamma_\beta = b_{\beta\gamma}\,b^\gamma_\alpha\,, \tag{2.118}$$

which expresses the fact that the tensor $\boldsymbol{b}^2$ is symmetric and represents no additional restriction. Further, if we equate the coefficients of $\boldsymbol{a}^\gamma$ we derive

$$b_{\alpha\gamma,\beta} - b_{\alpha\lambda}\Gamma^{\lambda}_{\gamma\beta} = b_{\beta\gamma,\alpha} - b_{\beta\lambda}\Gamma^{\lambda}_{\gamma\alpha}$$

and adding $-b_{\lambda\gamma}\Gamma^{\lambda}_{\alpha\beta}$ to both sides we obtain the relations

$$b_{\alpha\gamma|\beta} = b_{\beta\gamma|\alpha}\,, \tag{2.119}$$

which are called the *Mainardi-Codazzi equations*. Note that only two of these compatibility conditions are non-trivial, namely

$$b_{12|1} = b_{11|2} \quad \text{and} \quad b_{21|2} = b_{22|1}\,. \tag{2.120}$$

If we multiply the Mainardi-Codazzi equations (2.119) with $a^{\gamma\delta}$ we can raise the subscript γ and deduce the alternative form

$$b^{\delta}_{\alpha|\beta} = b^{\delta}_{\beta|\alpha}\,. \tag{2.121}$$

From the Gauss and Weingarten equations (2.115) we also have

$$\begin{aligned} \boldsymbol{a}_{\alpha,\beta\gamma} &= \left(\Gamma^{\lambda}_{\alpha\beta}\,\boldsymbol{a}_{\lambda} + b_{\alpha\beta}\,\boldsymbol{n}\right)_{,\gamma} \\ &= \left(\Gamma^{\lambda}_{\alpha\beta,\gamma} + \Gamma^{\lambda}_{\mu\gamma}\,\Gamma^{\mu}_{\alpha\beta}\right)\boldsymbol{a}_{\lambda} + \left(b_{\alpha\beta,\gamma} + \Gamma^{\lambda}_{\alpha\beta}b_{\lambda\gamma}\right)\boldsymbol{n} - b_{\alpha\beta}\,b^{\lambda}_{\gamma}\,\boldsymbol{a}_{\lambda} \end{aligned} \tag{2.122}$$

and simultaneously

$$\begin{aligned} \boldsymbol{a}_{\alpha,\gamma\beta} &= \left(\Gamma^{\lambda}_{\alpha\gamma}\,\boldsymbol{a}_{\lambda} + b_{\alpha\gamma}\,\boldsymbol{n}\right)_{,\beta} \\ &= \left(\Gamma^{\lambda}_{\alpha\gamma,\beta} + \Gamma^{\lambda}_{\mu\beta}\,\Gamma^{\mu}_{\alpha\gamma}\right)\boldsymbol{a}_{\lambda} + \left(b_{\alpha\gamma,\beta} + \Gamma^{\lambda}_{\alpha\gamma}b_{\lambda\beta}\right)\boldsymbol{n} - b_{\alpha\gamma}\,b^{\lambda}_{\beta}\,\boldsymbol{a}_{\lambda}\,. \end{aligned} \tag{2.123}$$

Let us compare the relations (2.122) and (2.123). If we equate the coefficients of $\boldsymbol{n}$, then we obtain again the Mainardi-Codazzi equations (2.119). By equating the coefficients of $\boldsymbol{a}_{\lambda}$ we arrive at

$$R^{\lambda}_{\cdot\alpha\beta\gamma} = b_{\alpha\gamma}\,b^{\lambda}_{\beta} - b_{\alpha\beta}\,b^{\lambda}_{\gamma}\,, \tag{2.124}$$

where

$$R^{\lambda}_{\cdot\alpha\beta\gamma} = \Gamma^{\lambda}_{\alpha\gamma,\beta} - \Gamma^{\lambda}_{\alpha\beta,\gamma} + \Gamma^{\lambda}_{\mu\beta}\,\Gamma^{\mu}_{\alpha\gamma} - \Gamma^{\lambda}_{\mu\gamma}\,\Gamma^{\mu}_{\alpha\beta} \tag{2.125}$$

is the so-called *Riemann tensor*. (Note that the counterpart of the condition (2.124) in the three-dimensional Euclidean space reads $R^{i}_{\cdot jkl} = 0$, where $R^{i}_{\cdot jkl}$ is the corresponding Riemann tensor, see e.g. [1, Sect. 1.5] or [5, Sect. 9.7].)

We can lower the superscript λ in (2.125) and define the covariant components of the Riemann tensor through

$$R_{\delta\alpha\beta\gamma} = a_{\delta\lambda}\,R^{\lambda}_{\cdot\alpha\beta\gamma}\,. \tag{2.126}$$

By a straightforward calculation we find that

$$R_{2112} = -R_{1212}\,, \qquad R_{2121} = R_{1212}\,, \qquad R_{1221} = -R_{1212}\,, \tag{2.127}$$

and all other covariant components are zero. Thus, R_{1212} is the only independent component of the Riemann tensor.

In view of (2.126), the compatibility conditions (2.124) can be written in the equivalent form

$$R_{\delta\alpha\beta\gamma} = b_{\alpha\gamma}\, b_{\beta\delta} - b_{\alpha\beta}\, b_{\gamma\delta}\,. \tag{2.128}$$

In order to rewrite the Eq. (2.128) in a convenient form, we use the relations

$$\varepsilon_{\alpha\beta}\,\varepsilon_{\gamma\delta} = a_{\alpha\gamma}\, a_{\beta\delta} - a_{\alpha\delta}\, a_{\beta\gamma} \qquad \text{and} \qquad K\,\varepsilon_{\delta\alpha} = b^{\mu}_{\delta}\, b^{\lambda}_{\alpha}\,\varepsilon_{\mu\lambda}\,,$$

see Exercise 2.9. Thus, (2.128) becomes

$$R_{\delta\alpha\beta\gamma} = b^{\lambda}_{\alpha}\, b^{\mu}_{\delta}\,(a_{\beta\mu}\, a_{\gamma\lambda} - a_{\beta\lambda}\, a_{\gamma\mu}) = b^{\lambda}_{\alpha}\, b^{\mu}_{\delta}\,\varepsilon_{\beta\gamma}\,\varepsilon_{\mu\lambda} = \varepsilon_{\beta\gamma}\,(b^{\lambda}_{\alpha}\, b^{\mu}_{\delta}\,\varepsilon_{\mu\lambda}) = \varepsilon_{\beta\gamma}\,(K\,\varepsilon_{\delta\alpha}),$$

i.e. we obtain the renowned formula

$$R_{\alpha\beta\gamma\delta} = K\,\varepsilon_{\alpha\beta}\,\varepsilon_{\gamma\delta}\,, \tag{2.129}$$

also called the *Theorema Egregium* due to Gauss, see e.g. [1, Sect. 2.7] or [5, Sect. 6.11]. Thus, Eq. (2.129) yields

$$R_{1212} = K\big(\sqrt{a}\,\big)^{2} e_{12}\, e_{12} = K\,a\,, \qquad \text{or} \qquad K = \frac{1}{a}\,R_{1212}\,. \tag{2.130}$$

From the last relation we can deduce the surprising result that the Gaussian curvature K is determined entirely by the metric tensor $a_{\alpha\beta}(\theta^1, \theta^2)$.

Remark As a consequence of this result, for shells we see that bending deformations that induce change in K must be accompanied by change in $a_{\alpha\beta}$, i.e. by *in-plane strain*. Hence, a given curvature confers a high degree of stiffness to the shell. On the contrary, in the case of a space curve, we can change its curvature without changing its arc length. Therefore, *rods* are relatively flexible. □

In conclusion, the compatibility conditions are the Mainardi-Codazzi equations (2.119) and the Theorema Egregium (2.129). We have shown the *necessity* of these conditions for the existence of surfaces $\boldsymbol{r}(\theta^\alpha)$ of class C^3. Sufficiency of the compatibility conditions for local existence and uniqueness of $\boldsymbol{r}$, appart from orientation ($a_{\alpha\beta}$ and $b_{\alpha\beta}$ are invariant under superposed rigid body motions) may also be shown, see e.g. [1, p. 80] or [5, p. 146].

2.6 Green-Stokes Formula

Let $\mathscr{S}$ be an oriented surface with parametrization $\boldsymbol{r}(\theta^\alpha)$, such that the unit normal vector $\boldsymbol{n}(\theta^\alpha)$ is piecewise continuous on $\mathscr{S}$, see Fig. 2.11. We assume that $\mathscr{S}$ is simply-connected and designate by $\partial\mathscr{S}$ its boundary curve. We consider a domain $\mathscr{R} \supset \mathscr{S}$ in the Euclidean 3-space, which include the surface $\mathscr{S}$, and denote with $\boldsymbol{x}$ the position vector of a generic point in $\mathscr{R}$. Let $\boldsymbol{v}(\boldsymbol{x})$ be a smooth vector field in $\mathscr{R}$. Then, the following relation, called the *Green-Stokes formula*, holds (see, e.g. [6, Chap. 12])

$$\int_{\partial\mathscr{S}} \boldsymbol{v} \cdot \mathrm{d}\boldsymbol{r} = \int_{\mathscr{S}} (\mathrm{curl}\, \boldsymbol{v}) \cdot \boldsymbol{n}\, \mathrm{d}a\,. \tag{2.131}$$

Here, the surface integral is defined as

$$\int_{\mathscr{S}} f(\theta^\alpha)\, \mathrm{d}a = \sum_{k=1}^{m} \iint_{\mathscr{S}\cap\mathscr{S}_k} f(\theta^\alpha)\, \sqrt{a}\, \mathrm{d}\theta^1 \mathrm{d}\theta^2, \qquad \text{for scalar functions } f,$$

where $\mathscr{S}_k$ are non-overlapping coordinate patches, which together comprise $\mathscr{S}$. The line integral over the boundary curve in (2.131) must have positive orientation, meaning that the parametrization along $\partial\mathscr{S}$ runs counterclockwise when the surface normal $\boldsymbol{n}$ points toward the viewer.

Let s be the arclength parameter along $\partial\mathscr{S}$ and $\boldsymbol{t} = t^\alpha \boldsymbol{a}_\alpha$ be the unit tangent vector to the boundary curve such that

$$\mathrm{d}\boldsymbol{r} = \boldsymbol{t}\, \mathrm{d}s = \boldsymbol{n} \times \boldsymbol{\nu}\, \mathrm{d}s = \boldsymbol{n} \times \nu_\alpha \boldsymbol{a}^\alpha\, \mathrm{d}s = \varepsilon^{\alpha\beta} \nu_\alpha \boldsymbol{a}_\beta\, \mathrm{d}s\,, \tag{2.132}$$

where $\boldsymbol{\nu} = \nu_\alpha \boldsymbol{a}^\alpha = \boldsymbol{t} \times \boldsymbol{n}$ is the normal vector to the boundary curve, lying in the tangent plane T_P. So, it follows that

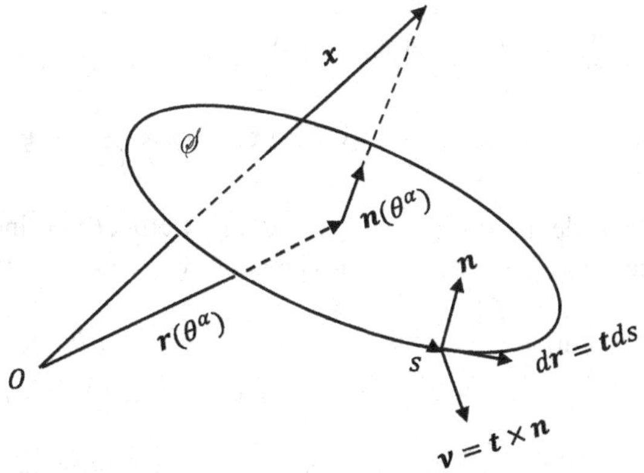

Fig. 2.11 The oriented surface $\mathscr{S}$, for which the Green-Stokes formula holds

$$t^\beta = \varepsilon^{\alpha\beta}\, \nu_\alpha\,. \tag{2.133}$$

Let us put the Green-Stokes formula (2.131) in another useful form. In view of (2.132), we have for any vector field $\boldsymbol{v} = v_\alpha\, \boldsymbol{a}^\alpha + v\, \boldsymbol{n}$ that

$$\boldsymbol{v}\cdot \mathrm{d}\boldsymbol{r} = \varepsilon^{\alpha\beta}\, \nu_\alpha (\boldsymbol{v}\cdot \boldsymbol{a}_\beta)\, \mathrm{d}s = \varepsilon^{\alpha\beta}\, \nu_\alpha\, v_\beta\, \mathrm{d}s\,. \tag{2.134}$$

Locally (in the neighborhood of the surface), we can represent the position vector $\boldsymbol{x} = x^i\, \boldsymbol{e}_i$ in the form

$$\boldsymbol{x}(\theta^1,\theta^2,\theta^3) = \boldsymbol{r}(\theta^1,\theta^2) + \theta^3\, \boldsymbol{n}(\theta^1,\theta^2). \tag{2.135}$$

Then, we can use $\boldsymbol{g}_i = \boldsymbol{x}_{,i}$ and (2.69) to get

$$\boldsymbol{g}_\alpha = \boldsymbol{a}_\alpha + \theta^3 \boldsymbol{n}_{,\alpha} = \mu^\beta_\alpha\, \boldsymbol{a}_\beta\,, \qquad \boldsymbol{g}_3 = \boldsymbol{n}\,, \qquad \text{where } \mu^\beta_\alpha = \delta^\beta_\alpha - \theta^3\, b^\beta_\alpha\,. \tag{2.136}$$

Also, for the dual basis vectors $\boldsymbol{g}^i$ we obtain

$$\boldsymbol{g}^\alpha = \frac{1}{\mu}\big[\boldsymbol{a}^\alpha + \theta^3 (b^\alpha_\beta\, \boldsymbol{a}^\beta - 2H\, \boldsymbol{a}^\alpha)\big], \qquad \boldsymbol{g}^3 = \boldsymbol{n}\,, \tag{2.137}$$

where

$$\mu = \det\left(\mu^\alpha_\beta\right) = 1 - 2H\theta^3 + K(\theta^3)^2 = (1-\theta^3\kappa_1)(1-\theta^3\kappa_2) > 0 \tag{2.138}$$

in a neighborhood of the surface for which $|\theta^3| < |\kappa_\alpha|^{-1}$. Indeed, in order to prove (2.137) we can verify that $\boldsymbol{g}^i\cdot \boldsymbol{g}_j = \delta^i_j$ using the relation (2.109). Thus, from (2.136) and (2.137) we see that *on* the surface $\mathscr{S}$ (i.e. for $\theta^3 = 0$) we have

$$\boldsymbol{g}_\alpha\big|_{\mathscr{S}} = \boldsymbol{a}_\alpha \qquad \text{and} \qquad \boldsymbol{g}^\alpha\big|_{\mathscr{S}} = \boldsymbol{a}^\alpha\,. \tag{2.139}$$

The curl operator is given by

$$\operatorname{curl} \boldsymbol{v}(\boldsymbol{x}) = \boldsymbol{e}_i \times \frac{\partial \boldsymbol{v}}{\partial x^i} = \boldsymbol{g}^i \times \boldsymbol{v}_{,i}\,. \tag{2.140}$$

Consider now that the vector field $\boldsymbol{v} = \boldsymbol{v}(\theta^1,\theta^2)$ is independent of θ^3. Then, we have $\operatorname{curl} \boldsymbol{v} = \boldsymbol{g}^\alpha \times \boldsymbol{v}_{,\alpha}$ and, in view of (2.82) and (2.139), we obtain *on* the surface

$$\begin{aligned}\operatorname{curl} \boldsymbol{v}\big|_{\mathscr{S}} &= \boldsymbol{a}^\alpha \times \boldsymbol{v}_{,\alpha} = \boldsymbol{a}^\alpha \times \big[\big(v_{\beta|\alpha} - v\, b_{\beta\alpha}\big)\boldsymbol{a}^\beta + \big(v_{,\alpha} + v_\beta b^\beta_\alpha\big)\boldsymbol{n}\big] \\ &= v_{\beta|\alpha}\, \boldsymbol{a}^\alpha \times \boldsymbol{a}^\beta + \big(v_{,\alpha} + v_\beta b^\beta_\alpha\big)\boldsymbol{a}^\alpha \times \boldsymbol{n}\,,\end{aligned} \tag{2.141}$$

since $b_{\beta\alpha}\, \boldsymbol{a}^\alpha \times \boldsymbol{a}^\beta = b_{\alpha\beta}\, \varepsilon^{\alpha\beta} \boldsymbol{n} = \boldsymbol{0}$. From (2.60) and (2.141) it follows that

$$\boldsymbol{n} \cdot \operatorname{curl} \boldsymbol{v}_{|\mathscr{S}} = v_{\beta|\alpha}\, \boldsymbol{n} \cdot \left(\boldsymbol{a}^{\alpha} \times \boldsymbol{a}^{\beta} \right) = \varepsilon^{\alpha\beta}\, v_{\beta|\alpha} \,. \tag{2.142}$$

We combine the relations (2.134) and (2.142) in the Green-Stokes formula (2.131) and deduce

$$\int_{\partial\mathscr{S}} \varepsilon^{\alpha\beta} v_{\beta}\, \nu_{\alpha}\, \mathrm{d}s = \int_{\mathscr{S}} \varepsilon^{\alpha\beta}\, v_{\beta|\alpha}\, \mathrm{d}a \,.$$

If we use the identity ${\varepsilon^{\alpha\beta}}_{|\gamma} = 0$ (see Exercise 2.9) and the notation $u^{\alpha} = \varepsilon^{\alpha\beta}\, v_{\beta}$ we can write this in the form

$$\int_{\partial\mathscr{S}} u^{\alpha} \nu_{\alpha}\, \mathrm{d}s = \int_{\mathscr{S}} {u^{\alpha}}_{|\alpha}\, \mathrm{d}a \,, \tag{2.143}$$

where u^{α} are arbitrary smooth functions of (θ^1, θ^2). Note that (see (2.90))

$${u^{\alpha}}_{|\alpha} = \frac{1}{\sqrt{a}} \left(\sqrt{a}\, u^{\alpha} \right)_{,\alpha} \,. \tag{2.144}$$

Next, let us consider the vector field $\boldsymbol{u}(\theta^1, \theta^2) = u^{\alpha} \boldsymbol{a}_{\alpha} + u\, \boldsymbol{n}$ where $u = u(\theta^1, \theta^2)$ is a scalar smooth function. Then, the formula (2.143) can be written as

$$\int_{\partial\mathscr{S}} \boldsymbol{u} \cdot \boldsymbol{\nu}\, \mathrm{d}s = \int_{\mathscr{S}} \left(\boldsymbol{u} \cdot \boldsymbol{a}^{\alpha} \right)_{|\alpha} \mathrm{d}a \,. \tag{2.145}$$

Consider now three vectors $\boldsymbol{u}^i(\theta^1, \theta^2) = u^{i\alpha} \boldsymbol{a}_{\alpha} + u^i \boldsymbol{n}$ with $i = 1, 2, 3$. For each fixed i we have, according to (2.143) and (2.144),

$$\int_{\partial\mathscr{S}} u^{i\alpha} \nu_{\alpha}\, \mathrm{d}s = \int_{\mathscr{S}} {u^{i\alpha}}_{|\alpha}\, \mathrm{d}a \quad \text{and} \quad {u^{i\alpha}}_{|\alpha} = \frac{1}{\sqrt{a}} \left(\sqrt{a}\, u^{i\alpha} \right)_{,\alpha} \,. \tag{2.146}$$

Since the basis vectors $\boldsymbol{e}_i$ are constant, we can then write

$$\int_{\partial\mathscr{S}} u^{i\alpha} \nu_{\alpha}\, \boldsymbol{e}_i\, \mathrm{d}s = \int_{\mathscr{S}} {u^{i\alpha}}_{|\alpha}\, \boldsymbol{e}_i\, \mathrm{d}a \quad \text{and} \quad {u^{i\alpha}}_{|\alpha}\, \boldsymbol{e}_i = \frac{1}{\sqrt{a}} \left(\sqrt{a}\, u^{i\alpha} \boldsymbol{e}_i \right)_{,\alpha} \,.$$

Thus,

$$\int_{\partial\mathscr{S}} (u^{i\alpha} \boldsymbol{e}_i) \nu_{\alpha}\, \mathrm{d}s = \int_{\mathscr{S}} \left(u^{i\alpha} \boldsymbol{e}_i \right)_{|\alpha} \mathrm{d}a \,, \tag{2.147}$$

where we have denoted $\left(u^{i\alpha} \boldsymbol{e}_i \right)_{|\alpha} = {u^{i\alpha}}_{|\alpha}\, \boldsymbol{e}_i = \frac{1}{\sqrt{a}} \left(\sqrt{a}\, u^{i\alpha} \boldsymbol{e}_i \right)_{,\alpha}$, in view of (2.144).

If we define the two vector fields $\boldsymbol{U}^{\alpha}(\theta^1, \theta^2) = u^{i\alpha}\, \boldsymbol{e}_i = (\boldsymbol{u}^i \cdot \boldsymbol{a}^{\alpha})\, \boldsymbol{e}_i$, with $\alpha = 1, 2$, then the relation (2.147) yields the formula

$$\int_{\partial \mathscr{S}} U^{\alpha} \nu_{\alpha} \, ds = \int_{\mathscr{S}} U^{\alpha}{}_{|\alpha} \, da \,, \tag{2.148}$$

for *any* smooth 3-vectors $\boldsymbol{U}^1, \ \boldsymbol{U}^2$.

2.7 Surface Differential Operators

Using surface differential operators we can write some of the equations and formulas presented in the previous sections in a convenient condensed form.

In analogy to (1.34) and (1.37), the *surface gradient operator* grad_s is given by

$$\text{grad}_s \, f = f_{,\alpha} \, \boldsymbol{a}^{\alpha} \qquad \text{and} \qquad \text{grad}_s \, \boldsymbol{v} = \boldsymbol{v}_{,\alpha} \otimes \boldsymbol{a}^{\alpha} \,, \tag{2.149}$$

for any smooth scalar function $f(\theta^1, \theta^2)$ and any smooth vector field $\boldsymbol{v}(\theta^1, \theta^2)$. In view of (2.80) and (2.82) we obtain the following expressions in terms of the components of the vector $\boldsymbol{v} = v^{\alpha} \boldsymbol{a}_{\alpha} + v \, \boldsymbol{n} = v_{\alpha} \boldsymbol{a}^{\alpha} + v \, \boldsymbol{n}$:

$$\begin{aligned} \text{grad}_s \, \boldsymbol{v} &= \big(v^{\alpha}{}_{|\beta} - b^{\alpha}_{\beta} \, v \big) \boldsymbol{a}_{\alpha} \otimes \boldsymbol{a}^{\beta} + \big(v_{,\beta} + b_{\alpha\beta} \, v^{\alpha} \big) \, \boldsymbol{n} \otimes \boldsymbol{a}^{\beta} \\ &= \big(v_{\alpha|\beta} - b_{\alpha\beta} \, v \big) \boldsymbol{a}^{\alpha} \otimes \boldsymbol{a}^{\beta} + \big(v_{,\beta} + b^{\alpha}_{\beta} \, v_{\alpha} \big) \, \boldsymbol{n} \otimes \boldsymbol{a}^{\beta} \, . \end{aligned} \tag{2.150}$$

Note that $\boldsymbol{a}^{\alpha} = \text{grad}_s \, \theta^{\alpha}$. Moreover, since $\boldsymbol{a}_{\alpha} = \boldsymbol{r}_{,\alpha}$, the unit tensor $\mathbf{1}$ in the tangent plane can be written as (see (2.51) and (2.66))

$$\mathbf{1} = \boldsymbol{a}_{\alpha} \otimes \boldsymbol{a}^{\alpha} = \text{grad}_s \, \boldsymbol{r} \,, \tag{2.151}$$

whereas the curvature tensor $\boldsymbol{b}$ has the expression (see (2.69) and (2.71))

$$\boldsymbol{b} = -\boldsymbol{n}_{,\alpha} \otimes \boldsymbol{a}^{\alpha} = -\text{grad}_s \, \boldsymbol{n} \, . \tag{2.152}$$

Further, in analogy to (1.46) and (1.49), the *surface divergence* div_s is given by

$$\text{div}_s \, \boldsymbol{v} = \text{tr}(\text{grad}_s \, \boldsymbol{v}) = \boldsymbol{v}_{,\alpha} \cdot \boldsymbol{a}^{\alpha} \qquad \text{and} \qquad \text{div}_s \, \boldsymbol{A} = (\boldsymbol{A}_{,\alpha}) \boldsymbol{a}^{\alpha} \,, \tag{2.153}$$

for any smooth vector field $\boldsymbol{v}(\theta^1, \theta^2)$ and tensor field $\boldsymbol{A}(\theta^1, \theta^2)$. Then, from (2.80) and (2.90) we obtain

$$\text{div}_s \, \boldsymbol{v} \;=\; v^{\alpha}{}_{|\alpha} - b^{\alpha}_{\alpha} \, v \;=\; \frac{1}{\sqrt{a}} \big(\sqrt{a} \, v^{\alpha} \big)_{,\alpha} - 2H \, v \,, \tag{2.154}$$

since $2H = \text{tr} \, \boldsymbol{b} = b^{\alpha}_{\alpha}$. So, using (2.154) the relation (2.145) can be written in the equivalent form

$$\int_{\partial\mathscr{S}} \boldsymbol{u}\cdot\boldsymbol{\nu}\,\mathrm{d}s = \int_{\mathscr{S}} \big(\mathrm{div}_s\,\boldsymbol{u} + 2H\,\boldsymbol{u}\cdot\boldsymbol{n}\big)\,\mathrm{d}a\,, \tag{2.155}$$

which also called the *divergence formula* on surfaces.

For any smooth tensor field $\boldsymbol{A}(\theta^1,\theta^2)$ of the form (2.65) we have the following relation (which generalizes Eq. (2.86))

$$\begin{aligned}\boldsymbol{A}_{,\gamma} = {} & (A^{\alpha\beta}{}_{|\gamma} - b^{\beta}_{\gamma}\,A^{\alpha 3} - b^{\alpha}_{\gamma}\,A^{3\beta})\boldsymbol{a}_\alpha\otimes\boldsymbol{a}_\beta + (A^{\alpha 3}{}_{|\gamma} + b_{\beta\gamma}\,A^{\alpha\beta} - b^{\alpha}_{\gamma}\,A^{33})\boldsymbol{a}_\alpha\otimes\boldsymbol{n}\\ & +(A^{3\beta}{}_{|\gamma} + b_{\alpha\gamma}\,A^{\alpha\beta} - b^{\beta}_{\gamma}\,A^{33})\boldsymbol{n}\otimes\boldsymbol{a}_\beta + (A^{33}{}_{,\gamma} + b_{\alpha\gamma}\,A^{\alpha 3} + b_{\beta\gamma}\,A^{3\beta})\boldsymbol{n}\otimes\boldsymbol{n}\,,\end{aligned}$$

where

$$A^{\alpha 3}{}_{|\gamma} = A^{\alpha 3}{}_{,\gamma} + A^{\lambda 3}\,\Gamma^{\alpha}_{\gamma\lambda} \qquad \text{and} \qquad A^{3\beta}{}_{|\gamma} = A^{3\beta}{}_{,\gamma} + A^{3\lambda}\,\Gamma^{\beta}_{\lambda\gamma}\,.$$

Then, from $(2.153)_2$ we get

$$\mathrm{div}_s\,\boldsymbol{A} = (A^{\alpha\beta}{}_{|\beta} - b^{\beta}_{\beta}\,A^{\alpha 3} - b^{\alpha}_{\beta}\,A^{3\beta})\boldsymbol{a}_\alpha + (A^{3\beta}{}_{|\beta} + b_{\alpha\beta}\,A^{\alpha\beta} - b^{\beta}_{\beta}\,A^{33})\boldsymbol{n}\,. \tag{2.156}$$

Hence, in view of (2.76) we have $\boldsymbol{a}_{\alpha|\beta} = \boldsymbol{a}_{\alpha,\beta} - \Gamma^{\gamma}_{\alpha\beta}\,\boldsymbol{a}_\gamma = b_{\alpha\beta}\boldsymbol{n}$ and we find

$$\begin{aligned}\mathrm{div}_s\,\boldsymbol{A} &= \big(A^{\alpha\beta}\boldsymbol{a}_\alpha\big)_{|\beta} + \big(A^{3\beta}\boldsymbol{n}\big)_{|\beta} - b^{\beta}_{\beta}\,A^{\alpha 3}\boldsymbol{a}_\alpha - b^{\beta}_{\beta}\,A^{33}\boldsymbol{n}\\ &= \big(\boldsymbol{A}\boldsymbol{a}^{\beta}\big)_{|\beta} - b^{\alpha}_{\alpha}\,(A^{\alpha 3}\boldsymbol{a}_\alpha + A^{33}\boldsymbol{n}) \;=\; \big(\boldsymbol{A}\boldsymbol{a}^{\alpha}\big)_{|\alpha} - 2H\,\boldsymbol{A}\boldsymbol{n}\,,\end{aligned} \tag{2.157}$$

so we have

$$\big(\boldsymbol{A}\boldsymbol{a}^{\alpha}\big)_{|\alpha} = \mathrm{div}_s\,\boldsymbol{A} + 2H\,\boldsymbol{A}\boldsymbol{n}\,. \tag{2.158}$$

Denoting by $\boldsymbol{U}^\alpha = \boldsymbol{A}\boldsymbol{a}^\alpha$, we see that $\boldsymbol{U}^\alpha\,\nu_\alpha = \boldsymbol{A}(\nu_\alpha\,\boldsymbol{a}^\alpha) = \boldsymbol{A}\boldsymbol{\nu}$ and the relation (2.148) can be written with the help of (2.158) in the form

$$\int_{\partial\mathscr{S}} \boldsymbol{A}\boldsymbol{\nu}\,\mathrm{d}s = \int_{\mathscr{S}} \big(\mathrm{div}_s\,\boldsymbol{A} + 2H\,\boldsymbol{A}\boldsymbol{n}\big)\,\mathrm{d}a\,, \tag{2.159}$$

which is also called the *divergence theorem for surfaces* and holds for any smooth tensor field $\boldsymbol{A}(\theta^1,\theta^2)$ of the general form (2.65).

The *surface curl* is given for any smooth 3-vector field $\boldsymbol{v}(\theta^1,\theta^2)$ by

$$\mathrm{curl}_s\,\boldsymbol{v} = \boldsymbol{a}^\alpha \times \boldsymbol{v}_{,\alpha}\,. \tag{2.160}$$

Then, substituting (2.82) here and using (2.60) and (2.62) we get the following expression in terms of vector components $v_\alpha = \boldsymbol{v}\cdot\boldsymbol{a}_\alpha$ and $v = \boldsymbol{v}\cdot\boldsymbol{n}$:

$$\mathrm{curl}_s\,\boldsymbol{v} = \varepsilon^{\alpha\beta}\,v_{\beta|\alpha}\,\boldsymbol{n} + \varepsilon^{\alpha\beta}\big(v_{,\beta} + b^{\gamma}_{\beta}\,v_\gamma\big)\boldsymbol{a}_\alpha\,, \tag{2.161}$$

since $\varepsilon^{\alpha\beta} b_{\alpha\beta} = 0$. In view of the first equation in (2.141), the Green-Stokes formula (2.131) can be written in the following manner

$$\int_{\partial\mathscr{S}} \boldsymbol{v} \cdot \mathrm{d}\boldsymbol{r} = \int_{\mathscr{S}} (\mathrm{curl}_s\, \boldsymbol{v}) \cdot \boldsymbol{n}\, \mathrm{d}a\,, \tag{2.162}$$

which holds for any smooth 3-vector field $\boldsymbol{v}(\theta^1, \theta^2)$ given *on the surface* $\mathscr{S}$.

Finally, we can express the geodesic curvature of a curve on $\mathscr{S}$ with the help of the operators curl_s and div_s. Let $\mathscr{C}$ be a curve on the surface with geodesic curvature κ_g, let $\boldsymbol{\lambda}$ be the unit tangent field and $\boldsymbol{\mu} = \boldsymbol{n} \times \boldsymbol{\lambda}$, see Sect. 2.4.1. Then, one can establish the relations

$$\kappa_g = (\mathrm{curl}_s\, \boldsymbol{\lambda}) \cdot \boldsymbol{n} \qquad \text{and} \qquad -\kappa_g = \mathrm{div}_s\, \boldsymbol{\mu}\,, \tag{2.163}$$

i.e. the geodesic curvature κ_g equals the normal component of the surface curl of λ, where as $-\kappa_g$ is equal to the surface divergence of $\boldsymbol{\mu}$. To prove (2.163), let us remark that

$$\boldsymbol{a}^\alpha = (\boldsymbol{a}^\alpha \cdot \boldsymbol{\lambda})\boldsymbol{\lambda} + (\boldsymbol{a}^\alpha \cdot \boldsymbol{\mu})\boldsymbol{\mu} = \lambda^\alpha \boldsymbol{\lambda} + \mu^\alpha \boldsymbol{\mu}$$

and, hence,

$$\mathrm{curl}_s\, \boldsymbol{\lambda} = \boldsymbol{a}^\alpha \times \boldsymbol{\lambda}_{,\alpha} = (\lambda^\alpha \boldsymbol{\lambda} + \mu^\alpha \boldsymbol{\mu}) \times \boldsymbol{\lambda}_{,\alpha}\,. \tag{2.164}$$

For the curvature vector $\boldsymbol{\kappa}$ we have by definition

$$\boldsymbol{\kappa} = \frac{\mathrm{d}\boldsymbol{\lambda}}{\mathrm{d}u} = \boldsymbol{\lambda}_{,\alpha} \frac{\mathrm{d}\theta^\alpha}{\mathrm{d}u} = \boldsymbol{\lambda}_{,\alpha} \lambda^\alpha\,,$$

in view of (2.96). Then, for the geodesic curvature we obtain

$$\kappa_g = \boldsymbol{\mu} \cdot \boldsymbol{\kappa} = (\boldsymbol{n} \times \boldsymbol{\lambda}) \cdot (\lambda^\alpha \boldsymbol{\lambda}_{,\alpha}) = (\lambda^\alpha \boldsymbol{\lambda} \times \boldsymbol{\lambda}_{,\alpha}) \cdot \boldsymbol{n}\,. \tag{2.165}$$

Taking into account (2.164) and (2.165), the relation $\kappa_g = (\mathrm{curl}_s\, \boldsymbol{\lambda}) \cdot \boldsymbol{n}$ is equivalent to

$$(\lambda^\alpha \boldsymbol{\lambda} \times \boldsymbol{\lambda}_{,\alpha}) \cdot \boldsymbol{n} = \big[(\lambda^\alpha \boldsymbol{\lambda} + \mu^\alpha \boldsymbol{\mu}) \times \boldsymbol{\lambda}_{,\alpha}\big] \cdot \boldsymbol{n}\,,$$

or

$$(\mu^\alpha \boldsymbol{\mu} \times \boldsymbol{\lambda}_{,\alpha}) \cdot \boldsymbol{n} = 0\,. \tag{2.166}$$

The last relation (2.166) holds indeed, since

$$(\boldsymbol{\mu} \times \boldsymbol{\lambda}_{,\alpha}) \cdot \boldsymbol{n} = (\boldsymbol{n} \times \boldsymbol{\mu}) \cdot \boldsymbol{\lambda}_{,\alpha} = -\boldsymbol{\lambda} \cdot \boldsymbol{\lambda}_{,\alpha} = 0.$$

Thus, the first equation in (2.163) is proved. To verify the second equation in (2.163), we write directly

$$-\mathrm{div}_s\,\boldsymbol{\mu} = -\boldsymbol{\mu}_{,\alpha}\cdot\boldsymbol{a}^\alpha = (\boldsymbol{\lambda}\times\boldsymbol{n})_{,\alpha}\cdot\boldsymbol{a}^\alpha = (\boldsymbol{\lambda}_{,\alpha}\times\boldsymbol{n})\cdot\boldsymbol{a}^\alpha + (\boldsymbol{\lambda}\times\boldsymbol{n}_{,\alpha})\cdot\boldsymbol{a}^\alpha$$
$$= (\boldsymbol{a}^\alpha\times\boldsymbol{\lambda}_{,\alpha})\cdot\boldsymbol{n} = (\mathrm{curl}_s\,\boldsymbol{\lambda})\cdot\boldsymbol{n}\,,$$

since $\boldsymbol{\lambda}\times\boldsymbol{n}_{,\alpha}$ is orthogonal to the tangent plane. The relations (2.163) are proved.

2.8 Exercises

2.1 Consider the torsional deformation $\boldsymbol{x}\to\boldsymbol{y}$ of a right circular cylinder, where

$$\boldsymbol{x} = r\boldsymbol{e}_r(\theta) + z\boldsymbol{e}_3, \qquad \boldsymbol{y} = r\boldsymbol{e}_r(\theta+\tau z) + z\boldsymbol{e}_3$$

Here, $\boldsymbol{e}_r(\theta) = \cos\theta\,\boldsymbol{e}_1 + \sin\theta\,\boldsymbol{e}_2$ and τ is the constant twist per unit axial length. Set $\{\theta^i\} = \{r,\theta,z\}$ and obtain $\{\boldsymbol{G}_i\}$, $\{\boldsymbol{g}_i\}$, the associated metric components, and $\boldsymbol{F}$. What are the physical components of $\boldsymbol{C} = \boldsymbol{F}^T\boldsymbol{F}$ with respect to $\{\boldsymbol{e}_r, \boldsymbol{e}_\theta, \boldsymbol{e}_3\}$? Sketch the curve defined by $r = r_0$, $\theta = \theta_0$ (constants) in the deformed body.

2.2 Let $\boldsymbol{P}$ be a tensor field and define $\boldsymbol{P}^i = \boldsymbol{P}\boldsymbol{G}^i$, where $\boldsymbol{G}^i$ are the dual vectors induced by coordinates $\{\theta^i\}$ in a reference configuration. Show that the differential operator $G^{-1/2}\left(G^{1/2}\boldsymbol{P}^i\right)_{,i}$, with $G = \det\left(G_{ij}\right)$, is invariant under coordinate transformation $\theta^i \to \bar{\theta}^i$.

2.3 The parametric equation of a plane with unit normal $\boldsymbol{n}$ is given by $\boldsymbol{n}\cdot\boldsymbol{r}(\theta^\alpha) = c$, a constant (the perpendicular distance from the origin to the plane). Show that a surface is a plane *if and only if* $b_{\alpha\beta} = 0$.

2.4 Consider the hyperbolic paraboloid described (in Cartesians) by $z(x,y) = xy/c$ with c a non-zero constant. Let $\{\theta^\alpha\} = \{x,y\}$ and find $a_{\alpha\beta}$, $b_{\alpha\beta}$, $\Gamma^\alpha_{\beta\gamma}$ in terms of x, y.

2.5 Consider a cone with vertex angle γ and parametric representation $\boldsymbol{r}(\theta^\alpha) = \theta^1\left[\sin\gamma\left(\sin\theta^2\boldsymbol{e}_1 + \cos\theta^2\boldsymbol{e}_2\right) + \cos\gamma\,\boldsymbol{e}_3\right]$. Sketch the cone and the curve of constant θ^1, θ^2. Find the functions $a_{\alpha\beta}$, $b_{\alpha\beta}$, $\Gamma^\alpha_{\beta\gamma}$.

2.6 The equation of a paraboloid of revolution is $\boldsymbol{r} = r\boldsymbol{e}_r(\theta) + z(r)\boldsymbol{e}_3$, where $\boldsymbol{e}_r(\theta) = \cos\theta\,\boldsymbol{e}_1 + \sin\theta\,\boldsymbol{e}_2$ and $z(r) = cr^2$, with c a positive constant. Find the normal curvature of a curve inclined at 45° to the meridian, as a function of r.

2.7 A surface is *ruled* if it is generated by a one-parameter family of straight lines. Such surfaces may be described by the parametric equation $\boldsymbol{r}(\phi,\psi) = \phi\boldsymbol{m}(\psi) + \boldsymbol{l}(\psi)$, where ϕ, ψ are the coordinate parameters and $\boldsymbol{m}$ is a field of unit vectors. Show that the curve $\psi =$ constant are both geodesics and 'asymptotic' lines (curves with zero normal curvature) and are therefore straight. Show that the Gaussian curvature is non-positive on any ruled surface.

2.8 Prove that the components of the surface metric tensor $a_{\alpha\beta}$ and the surface alternator $\varepsilon_{\alpha\beta}$ satisfy the following relations

(i) $\varepsilon_{\gamma\delta}\,a^{\alpha\gamma}a^{\beta\delta} = \varepsilon^{\alpha\beta}$.

(ii) $\varepsilon_{\alpha\beta}\,\varepsilon_{\gamma\delta} = a_{\alpha\gamma}\,a_{\beta\delta} - a_{\alpha\delta}\,a_{\beta\gamma}\,.$
Solution:
(i) This relation is an example of raising the subscripts γ and δ. To prove it, let us consider the four possible cases for the values of α, β.
If $\alpha = \beta = 1$, then both sides vanish.
If $\alpha = 1$ and $\beta = 2$, then from (2.54) the left side is $\varepsilon^{12} = \left(\sqrt{a}\right)^{-1}$, whereas the right side is

$$\varepsilon_{\gamma\delta}\,a^{1\gamma}a^{2\delta} = \sqrt{a}\left(a^{11}a^{22} - a^{12}a^{21}\right) = \sqrt{a}\,\det\left(a^{\alpha\beta}\right) = \sqrt{a}\,\frac{1}{a} = \frac{1}{\sqrt{a}}\,,$$

so we have $\varepsilon_{\gamma\delta}\,a^{1\gamma}a^{2\delta} = \varepsilon^{12}$ and the relation holds.
Similarly, we can treat the other two cases: $\alpha = 2,\ \beta = 1$ and, respectively, $\alpha = \beta = 2$.

(ii) If $\alpha = \beta$ or $\gamma = \delta$, then both sides vanish.
If $\alpha \neq \beta$ and $\gamma \neq \delta$, let us assume for example that $\alpha = \gamma = 1$ and $\beta = \delta = 2$. Then, the left side is

$$\varepsilon_{12}\,\varepsilon_{12} = \sqrt{a}\,e_{12}\,\sqrt{a}\,e_{12} = a,$$

whereas the right side is

$$a_{11}\,a_{22} - a_{12}\,a_{21} = \det(a_{\alpha\beta}) = a,$$

so both sides are equal. Similarly, we can treat the other cases ($\alpha = \delta = 1$, $\beta = \gamma = 2$), or ($\alpha = \gamma = 2$, $\beta = \delta = 1$), or ($\alpha = \delta = 2$, $\beta = \gamma = 1$).

2.9 Prove that the curvature tensor $\boldsymbol{b}$, and the surface alternator $\varepsilon_{\alpha\beta}$ satisfy the following relations
(i) $K\,\varepsilon_{\alpha\beta} = b^{\gamma}_{\alpha}\,b^{\delta}_{\beta}\,\varepsilon_{\gamma\delta}$, where K is the Gaussian curvature.
(ii) $\varepsilon^{\alpha\beta}{}_{|\gamma} = 0, \qquad \varepsilon_{\alpha\beta|\gamma} = 0\,.$
Solution:
(i) If $\alpha = \beta$, then both sides vanish.
If $\alpha \neq \beta$, let us asume for example that $\alpha = 1$ and $\beta = 2$. Then, the left side is

$$K\,\varepsilon_{12} = K\,\sqrt{a}\,e_{12} = K\,\sqrt{a}\,,$$

whereas the right side is

$$b^{\gamma}_{1}\,b^{\delta}_{2}\,\varepsilon_{\gamma\delta} = \sqrt{a}\,(b^{1}_{1}\,b^{2}_{2} - b^{2}_{1}\,b^{1}_{2}) = \sqrt{a}\,\det(b^{\alpha}_{\beta}) = \sqrt{a}\,K,$$

so both sides are equal. Similarly for the other case $\alpha = 2$ and $\beta = 1$.
(ii) If $\alpha = \beta$, then $\varepsilon^{\alpha\beta} = 0$ and the relation is obviously satisfied.
If $\alpha \neq \beta$, then we can write

$$\varepsilon^{\alpha\beta}{}_{|\gamma} = \varepsilon^{\alpha\beta}{}_{,\gamma} + \varepsilon^{\alpha\lambda}\Gamma^{\beta}_{\lambda\gamma} + \varepsilon^{\lambda\beta}\Gamma^{\alpha}_{\lambda\gamma}$$

and, in view of (2.54) and (2.89),

$$\varepsilon^{\alpha\beta}{}_{,\gamma} = \left(\frac{1}{\sqrt{a}}\right)_{,\gamma} e_{\alpha\beta} = -\frac{1}{a}\left(\sqrt{a}\right)_{,\gamma} e_{\alpha\beta} = -\frac{1}{\sqrt{a}}\Gamma^{\lambda}_{\lambda\gamma}\, e_{\alpha\beta}\,.$$

So, we obtain

$$\varepsilon^{\alpha\beta}{}_{|\gamma} = \frac{1}{\sqrt{a}}\left(-\, e_{\alpha\beta}\,\Gamma^{\lambda}_{\lambda\gamma} + e_{\alpha\lambda}\Gamma^{\beta}_{\lambda\gamma} + e_{\lambda\beta}\Gamma^{\alpha}_{\lambda\gamma}\right).$$

We can show that the left side of this equation vanishes: For example, when $\alpha = 1,\ \beta = 2$ we get

$$\varepsilon^{12}{}_{|\gamma} = \frac{1}{\sqrt{a}}\left(-\,\Gamma^{\lambda}_{\lambda\gamma} + \Gamma^{2}_{2\gamma} + \Gamma^{1}_{1\gamma}\right) = 0.$$

For $\alpha = 2,\ \beta = 1$ we have similarly

$$\varepsilon^{21}{}_{|\gamma} = \frac{1}{\sqrt{a}}\left(\Gamma^{\lambda}_{\lambda\gamma} - \Gamma^{1}_{1\gamma} - \Gamma^{2}_{2\gamma}\right) = 0.$$

The relation $\varepsilon_{\alpha\beta|\gamma} = 0$ can be proved analogously.

2.10 In 'lines of curvature' (l.o.c.) coordinates $\xi^1 = \xi$ and $\xi^2 = \eta$, we have $\boldsymbol{a}_1 = \alpha\boldsymbol{i}$ and $\boldsymbol{a}_2 = \beta\boldsymbol{j}$ ($\alpha = |\boldsymbol{a}_1|,\ \beta = |\boldsymbol{a}_2|$), where $\boldsymbol{i}$ and $\boldsymbol{j}$ are orthogonal unit vectors in the coordinate directions. Let the corresponding normal curvatures be κ_ξ and κ_η and let κ be the Gaussian curvature. Obtain the explicit forms of the Mainardi-Codazzi and Theorema-Egregium equations:

$$\begin{aligned}
&\frac{\partial}{\partial\eta}\left(\alpha\kappa_\xi\right) = \kappa_\eta\frac{\partial\alpha}{\partial\eta}\,,\\
&\frac{\partial}{\partial\xi}\left(\beta\kappa_\eta\right) = \kappa_\xi\frac{\partial\beta}{\partial\xi}\,,\\
&\frac{\partial}{\partial\xi}\left(\alpha^{-1}\frac{\partial\beta}{\partial\xi}\right) + \frac{\partial}{\partial\eta}\left(\beta^{-1}\frac{\partial\alpha}{\partial\eta}\right) + \alpha\beta\kappa = 0\,.
\end{aligned}$$

The third equation shows that the Gaussian curvature depends only on the surface metric (Gauss' famous theorem). Obtain formulas involving α and β for the geodesic curvatures of the coordinate curves, and show that if these vanish then the Gaussian curvature vanishes too. Use this to prove that the Gaussian curvature vanishes on any surface with orthogonal geodesics, whether or not the geodesics happen to be l.o.c.

2.11 Assume that we can write the vector $\boldsymbol{v}$ as

$$\boldsymbol{v} = \phi\,\boldsymbol{c}\,,$$

in which $\boldsymbol{c}$ is an arbitrary constant vector and ϕ is a scalar function. By using the definition of the surface divergence (Eq. (2.154)) and relation (2.155), prove the following relation

$$\int_{\mathscr{S}} (\text{grad}_s\phi)\,\text{d}a = \int_{\partial\mathscr{S}} \phi\boldsymbol{\nu}\,\text{d}s - \int_{\mathscr{S}} 2H\phi\boldsymbol{n}\,\text{d}a\,.$$

Now assume that ϕ is constant (not equal to zero) and derive the following relation

$$\int_{\partial\mathscr{S}} \boldsymbol{\nu}\,\text{d}s = \int_{\mathscr{S}} 2H\boldsymbol{n}\,\text{d}a\,.$$

Note: From this relation, we can conclude that for a closed surface, we have

$$\int_{\mathscr{S}} 2H\boldsymbol{n}\,\text{d}a = \boldsymbol{0}\,.$$

Solution:
We have

$$\boldsymbol{v} = \phi\boldsymbol{c}\,, \qquad \text{div}_s\boldsymbol{v} = \boldsymbol{v}_{,\beta}\cdot\boldsymbol{a}^{\beta}\,.$$

Since the vector $\boldsymbol{c}$ is constant, we find

$$\text{div}_s\boldsymbol{v} = \boldsymbol{c}\cdot\phi_{,\beta}\boldsymbol{a}^{\beta} = \boldsymbol{c}\cdot\text{grad}_s\phi\,.$$

By using relation (2.155) we find

$$\int_{\mathscr{S}} \boldsymbol{c}\cdot(\text{grad}_s\phi)\,\text{d}a = \int_{\partial\mathscr{S}} \phi\boldsymbol{c}\cdot\boldsymbol{\nu}\,\text{d}s - \int_{\mathscr{S}} 2H\phi\boldsymbol{c}\cdot\boldsymbol{n}\,\text{d}a\,.$$

Because $\boldsymbol{c}$ is an arbitrary constant vector, it follows

$$\boldsymbol{c}\cdot\left[\int_{\mathscr{S}} (\text{grad}_s\phi)\,\text{d}a - \int_{\partial\mathscr{S}} \phi\boldsymbol{\nu}\,\text{d}s + \int_{\mathscr{S}} 2H\phi\boldsymbol{n}\,\text{d}a\right] = 0\,,$$

so we conclude that

$$\int_{\mathscr{S}} (\text{grad}_s\phi)\,\text{d}a = \int_{\partial\mathscr{S}} \phi\boldsymbol{\nu}\,\text{d}s - \int_{\mathscr{S}} 2H\phi\boldsymbol{n}\,\text{d}a\,.$$

If ϕ is constant, we conclude that $\text{grad}_s\phi = \boldsymbol{0}$ and from the above relation we find

$$\int_{\partial\mathscr{S}} \phi\boldsymbol{\nu}\,\text{d}s = \int_{\mathscr{S}} 2H\phi\boldsymbol{n}\,\text{d}a\,.$$

2.12 Assume that vector $\boldsymbol{v}$ is

$$\boldsymbol{v} = \boldsymbol{f} \times \boldsymbol{c}\,,$$

in which vector $\boldsymbol{c}$ is an arbitrary constant vector. Show that

$$\mathrm{div}_s \boldsymbol{v} = (\mathrm{curl}_s \boldsymbol{f}\,) \cdot \boldsymbol{c}\,.$$

Use this relation to conclude

$$\int_{\mathscr{S}} (\mathrm{curl}_s \boldsymbol{f})\,\mathrm{d}a = \int_{\partial\mathscr{S}} (\boldsymbol{\nu} \times \boldsymbol{f})\,\mathrm{d}s - \int_{\mathscr{S}} 2H\,(\boldsymbol{n} \times \boldsymbol{f})\,\mathrm{d}a\,.$$

Solution:
From the definition of the surface divergence we find

$$\mathrm{div}_s \boldsymbol{v} = \boldsymbol{v}_{,\beta} \cdot \boldsymbol{a}^\beta = \left(\boldsymbol{a}^\beta \times \boldsymbol{f}_{,\beta}\right) \cdot \boldsymbol{c} = (\mathrm{curl}_s \boldsymbol{f}) \cdot \boldsymbol{c}\,.$$

By using relation (2.155) and given that vector $\boldsymbol{c}$ is an arbitrary constant vector, we find

$$\int_{\mathscr{S}} (\mathrm{curl}_s \boldsymbol{f})\,\mathrm{d}a = \int_{\partial\mathscr{S}} (\boldsymbol{\nu} \times \boldsymbol{f})\,\mathrm{d}s - \int_{\mathscr{S}} 2H\,(\boldsymbol{n} \times \boldsymbol{f})\,\mathrm{d}a.$$

References

1. Ciarlet, P.G.: An Introduction to Differential Geometry with Applications to Elasticity. Springer, Dordrecht (2005)
2. Dickmen, M.: Theory of Thin Elastic Shells. Pitman, Boston (1982)
3. Green, A.E., Zerna, W.: Theoretical Elasticity. Clarendon Press, Oxford (1968)
4. Naghdi, P.M.: The theory of shells and plates. In: Flügge, W. (ed.) Handbuch der Physik, vol. VIa/2, pp. 425–640. Springer, Berlin (1972)
5. Stoker, J.J.: Differential Geometry. Wiley, New York (1989)
6. Weatherburn, C.E.: Differential Geometry of Three Dimensions, vol. I. Cambridge University Press, London (1955)

Chapter 3
Hyperelastic Solids: Purely Mechanical Theory

Abstract The background on tensor analysis acquired in the first two chapters is used in the present chapter to cast the three-dimensional theory of nonlinear elasticity in a curvilinear-coordinate setting. This furnishes an immediate application of these ideas to a topic of mechanical significance and sets the stage for our subsequent work on elastic shells.

3.1 Constitutive Relations

Consider three-dimensional deformable bodies as decribed in Sect. 2.1.

The hyperelastic solids are characterized by the property that the second Piola-Kirchhoff stress tensor can be expressed as a function of the Cauchy-Green deformation tensor, i.e. $\boldsymbol{S} = \hat{\boldsymbol{S}}(\boldsymbol{C})$, and there is a function $U(\boldsymbol{C})$, the strain energy per unit reference volume, such that

$$\dot{U} = \frac{1}{2}\,\boldsymbol{S}(\boldsymbol{C}) \cdot \dot{\boldsymbol{C}}\,, \tag{3.1}$$

for any motion $\boldsymbol{y}(\theta^i, t)$. The superposed dot in the above relation denotes the time derivative, while the multiplication dot designates the scalar product between second order tensors. Taking into account that the symmetry of the second Piola-Kirchhoff stress tensor (cf. (2.46)), the relation (3.1) can be written equivalently in the following form

$$\frac{\partial U}{\partial g_{ij}}\,\dot{g}_{ij} = \frac{1}{2}\,(S^{ij}\boldsymbol{G}_i \otimes \boldsymbol{G}_j) \cdot (\dot{g}_{kl}\boldsymbol{G}^k \otimes \boldsymbol{G}^l),$$

so

$$\frac{\partial U}{\partial g_{ij}}\,\dot{g}_{ij} = \frac{1}{2}\,S^{ij}\dot{g}_{kl}(\boldsymbol{G}_i \cdot \boldsymbol{G}^k)(\boldsymbol{G}_j \cdot \boldsymbol{G}^l) = \frac{1}{2}\,S^{ij}\dot{g}_{ij},$$

or

$$\frac{1}{2}\Big(\frac{\partial U}{\partial g_{ij}} + \frac{\partial U}{\partial g_{ji}}\Big)\dot{g}_{ij} = \frac{1}{2}\,S^{ij}\dot{g}_{ij}\,,$$

D. J. Steigmann et al., *Lecture Notes on the Theory of Plates and Shells*,
Solid Mechanics and Its Applications 274,
https://doi.org/10.1007/978-3-031-25674-5_3

i.e.

$$\Big[S^{ij} - \Big(\frac{\partial U}{\partial g_{ij}} + \frac{\partial U}{\partial g_{ji}}\Big)\Big]\dot{g}_{ij} = 0, \qquad \text{for all } \dot{g}_{ij} = \dot{g}_{ji}\,.$$

Since the terms in the square bracket and $\dot{g}_{ij}$ are symmetric, and the term in the square bracket is not a function of $\dot{g}_{ij}$, we require for hyperelastic solids

$$S^{ij} = \frac{\partial U}{\partial g_{ij}} + \frac{\partial U}{\partial g_{ji}}\,. \tag{3.2}$$

In the case of incompressible materials ($\det \boldsymbol{C} = 1$), the relation (3.2) takes the form

$$S^{ij} = p\, g^{ij} + \Big(\frac{\partial U}{\partial g_{ij}} + \frac{\partial U}{\partial g_{ji}}\Big), \qquad p = \hat{p}(\theta^i). \tag{3.3}$$

Moreover, we have

$$\boldsymbol{T} = J^{-1}\boldsymbol{F}\boldsymbol{S}\boldsymbol{F}^{-T} \tag{3.4}$$

with

$$\boldsymbol{S} = S^{ij}\boldsymbol{G}_i \otimes \boldsymbol{G}_j = \left[pg^{ij} + \left(\frac{\partial U}{\partial g_{ij}} + \frac{\partial U}{\partial g_{ji}}\right)\right]\boldsymbol{G}_i \otimes \boldsymbol{G}_j\,. \tag{3.5}$$

As a result by using the relation $\boldsymbol{F}\boldsymbol{G}_i = \boldsymbol{g}_i$ and with $J = 1$ (incompressiblity condition) we conclude

$$T^{ij} = S^{ij} = pg^{ij} + \left(\frac{\partial U}{\partial g_{ij}} + \frac{\partial U}{\partial g_{ji}}\right). \tag{3.6}$$

Example: Isotropic materials

Let us consider hyperelastic materials which are isotropic relative to the reference configuratiion $\mathscr{K}$. Then, the strain energy density can be written in the compressible case as a function

$$U = U(I_1\,,\, I_2\,,\, I_3)\,, \tag{3.7}$$

where $I_1\,,\ I_2\,,\ I_3$ are the principal invariants of the Cauchy-Green deformation tensor $\boldsymbol{C}$. In view of the relations (2.9), (2.10) and (2.15), we have

$$\begin{aligned} I_1 &= \operatorname{tr}\boldsymbol{C} = \operatorname{tr}(g_{ij}\,\boldsymbol{G}^i \otimes \boldsymbol{G}^j) = G^{ij}g_{ij}\,, \\ I_2 &= \tfrac{1}{2}\left[(\operatorname{tr}\boldsymbol{C})^2 - \operatorname{tr}(\boldsymbol{C}^2)\right], \qquad \text{with} \end{aligned} \tag{3.8}$$

$$\operatorname{tr}(\boldsymbol{C}^2) = \operatorname{tr}\big[(C^{i}_{.\,j}\boldsymbol{G}_i \otimes \boldsymbol{G}^j)(C^{k}_{.\,l}\boldsymbol{G}_k \otimes \boldsymbol{G}^l)\big] = C^{i}_{.\,j}\,C^{j}_{.\,i} = (G^{ik}g_{kj})(G^{jl}g_{li}) = G^{ik}G^{jl}g_{kj}g_{li}\,,$$

and

$$I_3 = \det\boldsymbol{C} = J^2 = \frac{g}{G}\,. \tag{3.9}$$

Then, from (3.2) and (3.7) we obtain

$$S^{ij} = \sum_{k=1}^{3} \frac{\partial \hat{U}}{\partial I_k} \Big(\frac{\partial I_k}{\partial g_{ij}} + \frac{\partial I_k}{\partial g_{ji}} \Big) \tag{3.10}$$

In view of (3.8) and (3.9), we get by differentiation

$$\begin{aligned} \frac{\partial I_1}{\partial g_{ij}} &= G^{ij} , \\ \frac{\partial I_2}{\partial g_{ij}} &= \frac{1}{2} \left[2I_1 \frac{\partial I_1}{\partial g_{ij}} - \frac{\partial \, \mathrm{tr}(\boldsymbol{C}^2)}{\partial g_{ij}} \right] = I_1 \, G^{ij} - G^{ik} G^{jl} g_{kl} , \\ \frac{\partial I_3}{\partial g_{ij}} &= \frac{1}{G} \frac{\partial g}{\partial g_{ij}} = \frac{1}{G} \, g \, g^{ij} = I_3 \, g^{ij} , \end{aligned} \tag{3.11}$$

where we have used the relation (1.60). If we introduce the notations

$$\hat{U}_k := \frac{\partial \hat{U}}{\partial I_k} \quad \text{and} \quad H^{ij} := \frac{\partial I_2}{\partial g_{ij}} = I_1 \, G^{ij} - G^{ik} G^{jl} g_{kl} = H^{ji}, \tag{3.12}$$

then the Eq. (3.10) can be written in the form

$$S^{ij} = 2\big(\hat{U}_1 \, G^{ij} + \hat{U}_2 \, H^{ij} + I_3 \, \hat{U}_3 \, g^{ij} \big) . \tag{3.13}$$

Compare this relation with the results from [3], Sect. 1.15.

In the incompressible case, we have $I_3 = 1$. If we designate by $\tilde{U}(I_1, I_2) := \hat{U}(I_1, I_2, 1)$ and take into account (3.3), then we obtain similarly in this case

$$S^{ij} = 2\hat{U}_1 \, G^{ij} + 2\hat{U}_2 \, H^{ij} + p \, g^{ij} . \tag{3.14}$$

□

By virtue of the relations (2.10) and (2.23) we have $g_{ij} = C_{ij} = 2E_{ij} + G_{ij}$. Then, we can regard the strain energy density as a function $U = U(C_{ij}) = U(\frac{1}{2}(C_{ij} + C_{ji}))$ and from (3.2) we deduce

$$S^{ij} = \frac{\partial U}{\partial C_{ij}} + \frac{\partial U}{\partial C_{ji}} = 2 \frac{\partial U}{\partial C_{ij}} = \frac{\partial U}{\partial E_{ij}} . \tag{3.15}$$

The last relation can be written in condensed form

$$\boldsymbol{S} = U_{\boldsymbol{E}} , \tag{3.16}$$

where we have denoted the derivative of U with respect to the tensor $\boldsymbol{E} = E_{ij}\boldsymbol{G}^i \otimes \boldsymbol{G}^j$ by

$$U_{\boldsymbol{E}} := \frac{\partial U}{\partial E_{ij}} \boldsymbol{G}_i \otimes \boldsymbol{G}_j \,.$$

3.2 Small Strains

Let us consider first the case of small strains (we will generalize later to the case of finite strains). Thus, we assume that $|\boldsymbol{E}| = O(\epsilon)$, where $\epsilon \ll 1$ is a non-dimensional parameter.

We seek a stress-strain relation valid to order $O(\epsilon)$. In view of (3.16), we can write the expansion about $\boldsymbol{E} = \boldsymbol{0}$ and obtain

$$\boldsymbol{S}(\boldsymbol{E}) = U_{\boldsymbol{E}}(\boldsymbol{E}) = U_{\boldsymbol{E}}(\boldsymbol{0}) + U_{\boldsymbol{E}\boldsymbol{E}}(\boldsymbol{0})[\boldsymbol{E}] + o(\epsilon), \tag{3.17}$$

where we have designated by

$$U_{\boldsymbol{E}\boldsymbol{E}} := \frac{\partial^2 U}{\partial E_{ij}\, \partial E_{kl}} \boldsymbol{G}_i \otimes \boldsymbol{G}_j \otimes \boldsymbol{G}_k \otimes \boldsymbol{G}_l$$

the second order derivative with respect to $\boldsymbol{E}$. The relation (3.17) can be written as

$$\boldsymbol{S}(\boldsymbol{E}) = \underbrace{\boldsymbol{S}^0 + \underline{\boldsymbol{C}}[\boldsymbol{E}]}_{\text{linear part}} + o(\epsilon), \tag{3.18}$$

where we denote by

$$\boldsymbol{S}^0 := U_{\boldsymbol{E}}(\boldsymbol{0}) = S_0^{ij}\, \boldsymbol{G}_i \otimes \boldsymbol{G}_j \qquad \text{the “residual stress”}, \tag{3.19}$$

and by $\underline{\boldsymbol{C}}$ the fourth order tensor of *elastic moduli*

$$\underline{\boldsymbol{C}} := U_{\boldsymbol{E}\boldsymbol{E}}(\boldsymbol{0}) = C^{ijkl} \boldsymbol{G}_i \otimes \boldsymbol{G}_j \otimes \boldsymbol{G}_k \otimes \boldsymbol{G}_l \quad \text{with} \quad C^{ijkl} := \left.\frac{\partial^2 U}{\partial E_{ij}\, \partial E_{kl}}\right|_{\boldsymbol{E}=\boldsymbol{0}} . \tag{3.20}$$

From (3.20) we see that the elastic moduli satisfy the symmetry relations

$$C^{ijkl} = C^{klij} = C^{jikl} = C^{ijlk} \tag{3.21}$$

and we can write

$$U_{\boldsymbol{E}\boldsymbol{E}}(\boldsymbol{0})[\boldsymbol{E}] = \boldsymbol{G}_i \otimes \boldsymbol{G}_j \left.\frac{\partial^2 U}{\partial E_{ij}\, \partial E_{kl}}\right|_{\boldsymbol{E}=\boldsymbol{0}} E_{kl} = C^{ijkl}\, E_{kl}\, \boldsymbol{G}_i \otimes \boldsymbol{G}_j \,. \tag{3.22}$$

Integrating the equation $U_E(\boldsymbol{E}) = \boldsymbol{S}(\boldsymbol{E})$ with respect to $\boldsymbol{E}$ and taking into account (3.18), we obtain the associated strain energy function in the form

$$U(\boldsymbol{E}) = U(\boldsymbol{0}) + \boldsymbol{S}^0 \cdot \boldsymbol{E} + \frac{1}{2}\, \boldsymbol{E} \cdot \underline{\boldsymbol{C}}[\boldsymbol{E}] + o(\epsilon^2). \tag{3.23}$$

Further, we can set $U(\boldsymbol{0}) = 0$ (without loss of generality) and write the above relation with help of tensor comonents as

$$U \;=\; S_0^{ij}\, E_{ij} \;+\; \frac{1}{2}\, C^{ijkl}\, E_{ij}\, E_{kl} \;+ o(\epsilon^2). \tag{3.24}$$

These equations are valid for any kind of anisotropy.

As a typical situation, we have:

$$\boldsymbol{S}^0 = \boldsymbol{0} \qquad \text{and} \qquad U(\boldsymbol{E}) > 0 \;\; \text{for all} \;\; \boldsymbol{E} \neq \boldsymbol{0}. \tag{3.25}$$

Then, the tensor of elastic moduli $\underline{\boldsymbol{C}}$ is positive definite on the set of symmetric second order tensors.

3.2.1 Special Case: Isotropy

If the material is isotropic, then $U(\boldsymbol{E})$ is a function of $\operatorname{tr}\boldsymbol{E}$, $\operatorname{tr}(\boldsymbol{E}^2)$, $\operatorname{tr}(\boldsymbol{E}^3)$. In the case of small strains, retaining only homogeneous, quadratic terms, we obtain

$$U(\boldsymbol{E}) = \frac{1}{2}\,\lambda(\operatorname{tr}\boldsymbol{E})^2 \;+\; \mu \operatorname{tr}(\boldsymbol{E}^2), \tag{3.26}$$

where λ and μ are the Lamé moduli. Taking into account that $\operatorname{tr}\boldsymbol{E} = \boldsymbol{I} \cdot \boldsymbol{E}$ and $\operatorname{tr}(\boldsymbol{E}^2) = \boldsymbol{E} \cdot \boldsymbol{E} = |\boldsymbol{E}|^2$, we deduce from (3.16) and (3.26) the stress-strain relation for isotropic elastic materials

$$\boldsymbol{S} \;=\; U_{\boldsymbol{E}} \;=\; \lambda(\operatorname{tr}\boldsymbol{E})\,\boldsymbol{I} \;+\; 2\mu\,\boldsymbol{E}\,. \tag{3.27}$$

The following statement is a standard result from the classical theory of elasticity.

Theorem 3.1 *The strain energy density $U(\boldsymbol{E})$ defined by (3.26) is positive definite if and only if the shear modulus μ and the bulk modulus κ are positive, i.e.*

$$\mu > 0 \quad \textit{and} \quad \kappa = \lambda + \frac{2}{3}\,\mu > 0. \tag{3.28}$$

Proof For any second order tensor $\boldsymbol{T}$ let us denote its *deviatoric part* by

$$\operatorname{dev} \boldsymbol{T} := \boldsymbol{T} - \frac{1}{3}\,(\operatorname{tr} \boldsymbol{T})\,\boldsymbol{I}\,. \tag{3.29}$$

Then, we can decompose any second order tensor $\boldsymbol{T}$ in a direct sum as

$$\boldsymbol{T} = \operatorname{dev}(\operatorname{sym} \boldsymbol{T}) + \operatorname{skew} \boldsymbol{T} + \frac{1}{3}\,(\operatorname{tr} \boldsymbol{T})\,\boldsymbol{I}\,, \tag{3.30}$$

in the sense that all three terms in the right-hand side are mutually orthogonal:

$$\operatorname{dev}(\operatorname{sym} \boldsymbol{T}) \cdot \operatorname{skew} \boldsymbol{T} = 0, \quad \operatorname{dev}(\operatorname{sym} \boldsymbol{T}) \cdot \boldsymbol{I} = 0, \quad (\operatorname{skew} \boldsymbol{T}) \cdot \boldsymbol{I} = 0.$$

If we write the relation (3.30) for the symmetric tensor $\boldsymbol{E}$, we obtain

$$\boldsymbol{E} = \operatorname{dev} \boldsymbol{E} + \frac{1}{3}\,(\operatorname{tr} \boldsymbol{E})\,\boldsymbol{I} \tag{3.31}$$

and so

$$|\boldsymbol{E}|^2 = |\operatorname{dev} \boldsymbol{E}|^2 + \frac{1}{3}\,(\operatorname{tr} \boldsymbol{E})^2\,. \tag{3.32}$$

Substituting the last relation into (3.26), we get

$$U(\boldsymbol{E}) = \Big(\frac{1}{2}\,\lambda + \frac{1}{3}\,\mu\Big)(\operatorname{tr} \boldsymbol{E})^2 \;+\; \mu\,|\operatorname{dev} \boldsymbol{E}|^2 \;=\; \frac{1}{2}\,\kappa\,(\operatorname{tr} \boldsymbol{E})^2 \;+\; \mu\,|\operatorname{dev} \boldsymbol{E}|^2. \tag{3.33}$$

The last equation shows that $U(\boldsymbol{E})$ is positive definite if and only if $\mu > 0$ and $\kappa > 0$. □

Recall the strain-displacement relations (2.24) in the form

$$E_{ij} \;=\; \frac{1}{2}\,(u_{i|j} + u_{j|i} + u_{k|i}\,u^k{}_{|j}) \;=\; \varepsilon_{ij} + \frac{1}{2}\,u_{k|i}\,u^k{}_{|j}\,, \tag{3.34}$$

where we denote the small strain tensor by

$$\varepsilon_{ij} \;:=\; \frac{1}{2}\,(u_{i|j} + u_{j|i}). \tag{3.35}$$

Suppose that the displacement gradient is small, so that $|u_{i|j}| = O(\epsilon)$, with $\epsilon \ll 1$. Then, the Eq. (3.35) infer $|\boldsymbol{\varepsilon}| = O(\epsilon)$, so the above relations are valid, but now we can replace the tensor $\boldsymbol{E}$ by $\boldsymbol{\varepsilon}$ while retaining $O(\epsilon)$ accuracy. For instance, for isotropic materials we have

$$\begin{aligned} \boldsymbol{S} &= \lambda(\operatorname{tr} \boldsymbol{\varepsilon})\,\boldsymbol{I} \;+\; 2\mu\,\boldsymbol{\varepsilon} \;+\; O(\epsilon^2) \qquad \text{and} \\ U &= \tfrac{1}{2}\,\lambda(\operatorname{tr} \boldsymbol{\varepsilon})^2 \;+\; \mu\,\operatorname{tr}(\boldsymbol{\varepsilon}^2) \;+\; O(\epsilon^3). \end{aligned} \tag{3.36}$$

In the general case of anisotropy, these relations read

$$S = U_{\varepsilon} = \underline{C}[\varepsilon] + O(\epsilon^2) \quad \text{and} \quad U = \tfrac{1}{2}\, \varepsilon \cdot \underline{C}[\varepsilon] + O(\epsilon^3). \tag{3.37}$$

In the case of isotropy, the elastic moduli are

$$C^{ijkl} = \lambda G^{ij} G^{kl} + \mu\big(G^{ik} G^{jl} + G^{il} G^{kj}\big). \tag{3.38}$$

Indeed, we can verify the equality (3.38) by calculating

$$\begin{aligned}
\underline{C}[\varepsilon] = C^{ijkl} \varepsilon_{kl}\, G_i \otimes G_j &= \big(\lambda G^{ij} G^{kl} \varepsilon_{kl} + \mu G^{ik} G^{jl} \varepsilon_{kl} + \mu G^{il} G^{kj} \varepsilon_{kl}\big) G_i \otimes G_j \\
&= \lambda\big(G^{kl} \varepsilon_{kl}\big) G^j \otimes G_j + \big(\mu \varepsilon^{ij} + \mu \varepsilon^{ji}\big) G_i \otimes G_j \\
&= \lambda \operatorname{tr}\big(\varepsilon_{kl} G^k \otimes G^l\big) I + 2\mu\big(\varepsilon^{ij} G_i \otimes G_j\big) = \lambda(\operatorname{tr} \varepsilon)\, I + 2\mu\, \varepsilon\,,
\end{aligned}$$

which is in accordance with (3.36).

Let us denote with H the gradient of displacement. In view of (1.45), we have

$$H := \operatorname{Grad} u = u_{,i} \otimes G^i = u_{i|j}\, G^i \otimes G^j \tag{3.39}$$

and the deformation gradient can be written as

$$F = I + H. \tag{3.40}$$

Indeed, from (2.5) we have

$$F = g_i \otimes G^i = y_{,i} \otimes G^i = (x + u)_{,i} \otimes G^i = G_i \otimes G^i + u_{,i} \otimes G^i = I + H$$

and the relation (3.40) is proved.

By virtue of the relations (2.44) and (3.37), the Piola stress tensor satisfies

$$P = FS = (I + H)\, \underline{C}[\varepsilon]\,.$$

Thus,

$$P = \underline{C}[\varepsilon] + O(\varepsilon^2) = S + O(\varepsilon^2), \tag{3.41}$$

so, to leading order, the Piola stress tensors S and P coincide.

Further, taking into account the relation $\varepsilon = \operatorname{sym} H$ and the symmetries (3.21), we can write the Eq. (3.37) in the form

$$P = \underline{C}[H] = U_H \quad \text{and} \quad U = \tfrac{1}{2}\, H \cdot \underline{C}[H]\,. \tag{3.42}$$

Here, we have $U \geq 0$ and the equality $U = 0$ holds if and only if H is a skew-symmetric tensor.

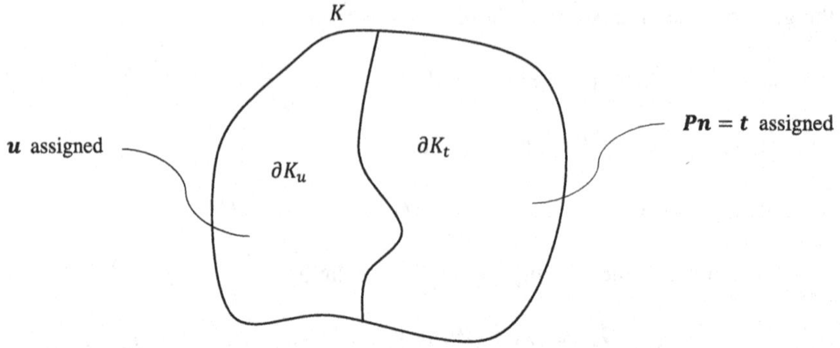

Fig. 3.1 Typical boundary conditions

With these notations, the local form of equations of motion (2.38) become

$$\operatorname{Div} \boldsymbol{P} + \rho_K \boldsymbol{b} = \rho_K \ddot{\boldsymbol{u}} , \qquad \text{or} \tag{3.43}$$

$$\operatorname{Div}\big(\underline{\mathbb{C}}[\boldsymbol{H}]\big) + \rho_K \boldsymbol{b} = \rho_K \ddot{\boldsymbol{u}} , \qquad \text{with} \qquad \boldsymbol{H} = \operatorname{Grad} \boldsymbol{u} . \tag{3.44}$$

This is a linear partial differential equation for the unknown $\boldsymbol{u}(\boldsymbol{x}, t)$, which is investigated in details in the so-called *linear elasticity* theory [5].

To the vectorial partial differential equation (3.44) we adjoin boundary conditions and initial conditions. Let $\partial\mathscr{K} = \partial\mathscr{K}_u \cup \partial\mathscr{K}_t$ be a disjoint partition of the boundary in the reference configuration. Then, the typical boundary conditions have the following form: the displacement $\boldsymbol{u}$ is assigned on the subset $\partial\mathscr{K}_u$, whereas the stress vector $\boldsymbol{t} = \boldsymbol{P}\boldsymbol{n} = \big(\underline{\mathbb{C}}[\boldsymbol{H}]\big)\boldsymbol{n}$ is assigned on the subset $\partial\mathscr{K}_t$ (Fig. 3.1). Also, the initial conditions prescribe the values of $\boldsymbol{u}(\boldsymbol{x}, 0)$ and $\dot{\boldsymbol{u}}(\boldsymbol{x}, 0)$. Thus, we see that the boundary conditions and initial conditions are linear too.

3.3 Potential Energy

We consider the potential energy of the deformable solid as

$$E[\boldsymbol{u}] := \int_{\mathscr{K}} U \,\mathrm{d}v - \int_{\mathscr{K}} \rho_K \boldsymbol{b} \cdot \boldsymbol{u} \,\mathrm{d}v - \int_{\partial\mathscr{K}_t} \boldsymbol{t} \cdot \boldsymbol{u} \,\mathrm{d}a \tag{3.45}$$

in the case of dead load. Let $\boldsymbol{b}$ denote the body load vector, which is prescribed. If we take the material time derivative $(\ )^{\cdot} = \frac{\partial}{\partial t}\Big|_{\boldsymbol{x}}$ of the relation (3.45), with $\boldsymbol{t}(\boldsymbol{x})$ assigned on $\partial\mathscr{K}_t$ and $\boldsymbol{u}(\boldsymbol{x})$ assigned on $\partial\mathscr{K}_u$, we obtain

$$\frac{\mathrm{d}E}{\mathrm{d}t} = \int_{\mathscr{K}} \dot{U} \,\mathrm{d}v - \int_{\mathscr{K}} \rho_K \boldsymbol{b} \cdot \dot{\boldsymbol{u}} \,\mathrm{d}v - \int_{\partial\mathscr{K}_t} \boldsymbol{t} \cdot \dot{\boldsymbol{u}} \,\mathrm{d}a. \tag{3.46}$$

For any tensor field $\boldsymbol{T}$ and vector field $\boldsymbol{v}$ we have the identity

$$\mathrm{Div}(\boldsymbol{T}^T\boldsymbol{v}) = \boldsymbol{T}\cdot\mathrm{Grad}\,\boldsymbol{v} + \big(\mathrm{Div}\,\boldsymbol{T}\big)\cdot\boldsymbol{v}, \tag{3.47}$$

which can be verified using the definitions (1.44), (1.46), and (1.51). Then, the derivative $\dot{U}$ appearing in (3.46) can be calculated as follows

$$\dot{U} = U_{\boldsymbol{H}}\cdot\dot{\boldsymbol{H}} = \boldsymbol{P}\cdot\mathrm{Grad}\,\dot{\boldsymbol{u}} = \mathrm{Div}(\boldsymbol{P}^T\dot{\boldsymbol{u}}) - \big(\mathrm{Div}\,\boldsymbol{P}\big)\cdot\dot{\boldsymbol{u}}\,. \tag{3.48}$$

If we insert (3.48) into (3.46) and use the divergence theorem, we obtain

$$\begin{aligned}\frac{\mathrm{d}E}{\mathrm{d}t} &= \int_{\partial\mathscr{K}}(\boldsymbol{P}^T\dot{\boldsymbol{u}})\cdot\boldsymbol{n}\,\mathrm{d}a - \int_{\mathscr{K}}\big(\mathrm{Div}\,\boldsymbol{P}+\rho_K\boldsymbol{b}\big)\cdot\dot{\boldsymbol{u}}\,\mathrm{d}v - \int_{\partial\mathscr{K}_t}\boldsymbol{t}\cdot\dot{\boldsymbol{u}}\,\mathrm{d}a\\ &= \int_{\partial\mathscr{K}_t}\dot{\boldsymbol{u}}\cdot\underbrace{\big(\boldsymbol{P}\boldsymbol{n}-\boldsymbol{t}\big)}_{=\boldsymbol{0}}\,\mathrm{d}a - \int_{\mathscr{K}}\rho_K\ddot{\boldsymbol{u}}\cdot\dot{\boldsymbol{u}}\,\mathrm{d}v\,,\end{aligned} \tag{3.49}$$

since $\dot{\boldsymbol{u}} = \boldsymbol{0}$ on $\partial\mathscr{K}_u$. If we denote the kinetic energy of the body by

$$K[\boldsymbol{u}] := \frac{1}{2}\int_{\mathscr{K}}\rho_K|\dot{\boldsymbol{u}}|^2\,\mathrm{d}v\,, \tag{3.50}$$

then from the relation (3.49) we derive the identity

$$\frac{\mathrm{d}}{\mathrm{d}t}\big(E+K\big) = 0, \tag{3.51}$$

which express the statement that "total energy is conserved" for purely elastic bodies.

Remark Real systems are always dissipative (due to heat flow, viscosity, etc.), i.e. they satisfy the inequality

$$\frac{\mathrm{d}}{\mathrm{d}t}\big(E+K\big) \le 0. \tag{3.52}$$

Consider that the body occupies at time $t = t_0$ the state with displacement vector $\boldsymbol{u}(\boldsymbol{x},t_0) =: \boldsymbol{u}_{t_0}$ and velocity $\dot{\boldsymbol{u}}(\boldsymbol{x},t_0) = \boldsymbol{0}$ (but $\ddot{\boldsymbol{u}}(\boldsymbol{x},t_0) \neq \boldsymbol{0}$). In the limit $t\to\infty$, we can assume that the body has reached an asymptotically stable state with displacement vector $\boldsymbol{u}(\boldsymbol{x},\infty) =: \boldsymbol{u}_\infty$, velocity $\dot{\boldsymbol{u}}(\boldsymbol{x},\infty) = \boldsymbol{0}$ and higher time derivatives $\boldsymbol{u}^{(n)}(\boldsymbol{x},\infty) = \boldsymbol{0}$ (Fig. 3.2). Then, the inequality (3.52) implies that

$$E[\boldsymbol{u}_\infty] + \underbrace{K[\boldsymbol{u}_\infty]}_{=0} \le E[\boldsymbol{u}_{t_0}] + \underbrace{K[\boldsymbol{u}_{t_0}]}_{=0}\,, \qquad \text{so} \qquad E[\boldsymbol{u}_\infty] \le E[\boldsymbol{u}_{t_0}]\,,$$

i.e., stable equilibria minimize the potential energy. □

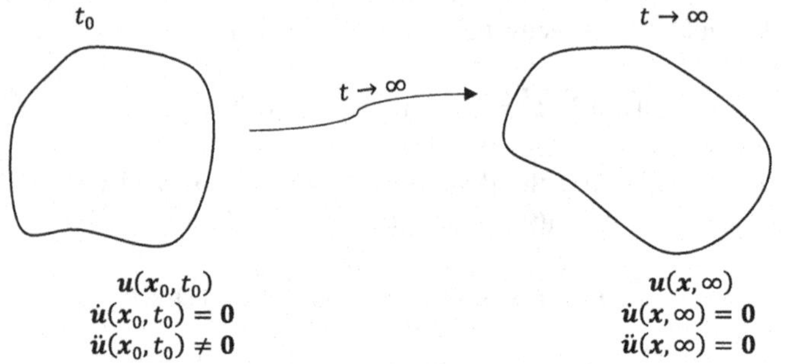

Fig. 3.2 Asymptotically stable state of the body for $t \to \infty$

Under the above hypothesis on U (namely $U[\boldsymbol{H}] \geq 0$ for all $\boldsymbol{H}$), the following classical result holds

Theorem 3.2 *Suppose $\boldsymbol{u}(\boldsymbol{x})$ is an equilibrium state. Then,*

$$E[\boldsymbol{u}] \leq E[\boldsymbol{u} + \boldsymbol{v}], \quad \textit{for all } \boldsymbol{v}(\boldsymbol{x}) \textit{ with } \boldsymbol{v} = \boldsymbol{0} \textit{ on } \partial\mathscr{K}_u\,, \tag{3.53}$$

and conversely.

Proof In the framework of linear elasticity, we can use the relations (3.42) and (3.45) to write

$$E[\boldsymbol{u}+\boldsymbol{v}] - E[\boldsymbol{u}] = \int_{\mathscr{K}} \left[U\big(\text{Grad}\,(\boldsymbol{u}+\boldsymbol{v})\big) - U\big(\text{Grad}\,\boldsymbol{u}\big)\right] \text{d}v - \int_{\mathscr{K}} \rho_K \boldsymbol{b}\cdot\boldsymbol{v}\,\text{d}v - \int_{\partial\mathscr{K}_t} \boldsymbol{t}\cdot\boldsymbol{v}\,\text{d}a,$$

where

$$\begin{aligned} U\big(\text{Grad}\,(\boldsymbol{u}+\boldsymbol{v})\big) = U\big(\text{Grad}\,\boldsymbol{u} + \text{Grad}\,\boldsymbol{v}\big) &= \tfrac{1}{2}\,(\text{Grad}\,\boldsymbol{u})\cdot \underline{\boldsymbol{C}}[\text{Grad}\,\boldsymbol{u}] \\ &+ \tfrac{1}{2}\,(\text{Grad}\,\boldsymbol{v})\cdot \underline{\boldsymbol{C}}[\text{Grad}\,\boldsymbol{v}] + (\text{Grad}\,\boldsymbol{v})\cdot \underline{\boldsymbol{C}}[\text{Grad}\,\boldsymbol{u}]. \end{aligned}$$

Substituting this relation, we obtain

$$\begin{aligned} E[\boldsymbol{u}+\boldsymbol{v}] - E[\boldsymbol{u}] &= \frac{1}{2}\int_{\mathscr{K}} (\text{Grad}\,\boldsymbol{v})\cdot \underline{\boldsymbol{C}}[\text{Grad}\,\boldsymbol{v}]\,\text{d}v \\ &+ \int_{\mathscr{K}} \left[(\text{Grad}\,\boldsymbol{v})\cdot \underline{\boldsymbol{C}}[\text{Grad}\,\boldsymbol{u}] - \rho_K \boldsymbol{b}\cdot\boldsymbol{v}\right] \text{d}v - \int_{\partial\mathscr{K}_t} \boldsymbol{t}\cdot\boldsymbol{v}\,\text{d}a. \end{aligned} \tag{3.54}$$

Further, on the basis of Eq. (3.47) and the divergence theorem, we see that the following relation holds for any tensor field $\boldsymbol{T}$ and vector field $\boldsymbol{v}$

$$\int_{\mathscr{K}} \boldsymbol{T}\cdot \text{Grad}\,\boldsymbol{v}\,\text{d}v = \int_{\partial\mathscr{K}} (\boldsymbol{T}\boldsymbol{n})\cdot\boldsymbol{v}\,\text{d}a - \int_{\mathscr{K}} \big(\text{Div}\,\boldsymbol{T}\big)\cdot\boldsymbol{v}\,\text{d}v\,. \tag{3.55}$$

Using the Eq. (3.55) in (3.54) for $\boldsymbol{T} = \underline{\boldsymbol{C}}[\text{Grad}\,\boldsymbol{u}]$, we derive

$$\begin{aligned} E[\boldsymbol{u}+\boldsymbol{v}] - E[\boldsymbol{u}] &= \frac{1}{2}\int_{\mathscr{K}} \underbrace{(\text{Grad}\,\boldsymbol{v})\cdot\underline{\boldsymbol{C}}[\text{Grad}\,\boldsymbol{v}]\,\mathrm{d}v}_{\geq 0} \\ &+ \int_{\partial\mathscr{K}_t} \big\{(\underline{\boldsymbol{C}}[\text{Grad}\,\boldsymbol{u}])\boldsymbol{n} - \boldsymbol{t}\big\}\cdot\boldsymbol{v}\,\mathrm{d}a - \int_{\mathscr{K}} \big\{\text{Div}\big(\underline{\boldsymbol{C}}[\text{Grad}\,\boldsymbol{u}]\big) + \rho_K\boldsymbol{b}\big\}\cdot\boldsymbol{v}\,\mathrm{d}v. \end{aligned} \tag{3.56}$$

1. Suppose that $\boldsymbol{u}(\boldsymbol{x})$ is an equilibrium state. Then, in view of the equilibrium equation (3.44) and the boundary condition on $\partial\mathscr{K}_t$, the second line of (3.56) is zero and we deduce

$$E[\boldsymbol{u}+\boldsymbol{v}] \;\geq\; E[\boldsymbol{u}],$$

so the inequality (3.53) is proved. This is consistent with the argument involving dissipation of energy in the above remark.

2. Conversely, suppose that the relation (3.53) holds and put $\boldsymbol{v} = \varepsilon\boldsymbol{w}$, where $\varepsilon \in (-\varepsilon_0, \varepsilon_0) \subset (-1, 1)$ is a parameter and $\boldsymbol{w}$ is an arbitrary vector field such that $\boldsymbol{w} = \boldsymbol{0}$ on $\partial\mathscr{K}_u$.

Let us define the function $F : (-\varepsilon_0, \varepsilon_0) \to \mathbb{R}$ with $F(\varepsilon) := E[\boldsymbol{u}+\varepsilon\boldsymbol{w}] - E[\boldsymbol{u}]$. We notice that it satisfies $F(\varepsilon) \geq 0$, $F(0) = 0$, and

$$\begin{aligned} F(\varepsilon) = \varepsilon &\left[\int_{\partial\mathscr{K}_t} \big(\boldsymbol{P}\boldsymbol{n} - \boldsymbol{t}\big)\cdot\boldsymbol{w}\,\mathrm{d}a - \int_{\mathscr{K}} \big(\text{Div}\,\boldsymbol{P} + \rho_K\boldsymbol{b}\big)\cdot\boldsymbol{w}\,\mathrm{d}v\right] \\ &+ \frac{1}{2}\,\varepsilon^2 \int_{\mathscr{K}} (\text{Grad}\,\boldsymbol{w})\cdot\underline{\boldsymbol{C}}[\text{Grad}\,\boldsymbol{w}]\,\mathrm{d}v, \end{aligned} \tag{3.57}$$

where we have denoted by $\boldsymbol{P} = \underline{\boldsymbol{C}}[\text{Grad}\,\boldsymbol{u}]$. Thus, the function $F(\varepsilon)$ is quadratic in ε and we can write

$$0 \;\leq\; F(\varepsilon) \;=\; \varepsilon\,\dot{F}(0) + \frac{1}{2}\,\varepsilon^2\ddot{F}(0), \tag{3.58}$$

where superposed dots designate derivatives with respect to ε and

$$\ddot{F}(0) = \int_{\mathscr{K}} (\text{Grad}\,\boldsymbol{w})\cdot\underline{\boldsymbol{C}}[\text{Grad}\,\boldsymbol{w}]\,\mathrm{d}v \;\geq\; 0. \tag{3.59}$$

Now, if we divide (3.58) by ε and let $\varepsilon \searrow 0$ (with $\varepsilon > 0$), then we deduce $\dot{F}(0) \geq 0$. Similarly, divide (3.58) by ε, let $\varepsilon \nearrow 0$ (with $\varepsilon < 0$), and find $\dot{F}(0) \leq 0$. Hence, we have necessarily $\dot{F}(0) = 0$, and from (3.57) we obtain

$$\int_{\mathscr{K}} \big(\text{Div}\,\boldsymbol{P} + \rho_K\boldsymbol{b}\big)\cdot\boldsymbol{w}\,\mathrm{d}v \;-\; \int_{\partial\mathscr{K}_t} \big(\boldsymbol{P}\boldsymbol{n} - \boldsymbol{t}\big)\cdot\boldsymbol{w}\,\mathrm{d}a \;=\; 0, \tag{3.60}$$

which holds for all vector fields $\boldsymbol{w}$ such that $\boldsymbol{w} = \boldsymbol{0}$ on $\partial\mathscr{K}_u$. Finally, using the fundamental lemma of the calculus of variations, we derive from (3.60) that

$$\mathrm{Div}\,\boldsymbol{P} + \rho_K \boldsymbol{b} = \boldsymbol{0} \quad \text{in } \mathscr{K} \qquad \text{and} \qquad \boldsymbol{P}\boldsymbol{n} - \boldsymbol{t} = \boldsymbol{0} \quad \text{on } \partial\mathscr{K}_t\,,$$

so $\boldsymbol{u}(\boldsymbol{x})$ is an equilibrium state. □

Notice that the relation (3.60) is also referred to as the "virtual work theorem" or the "weak form of the equilibrium equations".

In the above proof, we have shown the following result:

Let us denote by $\boldsymbol{u}(\boldsymbol{x}, \varepsilon) := \boldsymbol{u}(\boldsymbol{x}) + \varepsilon\boldsymbol{w}(\boldsymbol{x})$, where $\boldsymbol{w} = \boldsymbol{0}$ on $\partial\mathscr{K}_u$. Introduce the function $F(\varepsilon) := E[\boldsymbol{u}(\boldsymbol{x}, \varepsilon)]$, with $\varepsilon \in (-\varepsilon_0, \varepsilon_0) \subset (-1, 1)$. If the displacement field $\boldsymbol{u}(\boldsymbol{x}, 0)$ is an equilibrium state, then we have

$$\dot{F}(\varepsilon)\big|_{\varepsilon=0} = 0\,, \quad \text{and conversely;} \tag{3.61}$$

i.e., *equilibria yield stationary values of the total potential energy, and conversely, the stationary points of the energy are equilibrium states.*

Moreover, the following statement holds:

If the equilibrium is stable, then the equilibrium state is a minimizer of the energy; conversely, energy minimizers are equilibria (i.e., stationary states).

This statement, as required by physics, is assured by the assumption

$$\boldsymbol{H} \cdot \underline{\boldsymbol{C}}[\boldsymbol{H}] \geq 0\,, \qquad \text{for all} \quad \boldsymbol{H}\,.$$

Thus, the latter assumption is sufficient for the above property (but not necessary!).

Remark This interpretation of (stable) equilibria is especially useful for our purpose because the potential energy E of a shell-like body involves the shell (or plate) thickness h explicitly. Thus, e.g., for a reference configuration of the form $\mathscr{K} = \Omega \times [-h/2\,,\ h/2]$ we can write (see details in Fig. 3.3 and the next chapter)

$$\int_{\mathscr{K}} U\,\mathrm{d}v \;=\; \int_{\Omega} \Big(\int_{-h/2}^{h/2} U\,\mathrm{d}\zeta \Big)\,\mathrm{d}a \;=\; \int_{\Omega} W\,\mathrm{d}a\,,$$

where $W = \displaystyle\int_{-h/2}^{h/2} U\,\mathrm{d}\zeta$ is the areal strain energy density. □

3.4 Legendre-Hadamard Condition for Stability

The Legendre-Hadamard inequality proceeds from the notion that the second variation of the potential energy is necessarily non-negative if an equilibrium state furnishes a minimum of the potential energy (see also (3.59)).

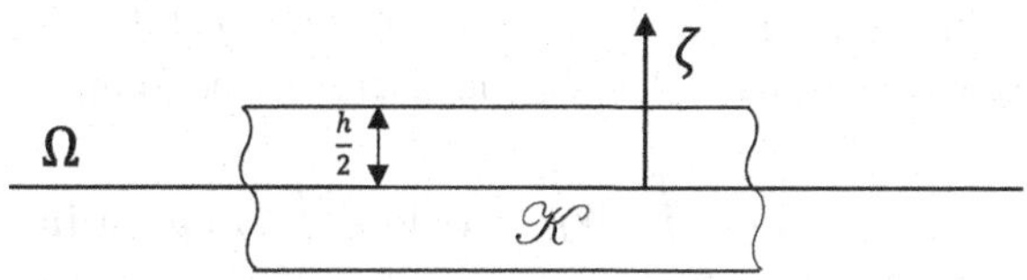

Fig. 3.3 Reference configuration of a thin plate

To present the Legendre-Hadamard condition in a general nonlinear setting, let us consider the deformation field $\boldsymbol{\chi}(\boldsymbol{X})$ of a solid, where $\boldsymbol{X}$ is the position of a material point in the reference configuration $\mathscr{K}$. The field $\boldsymbol{\chi}$ may also depend on time, but this dependence is not made explicit, since it is not important for our purpose. Let us denote by $\boldsymbol{F} = \nabla\boldsymbol{\chi}$ the deformation gradient and by $W(\boldsymbol{F};\boldsymbol{X})$ the strain energy density per unit volume of $\mathscr{K}$. We assume $W(\boldsymbol{F};\boldsymbol{X})$ to be a continuous function of $\boldsymbol{X}$ and twice continuously differentiable with respect to $\boldsymbol{F}$.

In this section, superposed dots stand for variational derivatives: these are ordinary derivatives of one-parameter families of the varied functions with respect to the parameter, evaluated at parameter value zero, which we identify with an equilibrium state. Then, we have

$$\dot{\boldsymbol{F}} = \nabla\boldsymbol{v}\,, \qquad \text{where} \quad \boldsymbol{v} = \dot{\boldsymbol{\chi}}\,. \tag{3.62}$$

Consider the case of dead load; the body force $\boldsymbol{b}$ and the stress vector $\boldsymbol{t}$ are fixed and are not functions of deformation. Concerning boundary conditions, we assume that position is assigned on the subset $\partial\mathscr{K}_u$ and the stress vector $\boldsymbol{t}$ is assigned on $\partial\mathscr{K}_t = \partial\mathscr{K} \setminus \partial\mathscr{K}_u$. Then, the admissibility condition on $\boldsymbol{v}$ is

$$\boldsymbol{v} = \boldsymbol{0} \quad \text{on} \quad \partial\mathscr{K}_u\,. \tag{3.63}$$

The potential energy is given by

$$E = \int_{\mathscr{K}} W\,\mathrm{d}v - \int_{\mathscr{K}} \rho_K \boldsymbol{b}\cdot\boldsymbol{\chi}\,\mathrm{d}v - \int_{\partial\mathscr{K}_t} \boldsymbol{t}\cdot\boldsymbol{\chi}\,\mathrm{d}a, \tag{3.64}$$

apart from an unimportant constant. Equilibria are thus seen to be those states that render the potential energy stationary, i.e.,

$$\dot{E} = 0, \tag{3.65}$$

for all admissible fields $\boldsymbol{v}$. Confining our attention to dead-load problems, we can write the first variation of the potential energy (3.64) in the form

$$\dot{E} = \int_{\mathscr{K}} W_{\boldsymbol{F}}\cdot\nabla\boldsymbol{v}\,\mathrm{d}v - \int_{\mathscr{K}} \rho_K \boldsymbol{b}\cdot\boldsymbol{v}\,\mathrm{d}v - \int_{\partial\mathscr{K}_t} \boldsymbol{t}\cdot\boldsymbol{v}\,\mathrm{d}a, \tag{3.66}$$

which vanishes if and only if the state $\boldsymbol{\chi}$ is equilibrated (see also (3.61)), since $\boldsymbol{b}$ and $\boldsymbol{t}$ are fixed and are not functions of deformation.

On taking a further variation of (3.66) and denoting by $\boldsymbol{w} = \ddot{\boldsymbol{\chi}}$, we obtain the second variation of the potential energy in the form

$$\begin{aligned}\ddot{E} = &\int_{\mathscr{K}} W_{\boldsymbol{F}} \cdot \nabla \boldsymbol{w} \, \mathrm{d}v - \int_{\mathscr{K}} \rho_K \boldsymbol{b} \cdot \boldsymbol{w} \, \mathrm{d}v - \int_{\partial \mathscr{K}_t} \boldsymbol{t} \cdot \boldsymbol{w} \, \mathrm{d}a \\ &+ \int_{\mathscr{K}} \left(W_{\boldsymbol{F}}\right)^{\cdot} \cdot \nabla \boldsymbol{v} \, \mathrm{d}v \,,\end{aligned} \tag{3.67}$$

where by the chain rule we have

$$\left(W_{\boldsymbol{F}}\right)^{\cdot} = W_{\boldsymbol{F}\boldsymbol{F}}\big[\dot{\boldsymbol{F}}\big] = W_{\boldsymbol{F}\boldsymbol{F}}\big[\nabla \boldsymbol{v}\big] \tag{3.68}$$

and $\boldsymbol{w} = \boldsymbol{0}$ on $\partial \mathscr{K}_u$.

If the state $\boldsymbol{\chi}$ is equilibrated, then the first line of relation (3.67) vanishes by virtue of (3.65) and (3.66). Thus, the second variation at equilibrium becomes

$$\ddot{E} = \int_{\mathscr{K}} \nabla \boldsymbol{v} \cdot W_{\boldsymbol{F}\boldsymbol{F}}\big[\nabla \boldsymbol{v}\big] \, \mathrm{d}v \,. \tag{3.69}$$

If the equilibrium state is an energy minimizer, it is necessary that

$$\ddot{E} \geq 0, \tag{3.70}$$

for all $\boldsymbol{v}$ which vanish on $\partial \mathscr{K}_u$.

Let $\mathscr{M}$ denote the tensor of elastic moduli, defined as the second gradient of W, i.e.

$$\mathscr{M} = W_{\boldsymbol{F}\boldsymbol{F}} \,.$$

The following result states the Legendre-Hadamard condition for stability.

Theorem 3.3 *If the second variation of the potential energy is non-negative at an energy-minimizing deformation, i.e.,*

$$\int_{\mathscr{K}} \nabla \boldsymbol{v} \cdot \mathscr{M}(\boldsymbol{F}; \boldsymbol{X})\big[\nabla \boldsymbol{v}\big] \, \mathrm{d}v \geq 0 \tag{3.71}$$

for all $\boldsymbol{v}$ *that vanish on* $\partial \mathscr{K}_u$*, then it is necessary that the Legendre-Hadamard inequality*

$$(\boldsymbol{a} \otimes \boldsymbol{n}) \cdot \mathscr{M}(\boldsymbol{F}; \boldsymbol{X})\big[\boldsymbol{a} \otimes \boldsymbol{n}\big] \geq 0 \tag{3.72}$$

be satisfied at every point $\boldsymbol{X} \in \mathscr{K}$ *and for all vectors* $\boldsymbol{a}$ *and* $\boldsymbol{n}$.

Proof Following the ideas of [2], we consider variations

$$\boldsymbol{v}(\boldsymbol{X}) = \epsilon \boldsymbol{\xi}(\boldsymbol{Y}) \qquad \text{with} \qquad \boldsymbol{Y} = \epsilon^{-1}(\boldsymbol{X} - \boldsymbol{X}_0), \tag{3.73}$$

where $\boldsymbol{X}_0$ is an interior point of $\mathscr{K}$, ϵ is a positive constant, and $\boldsymbol{\xi}$ is compactly supported in a region D, the image of a strictly interior neighborhood $\mathscr{K}' \subset \mathscr{K}$ of $\boldsymbol{X}_0$ under the map $\boldsymbol{Y}(\cdot)$. Accordingly, $\boldsymbol{v}$ vanishes on $\partial\mathscr{K}_u$ and is therefore admissible. For these variations (3.71) reduces, after dividing by ϵ^3, passing to the limit $\epsilon \to 0$ and invoking the Dominated Convergence Theorem, to

$$\int_D \nabla\boldsymbol{\xi} \cdot \mathscr{A}[\nabla\boldsymbol{\xi}]\, dv \geq 0, \tag{3.74}$$

where $\mathscr{A} = \mathscr{M}_{|\boldsymbol{X}_0} = \mathscr{A}^T$. Here and henceforth ∇ is the gradient with respect to $\boldsymbol{Y}$.

We extend $\boldsymbol{\xi}$ to complex-valued vector fields as

$$\boldsymbol{\xi} = \boldsymbol{\xi}_1 + i\,\boldsymbol{\xi}_2, \tag{3.75}$$

where $\boldsymbol{\xi}_1(\boldsymbol{Y})$ and $\boldsymbol{\xi}_2(\boldsymbol{Y})$ are real-valued. We use these fields to derive

$$\nabla\boldsymbol{\xi} \cdot \mathscr{A}[\nabla\bar{\boldsymbol{\xi}}] = \nabla\boldsymbol{\xi}_1 \cdot \mathscr{A}[\nabla\boldsymbol{\xi}_1] + \nabla\boldsymbol{\xi}_2 \cdot \mathscr{A}[\nabla\boldsymbol{\xi}_2], \tag{3.76}$$

in which an overbar denotes complex conjugate. The imaginary part of the above expression vanishes by virtue of the fact that $\mathscr{A}$ possesses major symmetry, so we have

$$\boldsymbol{S} \cdot \mathscr{A}[\boldsymbol{T}] = \boldsymbol{T} \cdot \mathscr{A}^T[\boldsymbol{S}] = \boldsymbol{T} \cdot \mathscr{A}[\boldsymbol{S}] \qquad \text{for any} \quad \boldsymbol{S}, \boldsymbol{T}.$$

Thus, if (3.74) holds for real-valued $\boldsymbol{\xi}$, then it follows that

$$\int_D \nabla\boldsymbol{\xi} \cdot \mathscr{A}[\nabla\bar{\boldsymbol{\xi}}]\, dv \geq 0 \tag{3.77}$$

for complex-valued $\boldsymbol{\xi}$.

Consider

$$\boldsymbol{\xi}(\boldsymbol{Y}) = \boldsymbol{a}\, \exp(ik\boldsymbol{n} \cdot \boldsymbol{Y})\, f(\boldsymbol{Y}), \tag{3.78}$$

where $\boldsymbol{a}$ and $\boldsymbol{n}$ are real fixed vectors, k is a non-zero real number and f is a real-valued differentiable function compactly supported in D. This yields

$$\nabla\boldsymbol{\xi} = \exp(ik\boldsymbol{n} \cdot \boldsymbol{Y})(ikf\, \boldsymbol{a} \otimes \boldsymbol{n} + \boldsymbol{a} \otimes \nabla f). \tag{3.79}$$

Substitution into (3.77) and division by k^2 results in

$$0 \leq (\boldsymbol{a} \otimes \boldsymbol{n}) \cdot \mathscr{A}[\boldsymbol{a} \otimes \boldsymbol{n}] \int_D f^2\, dv + k^{-2} \int_D (\boldsymbol{a} \otimes \nabla f) \cdot \mathscr{A}[\boldsymbol{a} \otimes \nabla f]\, dv. \tag{3.80}$$

Finally, as $k \to \infty$ we obtain

$$(\boldsymbol{a} \otimes \boldsymbol{n}) \cdot \mathscr{A}[\boldsymbol{a} \otimes \boldsymbol{n}] \geq 0, \tag{3.81}$$

which is just the inequality (3.72) on account of the arbitrariness of $\boldsymbol{X}_0$. □

Remark We have shown that the Legendre–Hadamard inequality (3.72) is necessary for (3.71). The converse is generally not true. □

References

1. Ciarlet, P.G.: Mathematical Elasticity: Three-Dimensional Elasticity, vol. I. Elsevier, Amsterdam (1988)
2. Giaquinta, M., Hildebrandt, S.: Calculus of Variations I. Springer, Berlin (2004)
3. Green, A.E., Adkins, J.E.: Large Elastic Deformations. Oxford University Press, Oxford (1970)
4. Green, A.E., Zerna, W.: Theoretical Elasticity. Courier Corporation (1992)
5. Gurtin, M.E.: The Linear Theory of Elasticity. In: Truesdell, C. (ed.) Handbuch der Physik, vol. VIa/2, pp. 1–296. Springer, Berlin (1972)
6. Ogden, R.W.: Nonlinear Elastic Deformations. Dover, Mineola (1997)
7. Steigmann, D.J.: Finite Elasticity Theory. Oxford University Press (2017)

Chapter 4
Linearly Elastic Plates

Abstract To emphasize the main aspects of our procedure in as simple a manner as possible, we start with the theory of thin flat plates. This is based on classical linear elasticity under the assumption that the three-dimensional body is generated by the parallel translation of a flat midsurface. Accordingly the complexities associated with nonlinear elasticity and the differential geometry of curved surfaces are deferred to later chapters. The derived areal energy density of the plate is used to obtain the relevant two-dimensional equilibrium equations via variational methods.

4.1 Three-dimensional Theory

Let us recall briefly the important equations of three-dimensional linear elasticity theory, which will be used in this chapter. For a more detailed presentation we refer to Chap. 3.

Let $\tilde{\boldsymbol{u}}$ be the displacement vector field of a three-dimensional body ($\tilde{\boldsymbol{u}}$ is assumed to be small) and let us designate by $\tilde{\boldsymbol{H}} = \text{Grad}\,\tilde{\boldsymbol{u}}$ its gradient. According to (3.42), the strain energy is given by

$$U = \frac{1}{2}\,\tilde{\boldsymbol{H}} \cdot \underline{\boldsymbol{C}}[\tilde{\boldsymbol{H}}] \tag{4.1}$$

and the Piola stress tensor is (when the reference configuration is stress free)

$$\tilde{\boldsymbol{P}} \;=\; \underline{\boldsymbol{C}}[\tilde{\boldsymbol{H}}] \;=\; U_{\tilde{\boldsymbol{H}}}\,, \tag{4.2}$$

where the tensor of elastic moduli $\underline{\boldsymbol{C}}$ satisfies the inequality

$$\boldsymbol{A} \cdot \underline{\boldsymbol{C}}[\boldsymbol{A}] > 0\,, \qquad \text{for all non-zero symmetric tensors } \boldsymbol{A}. \tag{4.3}$$

Consider tensors $\boldsymbol{A}$ of the form $\boldsymbol{A} = \boldsymbol{v} \otimes \boldsymbol{w}$. We note that

$$\text{sym}(\boldsymbol{v} \otimes \boldsymbol{w}) = \boldsymbol{0} \quad \text{ if and only if } \quad \boldsymbol{v} \otimes \boldsymbol{w} = \boldsymbol{0}.$$

D. J. Steigmann et al., *Lecture Notes on the Theory of Plates and Shells*,
Solid Mechanics and Its Applications 274,
https://doi.org/10.1007/978-3-031-25674-5_4

Then, from (4.3) we deduce

$$(\boldsymbol{v}\otimes\boldsymbol{w})\cdot\underline{\boldsymbol{C}}[\boldsymbol{v}\otimes\boldsymbol{w}] > 0\,, \qquad \text{for all} \quad \boldsymbol{v}\otimes\boldsymbol{w}\neq\boldsymbol{0}. \tag{4.4}$$

Thus, we have the "strong ellipticity" of the tensor of elastic moduli $\underline{\boldsymbol{C}}$.

The equation of equilibrium in the absence of body loads has the form

$$\operatorname{Div}\tilde{\boldsymbol{P}} = \boldsymbol{0}, \tag{4.5}$$

according to (3.43). In the case of a dead-traction/displacement boundary value problem, the potential energy given in (3.45) becomes

$$\mathscr{E}[\tilde{\boldsymbol{u}}] = \int_{\mathscr{K}} U\,\mathrm{d}v - \int_{\partial\mathscr{K}_t} \tilde{\boldsymbol{t}}\cdot\tilde{\boldsymbol{u}}\,\mathrm{d}a\,, \tag{4.6}$$

where the stress vector $\tilde{\boldsymbol{t}} = \tilde{\boldsymbol{P}}\boldsymbol{n}$ is assigned on the subset $\partial\mathscr{K}_t$ of the boundary.

4.2 Derivation of the Plate Model

Let us consider now the deformation of elastic plates (flat shells). Assume that the thin body occupies in the reference configuration a domain of the form (Fig. 4.1)

$$\mathscr{K} = \Big\{(x_1,x_2,x_3)\;\big|\;(x_1,x_2)\in\Omega\,,\;x_3\in\Big[-\frac{h}{2},\frac{h}{2}\Big]\Big\} = \Omega\times\Big[-\frac{h}{2},\frac{h}{2}\Big].$$

The position vector of a generic point in $\mathscr{K}$ is given by

$$\boldsymbol{x} = \boldsymbol{r} + \zeta\,\boldsymbol{k}\,, \qquad \text{with} \;\; \boldsymbol{r} = r_\alpha\,\boldsymbol{e}_\alpha \in \Omega \;\; \text{and} \;\; \boldsymbol{k} = \boldsymbol{e}_3\,, \tag{4.7}$$

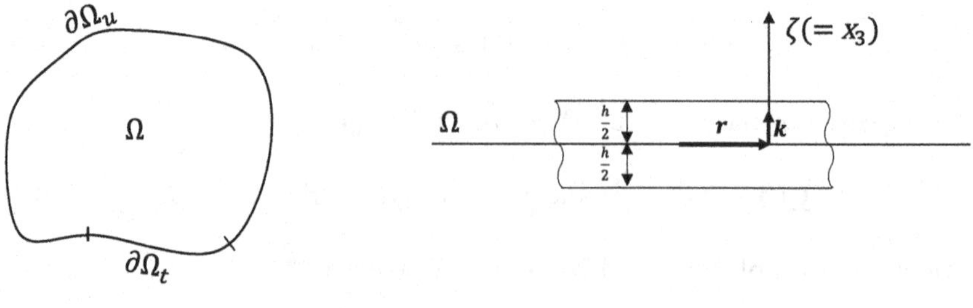

Fig. 4.1 Reference configuration of the plate (flat shell)

where we denote for convenience the coordinates by $x_\alpha = r_\alpha$ and $x_3 = \zeta$. Thus, the thickness coordinate is $\zeta \in \left[-\frac{h}{2}, \frac{h}{2}\right]$ and the thickness h is small.

If we designate by ℓ a typical spanwise dimension of the plate, then we assume that $h/\ell \ll 1$. Further, we take ℓ as the unit of length, so we have $\ell = 1$ and

$$h \ll 1. \tag{4.8}$$

Let us decompose the three-dimensional identity tensor as

$$\boldsymbol{I} = \boldsymbol{e}_\alpha \otimes \boldsymbol{e}_\alpha + \boldsymbol{k} \otimes \boldsymbol{k} = \mathbf{1} + \boldsymbol{k} \otimes \boldsymbol{k}, \tag{4.9}$$

where

$$\mathbf{1} = \boldsymbol{e}_\alpha \otimes \boldsymbol{e}_\alpha = \boldsymbol{I} - \boldsymbol{k} \otimes \boldsymbol{k} \tag{4.10}$$

designates the projection onto Ω (and the two–dimensional identity tensor on Ω). Then, in view of (4.7) and (4.9), we have

$$\mathrm{d}\boldsymbol{x} = \boldsymbol{I}\mathrm{d}\boldsymbol{x} = \mathbf{1}\mathrm{d}\boldsymbol{x} + \boldsymbol{k}(\boldsymbol{k} \cdot \mathrm{d}\boldsymbol{x}) = \mathrm{d}\boldsymbol{r} + \boldsymbol{k}\,\mathrm{d}\zeta \tag{4.11}$$

and, hence,

$$\mathrm{d}\tilde{\boldsymbol{u}} = \tilde{\boldsymbol{H}}\mathrm{d}\boldsymbol{x} = (\tilde{\boldsymbol{H}}\mathbf{1} + \tilde{\boldsymbol{H}}\boldsymbol{k} \otimes \boldsymbol{k})\mathrm{d}\boldsymbol{x} = (\tilde{\boldsymbol{H}}\mathbf{1})\mathrm{d}\boldsymbol{r} + (\tilde{\boldsymbol{H}}\boldsymbol{k})\mathrm{d}\zeta\,. \tag{4.12}$$

Let

$$\hat{\boldsymbol{u}}(\boldsymbol{r}, \zeta) := \tilde{\boldsymbol{u}}(\boldsymbol{r} + \zeta\,\boldsymbol{k}) \quad \text{and} \quad \hat{\boldsymbol{H}}(\boldsymbol{r}, \zeta) := \tilde{\boldsymbol{H}}(\boldsymbol{r} + \zeta\,\boldsymbol{k}),$$

and let $\nabla(\,\cdot\,)$ designate the two-dimensional gradient with respect to $\boldsymbol{r}$ and $(\,\cdot\,)' = \dfrac{\partial(\,\cdot\,)}{\partial\zeta}$ the derivative with respect to ζ at fixed $\boldsymbol{r}$. Then, we can write

$$\mathrm{d}\tilde{\boldsymbol{u}} = \mathrm{d}\hat{\boldsymbol{u}} = (\nabla\hat{\boldsymbol{u}})\mathrm{d}\boldsymbol{r} + \hat{\boldsymbol{u}}'\mathrm{d}\zeta\,. \tag{4.13}$$

Comparing (4.12) and (4.13), we deduce

$$\tilde{\boldsymbol{H}}\mathbf{1} = \nabla\hat{\boldsymbol{u}}\,, \qquad \tilde{\boldsymbol{H}}\boldsymbol{k} = \hat{\boldsymbol{u}}' \tag{4.14}$$

and

$$\tilde{\boldsymbol{H}} = \tilde{\boldsymbol{H}}\boldsymbol{I} = \tilde{\boldsymbol{H}}\mathbf{1} + \tilde{\boldsymbol{H}}\boldsymbol{k} \otimes \boldsymbol{k} = \nabla\hat{\boldsymbol{u}} + \hat{\boldsymbol{u}}' \otimes \boldsymbol{k}\,. \tag{4.15}$$

4.2.1 Integration over the Thickness

The strain energy of the plate is

$$\mathscr{S} := \int_{\mathscr{K}} U(\tilde{\boldsymbol{H}})\,\mathrm{d}v = \int_{\Omega}\Big(\int_{-h/2}^{h/2} U\big(\hat{\boldsymbol{H}}(\boldsymbol{r},\zeta)\big)\,\mathrm{d}\zeta\Big)\,\mathrm{d}a = \int_{\Omega} \bar{W}\,\mathrm{d}a\,, \tag{4.16}$$

where $\bar{W}$ is the strain energy density per unit area of Ω. We have

$$\bar{W} = \int_{-h/2}^{h/2} G(\zeta)\,\mathrm{d}\zeta =: I(h)\,, \qquad \text{with} \quad G(\zeta) := U\big(\hat{\boldsymbol{H}}(\boldsymbol{r},\zeta)\big), \tag{4.17}$$

and we have denoted the above integral by $I(h)$.

Now, we can expand the function $I(h)$ in the following form

$$I(h) = I(0) + h\,I'(0) + \frac{1}{2}\,h^2 I''(0) + \frac{1}{6}\,h^3 I'''(0) + \cdots\,, \tag{4.18}$$

where (in view of the definition $(4.17)_1$ and the Leibniz Rule)

$$\begin{aligned} I'(h) &= \tfrac{1}{2}\big[G(\tfrac{h}{2}) + G(-\tfrac{h}{2})\big], & I'(0) &= G(0),\\ I''(h) &= \tfrac{1}{4}\big[G'(\tfrac{h}{2}) - G'(-\tfrac{h}{2})\big], & I''(0) &= 0,\\ I'''(h) &= \tfrac{1}{8}\big[G''(\tfrac{h}{2}) + G''(-\tfrac{h}{2})\big], & I'''(0) &= \tfrac{1}{4}\,G''(0). \end{aligned} \tag{4.19}$$

Thus, on the basis of (4.17)–(4.19) we obtain

$$\bar{W} = h\,G(0) + \frac{1}{24}\,h^3 G''(0) + o(h^3). \tag{4.20}$$

In what follows, we consider the case of uniform materials (with no residual stress). Also, for any field $f = f(\boldsymbol{r},\zeta)$, let $f_0 := f(\boldsymbol{r},0)$. Then, in view of (4.1), (4.2), and $(4.17)_2$, we deduce

$$\begin{aligned} &G(0) = U(\hat{\boldsymbol{H}}_0) = \tfrac{1}{2}\,\boldsymbol{P}_0 \cdot \hat{\boldsymbol{H}}_0\,, \qquad G'(0) = \big(U_{\tilde{\boldsymbol{H}}} \cdot \hat{\boldsymbol{H}}'\big)_{|\zeta=0} = \boldsymbol{P}_0 \cdot \hat{\boldsymbol{H}}'_0\,,\\ &G''(0) = \big(\tilde{\boldsymbol{P}} \cdot \hat{\boldsymbol{H}}'\big)'_{|\zeta=0} = \boldsymbol{P}'_0 \cdot \hat{\boldsymbol{H}}'_0 + \boldsymbol{P}_0 \cdot \hat{\boldsymbol{H}}''_0\,, \end{aligned} \tag{4.21}$$

where

$$\boldsymbol{P}_0 = \underline{C}[\hat{\boldsymbol{H}}_0] \qquad \text{and} \qquad \boldsymbol{P}'_0 = \underline{C}[\hat{\boldsymbol{H}}'_0]. \tag{4.22}$$

In order to express the gradient of displacement appearing in the above relations, we write the expansion of $\tilde{\boldsymbol{u}}$ with respect to ζ in the form

$$\tilde{\boldsymbol{u}}(\boldsymbol{x}) = \hat{\boldsymbol{u}}(\boldsymbol{r},\zeta) = \boldsymbol{u}(\boldsymbol{r}) + \zeta\, \boldsymbol{a}(\boldsymbol{r}) + \frac{1}{2}\zeta^2\, \boldsymbol{b}(\boldsymbol{r}) + \frac{1}{6}\zeta^3\, \boldsymbol{c}(\boldsymbol{r}) + \cdots, \tag{4.23}$$

where

$$\begin{aligned} \boldsymbol{u}(\boldsymbol{r}) &:= \hat{\boldsymbol{u}}_0 = \hat{\boldsymbol{u}}(\boldsymbol{r},0), \quad \boldsymbol{a}(\boldsymbol{r}) := \hat{\boldsymbol{u}}_0' = \tfrac{\partial \hat{\boldsymbol{u}}}{\partial \zeta}\Big|_{\zeta=0}, \\ \boldsymbol{b}(\boldsymbol{r}) &:= \hat{\boldsymbol{u}}_0'' = \tfrac{\partial^2 \hat{\boldsymbol{u}}}{\partial \zeta^2}\Big|_{\zeta=0}, \quad \boldsymbol{c}(\boldsymbol{r}) := \hat{\boldsymbol{u}}_0''' = \tfrac{\partial^3 \hat{\boldsymbol{u}}}{\partial \zeta^3}\Big|_{\zeta=0}. \end{aligned} \tag{4.24}$$

Then, in view of (4.15) and (4.24) we find for the displacement gradient

$$\hat{\boldsymbol{H}}_0 = \nabla \boldsymbol{u} + \boldsymbol{a} \otimes \boldsymbol{k}, \qquad \hat{\boldsymbol{H}}_0' = \nabla \boldsymbol{a} + \boldsymbol{b} \otimes \boldsymbol{k}, \qquad \hat{\boldsymbol{H}}_0'' = \nabla \boldsymbol{b} + \boldsymbol{c} \otimes \boldsymbol{k}. \tag{4.25}$$

Thus, the relation (4.20) becomes

$$\begin{aligned} \bar{W} &= W + o(h^3), \qquad \text{with} \\ W(\nabla \boldsymbol{u}, \nabla \boldsymbol{a}, \nabla \boldsymbol{b}, \boldsymbol{a}, \boldsymbol{b}, \boldsymbol{c}) &= \tfrac{1}{2}\, h\, \boldsymbol{P}_0 \cdot \hat{\boldsymbol{H}}_0 + \tfrac{1}{24}\, h^3 \big(\boldsymbol{P}_0' \cdot \hat{\boldsymbol{H}}_0' + \boldsymbol{P}_0 \cdot \hat{\boldsymbol{H}}_0''\big) \end{aligned} \tag{4.26}$$

and the energy (4.16) can be written in the form

$$\mathscr{S} = S + o(h^3), \qquad \text{where} \qquad S := \int_{\Omega} W\, \mathrm{da}. \tag{4.27}$$

4.2.2 *Edge Loads*

Let $\partial\Omega = \partial\Omega_u \cup \partial\Omega_t$ be a disjoint partition of the boundary curve of Ω. As boundary conditions for the plate, we assume that the displacement $\tilde{\boldsymbol{u}}$ is assigned on

$$\partial\mathscr{K}_u := \partial\Omega_u \times \Big[-\frac{h}{2}, \frac{h}{2}\Big] \subset \partial\mathscr{K},$$

while the stress vector $\tilde{\boldsymbol{t}}$ is assigned on the rest of the boundary $\partial\mathscr{K}_t := \partial\mathscr{K} \setminus \partial\mathscr{K}_u$. Then, $\partial\mathscr{K}_t$ consists of a portion of edge boundary and of the lateral faces, i.e.

$$\partial\mathscr{K}_t = \partial\mathscr{K}_{tl} \times \Big\{(\boldsymbol{r},\zeta)\ \Big|\ \boldsymbol{r} \in \Omega,\ \zeta = \pm\frac{h}{2}\Big\}, \quad \text{with} \quad \partial\mathscr{K}_{tl} := \partial\Omega_t \times \Big[-\frac{h}{2}, \frac{h}{2}\Big].$$

Consider the following integral appearing in the potential energy (4.6)

$$\int_{\partial\mathscr{K}_t} \tilde{\boldsymbol{t}} \cdot \tilde{\boldsymbol{u}}\, \mathrm{da} = \underbrace{\int_{\partial\mathscr{K}_{tl}} \tilde{\boldsymbol{t}} \cdot \tilde{\boldsymbol{u}}\, \mathrm{da}}_{\text{edge loads}} + \underbrace{\int_{\partial\mathscr{K}_t \setminus \partial\mathscr{K}_{tl}} \tilde{\boldsymbol{t}} \cdot \tilde{\boldsymbol{u}}\, \mathrm{da}}_{\text{lateral loads}}. \tag{4.28}$$

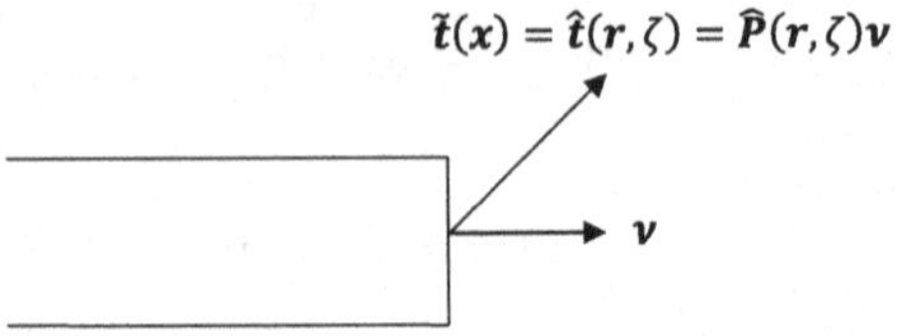

Fig. 4.2 The stress vector on the edge boundary

In this section we confine our attention to edge loads and we shall consider the lateral loads later. On the edge boundary, the stress vector is given by $\tilde{\boldsymbol{t}}(\boldsymbol{x}) = \hat{\boldsymbol{t}}(\boldsymbol{r},\zeta) = \hat{\boldsymbol{P}}(\boldsymbol{r},\zeta)\boldsymbol{v}$, where $\boldsymbol{v}$ is the unit outward normal vector to the plane curve $\partial\Omega_t$, see Fig. 4.2. Then, we have

$$\int_{\partial\mathscr{K}_{tI}} \tilde{\boldsymbol{t}}\cdot\tilde{\boldsymbol{u}}\,\mathrm{d}a = \int_{\partial\Omega_t}\Big(\int_{-h/2}^{h/2}\tilde{\boldsymbol{t}}\cdot\tilde{\boldsymbol{u}}\,\mathrm{d}\zeta\Big)\,\mathrm{d}s\,. \tag{4.29}$$

To estimate this integral, we expand $\tilde{\boldsymbol{t}}$ with respect to ζ

$$\tilde{\boldsymbol{t}}(\boldsymbol{x}) = \hat{\boldsymbol{t}}(\boldsymbol{r},\zeta) = \boldsymbol{t}_0 + \zeta\,\boldsymbol{t}_0' + \frac{1}{2}\zeta^2\,\boldsymbol{t}_0'' + \cdots, \tag{4.30}$$

where we denote the derivatives by $\boldsymbol{t}_0^{(n)} := \hat{\boldsymbol{t}}^{(n)}\big|_{\zeta=0}$. Then, in view of (4.23) we deduce

$$\tilde{\boldsymbol{t}}\cdot\tilde{\boldsymbol{u}} = \boldsymbol{t}_0\cdot\boldsymbol{u} + \zeta\,(\boldsymbol{t}_0'\cdot\boldsymbol{u} + \boldsymbol{t}_0\cdot\boldsymbol{a}) + \zeta^2\Big(\frac{1}{2}\boldsymbol{t}_0\cdot\boldsymbol{b} + \boldsymbol{t}_0'\cdot\boldsymbol{a} + \frac{1}{2}\boldsymbol{t}_0''\cdot\boldsymbol{u}\Big) + O(\zeta^3), \tag{4.31}$$

and inserting in (4.29) we obtain by integration

$$\int_{\partial\mathscr{K}_{tI}} \tilde{\boldsymbol{t}}\cdot\tilde{\boldsymbol{u}}\,\mathrm{d}a = \int_{\partial\Omega_t}(\boldsymbol{p}_u\cdot\boldsymbol{u} + \boldsymbol{p}_a\cdot\boldsymbol{a} + \boldsymbol{p}_b\cdot\boldsymbol{b})\,\mathrm{d}s + o(h^3), \tag{4.32}$$

where we have introduced the notations

$$\boldsymbol{p}_u := h\,\boldsymbol{t}_0 + \frac{1}{24}h^3\boldsymbol{t}_0'', \qquad \boldsymbol{p}_a := \frac{1}{12}h^3\boldsymbol{t}_0', \qquad \boldsymbol{p}_b := \frac{1}{24}h^3\boldsymbol{t}_0\,. \tag{4.33}$$

Altogether, on the basis of Eqs. (4.6), (4.27) and (4.32), the potential energy of an edge-loaded plate is

$$\mathscr{E} = E + o(h^3), \qquad \text{with}$$
$$E = \int_\Omega W\,\mathrm{d}a - \int_{\partial\Omega_t}(\boldsymbol{p}_u\cdot\boldsymbol{u} + \boldsymbol{p}_a\cdot\boldsymbol{a} + \boldsymbol{p}_b\cdot\boldsymbol{b})\,\mathrm{d}s\,. \tag{4.34}$$

The contribution of lateral loads to the potential energy will be added later, see Sect. 4.2.4.

4.2.3 Optimal Expressions

We seek optimal expressions for the vectors $\boldsymbol{a}, \boldsymbol{b}, \boldsymbol{c}$, for a given midplane displacement $\boldsymbol{u}(\boldsymbol{r})$. To this end, we use results from the (exact) three-dimensional theory.

Let $\tilde{\boldsymbol{t}}^{+}$ and $\tilde{\boldsymbol{t}}^{-}$ be the stress vectors on the lateral faces, i.e. $\tilde{\boldsymbol{t}}^{\pm} := \hat{\boldsymbol{t}}(\boldsymbol{r}, \pm\frac{h}{2})$. In view of Fig. 4.3, we have $\tilde{\boldsymbol{t}}^{\pm} = \pm\tilde{\boldsymbol{P}}^{\pm}\boldsymbol{k}$, where $\tilde{\boldsymbol{P}}^{\pm} := \hat{\boldsymbol{P}}(\boldsymbol{r}, \pm\frac{h}{2})$.

We employ the expansion of the Piola stress tensor with respect to ζ

$$\tilde{\boldsymbol{P}}(\boldsymbol{x}) = \hat{\boldsymbol{P}}(\boldsymbol{r}, \zeta) = \boldsymbol{P}_0 + \zeta\, \boldsymbol{P}_0' + \frac{1}{2}\zeta^2\, \boldsymbol{P}_0'' + \cdots ,$$

written for the values $\zeta = \pm\frac{h}{2}$. Then, we deduce

$$\begin{aligned} \tilde{\boldsymbol{t}}^{+} &= \tilde{\boldsymbol{P}}^{+}\boldsymbol{k} = \boldsymbol{P}_0\boldsymbol{k} + \tfrac{h}{2}\, \boldsymbol{P}_0'\boldsymbol{k} + \tfrac{h^2}{8}\, \boldsymbol{P}_0''\boldsymbol{k} + O(h^3), \\ \tilde{\boldsymbol{t}}^{-} &= -\tilde{\boldsymbol{P}}^{-}\boldsymbol{k} = -\boldsymbol{P}_0\boldsymbol{k} + \tfrac{h}{2}\, \boldsymbol{P}_0'\boldsymbol{k} - \tfrac{h^2}{8}\, \boldsymbol{P}_0''\boldsymbol{k} + O(h^3), \end{aligned} \tag{4.35}$$

and, hence,

$$\begin{aligned} \tilde{\boldsymbol{t}}^{+} + \tilde{\boldsymbol{t}}^{-} &= h\, \boldsymbol{P}_0'\boldsymbol{k} + O(h^3), \\ \tilde{\boldsymbol{t}}^{+} - \tilde{\boldsymbol{t}}^{-} &= 2\boldsymbol{P}_0\boldsymbol{k} + O(h^2). \end{aligned} \tag{4.36}$$

Note that the sum $\tilde{\boldsymbol{t}}^{+} + \tilde{\boldsymbol{t}}^{-}$ represents the net lateral traction.

Also, recall that the equilibrium equation has the form $\mathrm{Div}\, \tilde{\boldsymbol{P}} = \boldsymbol{0}$ in $\mathscr{K}$, see (4.5). Using the decomposition $\tilde{\boldsymbol{P}} = \tilde{\boldsymbol{P}}\boldsymbol{I} = \tilde{\boldsymbol{P}}\boldsymbol{1} + \tilde{\boldsymbol{P}}\boldsymbol{k} \otimes \boldsymbol{k}$ we can express the divergence as

$$\mathrm{Div}\, \tilde{\boldsymbol{P}} = \mathrm{div}(\tilde{\boldsymbol{P}}\boldsymbol{1}) + \tilde{\boldsymbol{P}}'\boldsymbol{k}\,, \tag{4.37}$$

where $\mathrm{div}(\,\cdot\,) = \mathrm{tr}[\nabla(\,\cdot\,)]$ designates the two-dimensional divergence in Ω. Thus, we impose the equilibrium equation (4.5) and evaluate it on Ω (i.e., $\zeta = 0$):

$$\mathrm{div}(\boldsymbol{P}_0\boldsymbol{1}) + \boldsymbol{P}_0'\boldsymbol{k} = \boldsymbol{0} \quad \text{on} \;\; \Omega. \tag{4.38}$$

Notice that this equation is exact; no approximation is involved.

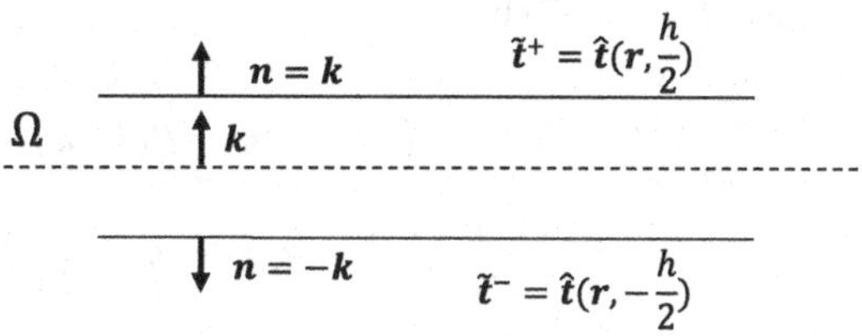

Fig. 4.3 The stress vector on the lateral faces

Suppose now that the stress vectors on the lateral faces vanish, i.e.

$$\tilde{\boldsymbol{t}}^{+} = \mathbf{0}, \qquad \tilde{\boldsymbol{t}}^{-} = \mathbf{0}. \tag{4.39}$$

Then, from (4.36) we obtain

$$\boldsymbol{P}_0\boldsymbol{k} = O(h^2) \qquad \text{and} \qquad \boldsymbol{P}_0'\boldsymbol{k} = O(h^2). \tag{4.40}$$

Further, in view of (4.38) we deduce that

$$\mathrm{div}(\boldsymbol{P}_0\mathbf{1}) = O(h^2). \tag{4.41}$$

Remark Using the same argument, one can show that the relations (4.40) and (4.41) are also valid if we replace the assumptions (4.39) by

$$\tilde{\boldsymbol{t}}^{\pm} = O(h^3). \tag{4.42}$$

□

If we insert the relations (4.25) and the decomposition $\boldsymbol{P}_0 = \boldsymbol{P}_0\mathbf{1} + \boldsymbol{P}_0\boldsymbol{k}\otimes\boldsymbol{k}$ into Eq. $(4.26)_2$, then we obtain the following form of the strain energy density

$$\begin{aligned} W &= \tfrac{1}{2}\,h\left(\boldsymbol{P}_0\mathbf{1}\cdot\nabla\boldsymbol{u} + \boldsymbol{P}_0\boldsymbol{k}\cdot\boldsymbol{a}\right) \\ &\quad + \tfrac{1}{24}\,h^3\left(\boldsymbol{P}_0'\mathbf{1}\cdot\nabla\boldsymbol{a} + \boldsymbol{P}_0'\boldsymbol{k}\cdot\boldsymbol{b} + \boldsymbol{P}_0\mathbf{1}\cdot\nabla\boldsymbol{b} + \boldsymbol{P}_0\boldsymbol{k}\cdot\boldsymbol{c}\right). \end{aligned} \tag{4.43}$$

Further, using Eq. (4.38) and a relation of the type (3.47), we get

$$\begin{aligned} \boldsymbol{P}_0'\boldsymbol{k}\cdot\boldsymbol{b} + \boldsymbol{P}_0\mathbf{1}\cdot\nabla\boldsymbol{b} &= \boldsymbol{P}_0'\boldsymbol{k}\cdot\boldsymbol{b} + \mathrm{div}\big[(\boldsymbol{P}_0\mathbf{1})^T\boldsymbol{b}\big] - \mathrm{div}(\boldsymbol{P}_0\mathbf{1})\cdot\boldsymbol{b} \\ &= \mathrm{div}\big[(\boldsymbol{P}_0\mathbf{1})^T\boldsymbol{b}\big] + 2\boldsymbol{P}_0'\boldsymbol{k}\cdot\boldsymbol{b}\,. \end{aligned} \tag{4.44}$$

Substituting (4.44) into (4.43) and taking into account (4.40), we find

$$\begin{aligned} W &= \tfrac{1}{2}\,h\left(\boldsymbol{P}_0\mathbf{1}\cdot\nabla\boldsymbol{u} + \boldsymbol{P}_0\boldsymbol{k}\cdot\boldsymbol{a}\right) + \tfrac{1}{24}\,h^3\left(\boldsymbol{P}_0'\mathbf{1}\cdot\nabla\boldsymbol{a}\right) \\ &\quad + \tfrac{1}{24}\,h^3\,\mathrm{div}\big[(\boldsymbol{P}_0\mathbf{1})^T\boldsymbol{b}\big] + o(h^3) \end{aligned} \tag{4.45}$$

and the energy (4.27) becomes

$$\begin{aligned} S = \int_\Omega W\,\mathrm{d}a &= \frac{1}{2}\,h\int_\Omega\left(\boldsymbol{P}_0\mathbf{1}\cdot\nabla\boldsymbol{u} + \boldsymbol{P}_0\boldsymbol{k}\cdot\boldsymbol{a}\right)\mathrm{d}a + \frac{1}{24}\,h^3\int_\Omega\left(\boldsymbol{P}_0'\mathbf{1}\cdot\nabla\boldsymbol{a}\right)\mathrm{d}a \\ &\quad + \frac{1}{24}\,h^3\int_{\partial\Omega}(\boldsymbol{P}_0\mathbf{1})\boldsymbol{\nu}\cdot\boldsymbol{b}\,\mathrm{d}s + o(h^3). \end{aligned} \tag{4.46}$$

Also, using $(4.33)_3$ and $\boldsymbol{t}_0 = \boldsymbol{P}_0\boldsymbol{\nu} = \boldsymbol{P}_0\mathbf{1}\boldsymbol{\nu}$ on $\partial\Omega_t$, we get for the contribution of edge loads

$$\int_{\partial\Omega_t} \boldsymbol{p}_b \cdot \boldsymbol{b}\, \mathrm{d}s = \frac{1}{24} h^3 \int_{\partial\Omega_t} \boldsymbol{P}_0 \mathbf{1}\boldsymbol{\nu} \cdot \boldsymbol{b}\, \mathrm{d}s \,. \tag{4.47}$$

If we insert the relations (4.46) and (4.47) into (4.34) and neglect the terms of order $o(h^3)$, we obtain

$$\begin{aligned} E &= \frac{1}{2} h \int_{\Omega} \big(\boldsymbol{P}_0 \mathbf{1} \cdot \nabla \boldsymbol{u} + \boldsymbol{P}_0 \boldsymbol{k} \cdot \boldsymbol{a}\big)\, \mathrm{d}a + \frac{1}{24} h^3 \int_{\Omega} \boldsymbol{P}'_0 \mathbf{1} \cdot \nabla \boldsymbol{a}\, \mathrm{d}a \\ &+ \frac{1}{24} h^3 \int_{\partial\Omega_u} \boldsymbol{P}_0 \mathbf{1}\boldsymbol{\nu} \cdot \boldsymbol{b}\, \mathrm{d}s - \int_{\partial\Omega_t} \big(\boldsymbol{p}_u \cdot \boldsymbol{u} + \boldsymbol{p}_a \cdot \boldsymbol{a}\big)\, \mathrm{d}s \,. \end{aligned} \tag{4.48}$$

Here, we notice that the first integrand (which determines the term of order $O(h)$) can be written as

$$\frac{1}{2} \big(\boldsymbol{P}_0 \mathbf{1} \cdot \nabla \boldsymbol{u} + \boldsymbol{P}_0 \boldsymbol{k} \cdot \boldsymbol{a}\big) = \frac{1}{2} \boldsymbol{P}_0 \cdot \boldsymbol{H}_0 = \frac{1}{2} \boldsymbol{H}_0 \cdot \underline{\boldsymbol{C}}[\boldsymbol{H}_0] = U(\boldsymbol{H}_0),$$

which is equal to the strain energy density U evaluated on the midplane Ω.

Remark In the course of the derivation (see (4.43)–(4.45)), we have suppressed $\boldsymbol{P}_0\boldsymbol{k}$ and $\boldsymbol{P}'_0\boldsymbol{k}$ in the term of order $O(h^3)$. That is, in the coefficient of h^3 we put

$$\boldsymbol{P}_0 \boldsymbol{k} = \mathbf{0} \ \text{(plane stress on } \Omega) \quad \text{and} \quad \boldsymbol{P}'_0 \boldsymbol{k} = \mathbf{0}. \tag{4.49}$$

Recall from (4.22) and (4.25) that

$$\boldsymbol{P}_0 = \underline{\boldsymbol{C}}[\nabla \boldsymbol{u} + \boldsymbol{a} \otimes \boldsymbol{k}] \quad \text{and} \quad \boldsymbol{P}'_0 = \underline{\boldsymbol{C}}[\nabla \boldsymbol{a} + \boldsymbol{b} \otimes \boldsymbol{k}] \tag{4.50}$$

for uniform materials. Then, we can write

$$\boldsymbol{P}_0 \boldsymbol{k} = \big(\underline{\boldsymbol{C}}[\nabla \boldsymbol{u}]\big)\boldsymbol{k} + \boldsymbol{A}\boldsymbol{a} \,, \tag{4.51}$$

where we define the second-order tensor $\boldsymbol{A}$ (also called "acoustic tensor") by the relation

$$\boldsymbol{A}\boldsymbol{a} := \big(\underline{\boldsymbol{C}}[\boldsymbol{a} \otimes \boldsymbol{k}]\big)\boldsymbol{k} \,, \tag{4.52}$$

which is a linear function of $\boldsymbol{a}$. Note that

$$\boldsymbol{a} \cdot \boldsymbol{A}\boldsymbol{a} = \boldsymbol{a} \cdot \big(\underline{\boldsymbol{C}}[\boldsymbol{a} \otimes \boldsymbol{k}]\big)\boldsymbol{k} = (\boldsymbol{a} \otimes \boldsymbol{k}) \cdot \underline{\boldsymbol{C}}[\boldsymbol{a} \otimes \boldsymbol{k}] > 0, \quad \text{for all } \boldsymbol{a} \neq \mathbf{0}, \tag{4.53}$$

by the strong ellipticity condition (4.4). Thus, $\boldsymbol{A}$ defined by (4.52) is positive definite. Let us show that the tensor $\boldsymbol{A}$ is also symmetric, due to the major symmetries of $\underline{\boldsymbol{C}}$. Indeed, in view of (4.7) we have for the case of plates $\boldsymbol{G}_i = \boldsymbol{G}^i = \boldsymbol{e}_i$ and $\boldsymbol{k} = \boldsymbol{G}_3$. Then, from (3.20) and $\boldsymbol{a} = a_j \boldsymbol{G}^j$ we get

$$\underline{\boldsymbol{C}}[\boldsymbol{a} \otimes \boldsymbol{k}] = C^{ilj3} a_j \boldsymbol{G}_i \otimes \boldsymbol{G}_l \quad \text{and} \quad \underline{\boldsymbol{C}}[\boldsymbol{a} \otimes \boldsymbol{k}]\boldsymbol{k} = C^{i3j3} a_j \boldsymbol{G}_i \,. \tag{4.54}$$

Hence, the tensor $\boldsymbol{A}$ defined in (4.52) is given by

$$\boldsymbol{A} = A^{ij}\boldsymbol{G}_i \otimes \boldsymbol{G}_j\,, \qquad \text{with} \qquad A^{ij} = C^{i3j3}\,, \tag{4.55}$$

which shows that $\boldsymbol{A}$ is symmetric, by virtue of (3.21).

Thus, the tensor $\boldsymbol{A}$ is invertible and the equation $\boldsymbol{P}_0\boldsymbol{k} = \boldsymbol{0}$ can be solved using (4.51) to obtain

$$\boldsymbol{a} = -\boldsymbol{A}^{-1}\big(\underline{\boldsymbol{C}}[\nabla\boldsymbol{u}]\big)\boldsymbol{k} =: \boldsymbol{f}(\nabla\boldsymbol{u}), \tag{4.56}$$

where $\boldsymbol{f}(\cdot)$ is the linear function determined by relation (4.56).

Similarly, we have $\boldsymbol{P}_0'\boldsymbol{k} = \big(\underline{\boldsymbol{C}}[\nabla\boldsymbol{a}]\big)\boldsymbol{k} + \big(\underline{\boldsymbol{C}}[\boldsymbol{b}\otimes\boldsymbol{k}]\big)\boldsymbol{k} = \big(\underline{\boldsymbol{C}}[\nabla\boldsymbol{a}]\big)\boldsymbol{k} + \boldsymbol{A}\boldsymbol{b}$ and the equation $\boldsymbol{P}_0'\boldsymbol{k} = \boldsymbol{0}$ yields

$$\boldsymbol{b} = -\boldsymbol{A}^{-1}\big(\underline{\boldsymbol{C}}[\nabla\boldsymbol{a}]\big)\boldsymbol{k} = \boldsymbol{f}(\nabla\boldsymbol{a}) =: \boldsymbol{g}(\nabla\nabla\boldsymbol{u}), \tag{4.57}$$

where $\boldsymbol{g}$ is the function determined by the relation $\boldsymbol{g}(\nabla\nabla\boldsymbol{u}) = \boldsymbol{f}\big(\nabla\boldsymbol{f}(\nabla\boldsymbol{u})\big)$.

Thus, according to (4.49), we impose that $\boldsymbol{a} = \boldsymbol{f}(\nabla\boldsymbol{u})$ and $\boldsymbol{b} = \boldsymbol{g}(\nabla\nabla\boldsymbol{u})$ in the coefficient of h^3. □

Now, we can write the third integral term in (4.48) as follows

$$\frac{1}{24}h^3\int_{\partial\Omega_u}\boldsymbol{P}_0\boldsymbol{1}\boldsymbol{\nu}\cdot\boldsymbol{b}\,\mathrm{d}s \;=\; \frac{1}{24}h^3\int_{\partial\Omega_u}\big(\underline{\boldsymbol{C}}[\nabla\boldsymbol{u}+\boldsymbol{a}\otimes\boldsymbol{k}]\big)\boldsymbol{1}\boldsymbol{\nu}\cdot\boldsymbol{b}\,\mathrm{d}s\,. \tag{4.58}$$

Let s designate the arclength parameter along the boundary curve $\partial\Omega$. Since $\hat{\boldsymbol{u}}(\boldsymbol{r},\zeta)$ is assigned on $\partial\Omega_u\times\big[-\frac{h}{2},\frac{h}{2}\big] = \partial\mathscr{K}_u$, so the derivatives $\hat{\boldsymbol{u}}^{(n)}(\boldsymbol{r},\zeta)$ are also assigned there. Hence, the fields

$$\boldsymbol{u}(s) = \hat{\boldsymbol{u}}(\boldsymbol{r},0),\quad \boldsymbol{a}(s) = \hat{\boldsymbol{u}}'(\boldsymbol{r},0),\quad \boldsymbol{b}(s) = \hat{\boldsymbol{u}}''(\boldsymbol{r},0),\quad \text{with } \boldsymbol{r} = \boldsymbol{r}(s), \tag{4.59}$$

are all assigned on $\partial\Omega_u$. Further, we can decompose the gradient $\nabla\boldsymbol{u}$ on $\partial\Omega$ as follows: if $\boldsymbol{\tau} := \dfrac{\mathrm{d}\boldsymbol{r}}{\mathrm{d}s}$ is the unit tangent vector and $\boldsymbol{\nu}$ the unit normal to $\partial\Omega$, as in Fig. 4.4, then we have

$$\begin{aligned}\nabla\boldsymbol{u} &= (\nabla\boldsymbol{u})\boldsymbol{1} = (\nabla\boldsymbol{u})(\boldsymbol{\tau}\otimes\boldsymbol{\tau}+\boldsymbol{\nu}\otimes\boldsymbol{\nu}) = \big[(\nabla\boldsymbol{u})\boldsymbol{\tau}\big]\otimes\boldsymbol{\tau} + \big[(\nabla\boldsymbol{u})\boldsymbol{\nu}\big]\otimes\boldsymbol{\nu}\\ &= \frac{\mathrm{d}\boldsymbol{u}(s)}{\mathrm{d}s}\otimes\boldsymbol{\tau} + \frac{\partial\boldsymbol{u}}{\partial\boldsymbol{\nu}}\otimes\boldsymbol{\nu} \;=\; \boldsymbol{u}_{,s}\otimes\boldsymbol{\tau} + \boldsymbol{u}_{,\nu}\otimes\boldsymbol{\nu}\,,\end{aligned} \tag{4.60}$$

where $\boldsymbol{u}_{,s} := \dfrac{\mathrm{d}\boldsymbol{u}(s)}{\mathrm{d}s}$, and $\boldsymbol{u}_{,\nu} := \dfrac{\partial\boldsymbol{u}}{\partial\boldsymbol{\nu}} = (\nabla\boldsymbol{u})\boldsymbol{\nu}$ is the normal derivative of $\boldsymbol{u}$ on $\partial\Omega$. So, the gradient $\nabla\boldsymbol{u}$ on $\partial\Omega$ is determined by $\boldsymbol{u}(s)$ and $\boldsymbol{u}_{,\nu}(s)$. In view of (4.56) and $(4.59)_2$, we have that $\boldsymbol{a} = \boldsymbol{f}(\nabla\boldsymbol{u}) = \boldsymbol{f}\big(\boldsymbol{u}_{,s}\otimes\boldsymbol{\tau} + \boldsymbol{u}_{,\nu}\otimes\boldsymbol{\nu}\big)$ is assigned on $\partial\Omega_u$. Then, $\boldsymbol{u}$ and $\boldsymbol{u}_{,\nu}$ are assigned on $\partial\Omega_u$, and the gradient $\nabla\boldsymbol{u}$ is assigned on $\partial\Omega_u$.

Fig. 4.4 Boundary curve of the midplane

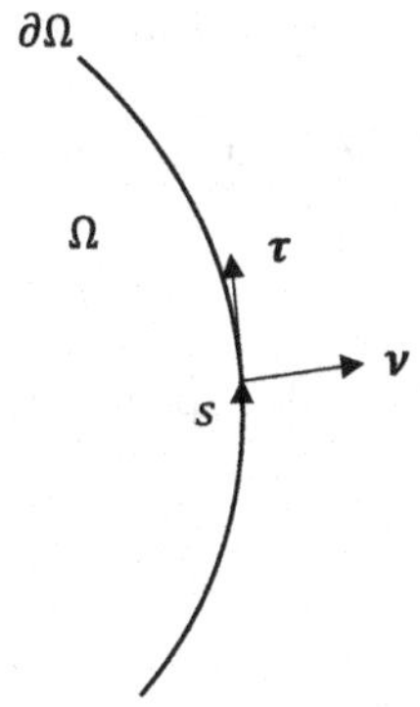

Altogether, we deduce that the integral (4.58) is a fixed constant, contributing a (disposable) constant to the overall energy; its variation vanishes identically and so does not affect the equilibrium equations.

We can therefore suppress the integral (4.58) in the expression of the energy (4.48) without loss of generality, obtaining

$$E = \frac{h}{2}\int_{\Omega} \underbrace{(\boldsymbol{P}_0\mathbf{1}\cdot\nabla\boldsymbol{u} + \boldsymbol{P}_0\boldsymbol{k}\cdot\boldsymbol{a})}_{=2U(\nabla\boldsymbol{u}+\boldsymbol{a}\otimes\boldsymbol{k})}\,\mathrm{d}a + \frac{h^3}{24}\int_{\Omega} \boldsymbol{P}_0'\mathbf{1}\cdot\nabla\boldsymbol{a}\,\mathrm{d}a - \int_{\partial\Omega_t} (\boldsymbol{p}_u\cdot\boldsymbol{u} + \boldsymbol{p}_a\cdot\boldsymbol{a})\mathrm{d}s\,, \tag{4.61}$$

in which

$$h^3\,\boldsymbol{P}_0'\mathbf{1}\cdot\nabla\boldsymbol{a} \;=\; h^3\big(\underline{\boldsymbol{C}}[\nabla\bar{\boldsymbol{a}} + \bar{\boldsymbol{b}}\otimes\boldsymbol{k}]\big)\mathbf{1}\cdot\nabla\boldsymbol{a}\,, \tag{4.62}$$

with $\bar{\boldsymbol{a}} := \boldsymbol{f}(\nabla\boldsymbol{u})$ and $\bar{\boldsymbol{b}} := \boldsymbol{g}(\nabla\nabla\boldsymbol{u}) = \boldsymbol{f}(\nabla\bar{\boldsymbol{a}})$.

Note that $\big(\underline{\boldsymbol{C}}[\nabla\bar{\boldsymbol{a}} + \bar{\boldsymbol{b}}\otimes\boldsymbol{k}]\big)\boldsymbol{k} = \mathbf{0}$ by virtue of $(4.49)_2$, so we obtain

$$\begin{aligned}\big(\underline{\boldsymbol{C}}[\nabla\bar{\boldsymbol{a}} + \bar{\boldsymbol{b}}\otimes\boldsymbol{k}]\big)\mathbf{1}\cdot\nabla\bar{\boldsymbol{a}} &= \underline{\boldsymbol{C}}[\nabla\bar{\boldsymbol{a}} + \bar{\boldsymbol{b}}\otimes\boldsymbol{k}]\cdot\big(\nabla\bar{\boldsymbol{a}} + \bar{\boldsymbol{b}}\otimes\boldsymbol{k}\big)\\ &= \underline{\boldsymbol{C}}[\bar{\boldsymbol{H}}_0']\cdot\bar{\boldsymbol{H}}_0' \;=\; 2U(\bar{\boldsymbol{H}}_0')\,,\end{aligned} \tag{4.63}$$

in which

$$\bar{\boldsymbol{H}}_0' \;:=\; \nabla\bar{\boldsymbol{a}} + \bar{\boldsymbol{b}}\otimes\boldsymbol{k} \;=\; \nabla\boldsymbol{f} + \boldsymbol{g}\otimes\boldsymbol{k} \tag{4.64}$$

is determined by $\nabla\nabla\boldsymbol{u}$. Thus, the term of order $O(h^3)$ in the energy includes the plate curvature, induced by deformation.

Comment We are *not* justified in using $\boldsymbol{a} = \boldsymbol{f}(\nabla\boldsymbol{u})$ in the $O(h)$ term, since $\boldsymbol{P}_0\boldsymbol{k} = O(h^2)$ contributes at $O(h^3)$, and those terms have been retained. However, let us show that we obtain the optimal $O(h^3)$ model by putting $\boldsymbol{a} = \boldsymbol{f}(\nabla\boldsymbol{u})$ in *all* terms. In this respect, we prove the following result □

Lemma 4.1 *With the above notations, the vector* $\boldsymbol{a} = \boldsymbol{f}(\nabla\boldsymbol{u})$ *minimizes the strain energy density* $U(\nabla\boldsymbol{u} + \boldsymbol{a}\otimes\boldsymbol{k})$ *for a given* $\boldsymbol{u}(\boldsymbol{r})$, *i.e. we have*

$$U(\nabla \boldsymbol{u} + \boldsymbol{a} \otimes \boldsymbol{k}) \; > \; U\big(\nabla \boldsymbol{u} + \boldsymbol{f}(\nabla \boldsymbol{u}) \otimes \boldsymbol{k}\big), \qquad \textit{for all} \;\; \boldsymbol{a} \neq \boldsymbol{f}(\nabla \boldsymbol{u}). \tag{4.65}$$

Proof Let us define the function

$$G(\boldsymbol{a}) := U(\nabla \boldsymbol{u} + \boldsymbol{a} \otimes \boldsymbol{k}) \quad \text{at fixed} \quad \boldsymbol{u}(\boldsymbol{r}).$$

Consider a path $\boldsymbol{a}(u)$ and let $\sigma(u) := G\big(\boldsymbol{a}(u)\big)$. Then, we have the derivatives

$$\sigma'(u) = \frac{\mathrm{d}}{\mathrm{d}u} G\big(\boldsymbol{a}(u)\big) = U_{\tilde{\boldsymbol{H}}}\big(\nabla \boldsymbol{u} + \boldsymbol{a}(u) \otimes \boldsymbol{k}\big) \cdot \big(\boldsymbol{a}'(u) \otimes \boldsymbol{k}\big) = \big(\boldsymbol{P}_0(u)\boldsymbol{k}\big) \cdot \boldsymbol{a}'(u). \tag{4.66}$$

and

$$\sigma''(u) = \big(\boldsymbol{P}_0(u)\boldsymbol{k}\big) \cdot \boldsymbol{a}''(u) + \big(\boldsymbol{P}_0'(u)\boldsymbol{k}\big) \cdot \boldsymbol{a}'(u), \tag{4.67}$$

where

$$\boldsymbol{P}_0'\boldsymbol{k} = \big(\underline{\boldsymbol{C}}[\boldsymbol{H}_0']\big)\boldsymbol{k} = \big(\underline{\boldsymbol{C}}[\boldsymbol{a}' \otimes \boldsymbol{k}]\big)\boldsymbol{k} = \boldsymbol{A}\boldsymbol{a}', \tag{4.68}$$

in view of (4.52). The derivative of $G(\boldsymbol{a})$ with respect to the vector $\boldsymbol{a} = a_i \boldsymbol{G}^i$ is

$$\begin{aligned} G_{\boldsymbol{a}} &= \frac{\partial G}{\partial a_i} \boldsymbol{G}_i = \left[U_{\tilde{\boldsymbol{H}}}\big(\nabla \boldsymbol{u} + \boldsymbol{a} \otimes \boldsymbol{k}\big) \cdot \frac{\partial}{\partial a_i}\big(\nabla \boldsymbol{u} + \boldsymbol{a} \otimes \boldsymbol{k}\big) \right] \boldsymbol{G}_i \\ &= \big[\boldsymbol{P}_0 \cdot \big(\boldsymbol{G}^i \otimes \boldsymbol{k}\big)\big]\boldsymbol{G}_i = \big[(\boldsymbol{P}_0\boldsymbol{k}) \cdot \boldsymbol{G}^i\big]\boldsymbol{G}_i = (\boldsymbol{P}_0\boldsymbol{k})(\boldsymbol{G}^i \otimes \boldsymbol{G}_i) = \boldsymbol{P}_0\boldsymbol{k}\,. \end{aligned} \tag{4.69}$$

Inserting (4.68), (4.69) into the relations (4.66) and (4.67), we obtain the derivatives

$$\sigma' = G_{\boldsymbol{a}} \cdot \boldsymbol{a}' \qquad \text{and} \qquad \sigma'' = G_{\boldsymbol{a}} \cdot \boldsymbol{a}'' + \underbrace{\boldsymbol{a}' \cdot \boldsymbol{A}\boldsymbol{a}'}_{>0}, \tag{4.70}$$

where we have $\boldsymbol{a}' \cdot \boldsymbol{A}\boldsymbol{a}' > 0$ for all $\boldsymbol{a}' \neq \boldsymbol{0}$, by virtue of (4.53).

Consider now the special path $\boldsymbol{a}(u) := (1-u)\boldsymbol{a}_1 + u\,\boldsymbol{a}_2$, with $\boldsymbol{a}_1 \neq \boldsymbol{a}_2$ and $u \in [0, 1]$. Then, we have

$$\boldsymbol{a}'(u) = \boldsymbol{a}_2 - \boldsymbol{a}_1 \neq \boldsymbol{0} \qquad \text{and} \qquad \boldsymbol{a}''(u) = \boldsymbol{0}, \tag{4.71}$$

and the relation $(4.70)_2$ becomes

$$\sigma''(u) = \boldsymbol{a}'(u) \cdot \boldsymbol{A}\boldsymbol{a}'(u) \; > \; 0\,, \tag{4.72}$$

so the function $\sigma(u)$ is convex on $[0, 1]$. Hence, we deduce

$$\sigma'(u) = \sigma'(0) + \int_0^u \sigma''(x)\,\mathrm{d}x \; > \; \sigma'(0) \qquad \text{and}$$

$$\sigma(1) - \sigma(0) = \int_0^1 \sigma'(u)\,\mathrm{d}u \; > \; \sigma'(0),$$

i.e., in view of $(4.70)_1$ and (4.71),

$$G(\boldsymbol{a}_2) - G(\boldsymbol{a}_1) \; > \; G_{\boldsymbol{a}}(\boldsymbol{a}_1)\cdot(\boldsymbol{a}_2 - \boldsymbol{a}_1), \qquad \text{for all} \;\; \boldsymbol{a}_1 \neq \boldsymbol{a}_2\,, \tag{4.73}$$

which expresses the convexity of the function $G(\boldsymbol{a})$.

Suppose now that $\boldsymbol{a}_1 = \boldsymbol{f}(\nabla\boldsymbol{u})$ and the vector $\boldsymbol{a}_2$ is arbitrary. Then, by virtue of (4.69) and $(4.49)_1$, we have

$$G_{\boldsymbol{a}}(\boldsymbol{a}_1) = \boldsymbol{P}_0\boldsymbol{k} = \underline{\boldsymbol{C}}[\nabla\boldsymbol{u} + \boldsymbol{f}(\nabla\boldsymbol{u})\otimes\boldsymbol{k}] = \boldsymbol{0}\,,$$

and the inequality (4.73) reduces to

$$G(\boldsymbol{a}) \; - \; G(\boldsymbol{f}(\nabla\boldsymbol{u})) \; > \; 0 \qquad \text{for all} \;\; \boldsymbol{a} \neq \boldsymbol{f}(\nabla\boldsymbol{u})\,,$$

i.e. the relation (4.65) is proved. □

Taking into account the above Lemma 4.1 and the relation (4.61), we shall put $\boldsymbol{a} = \boldsymbol{f}(\nabla\boldsymbol{u})$ in *all* terms of the strain energy density, in order to obtain an optimal model of order $O(h^3)$. Thus the Eqs. (4.61)–(4.63) become

$$\begin{aligned} E \; &= \int_{\Omega} W \,\mathrm{d}a - \int_{\partial\Omega_t} (\boldsymbol{p}_u\cdot\boldsymbol{u} + \boldsymbol{p}_a\cdot\boldsymbol{a})\,\mathrm{d}s\,, \qquad \text{where} \\ W \; &= \frac{1}{2}\,h\,\underline{\boldsymbol{C}}[\bar{\boldsymbol{H}}_0]\cdot\bar{\boldsymbol{H}}_0 + \frac{1}{24}\,h^3\underline{\boldsymbol{C}}[\bar{\boldsymbol{H}}_0']\cdot\bar{\boldsymbol{H}}_0' \qquad \text{with} \\ \bar{\boldsymbol{H}}_0 &= \nabla\boldsymbol{u} + \boldsymbol{f}(\nabla\boldsymbol{u})\otimes\boldsymbol{k}\,, \qquad \bar{\boldsymbol{H}}_0' = \nabla\boldsymbol{f}(\nabla\boldsymbol{u}) + \boldsymbol{g}(\nabla\nabla\boldsymbol{u})\otimes\boldsymbol{k}\,. \end{aligned} \tag{4.74}$$

So, the potential energy E is determined entirely by $\nabla\boldsymbol{u}$ and $\nabla\nabla\boldsymbol{u}$.

In the edge-load term $\boldsymbol{p}_a\cdot\boldsymbol{a}$ in (4.74) we put $\boldsymbol{a} = \boldsymbol{f}(\nabla\boldsymbol{u})$, since $\boldsymbol{p}_a$ is of order $O(h^3)$, cf. (4.33). Thus, we have the edge-load potential

$$\int_{\partial\Omega_t} (\boldsymbol{p}_u\cdot\boldsymbol{u} + \boldsymbol{p}_a\cdot\boldsymbol{a})\,\mathrm{d}s = \int_{\partial\Omega_t} \big(\boldsymbol{p}_u\cdot\boldsymbol{u} + \boldsymbol{p}_a\cdot\boldsymbol{f}(\nabla\boldsymbol{u})\big)\,\mathrm{d}s\,, \tag{4.75}$$

with

$$\boldsymbol{f}(\nabla\boldsymbol{u}) = -\boldsymbol{A}^{-1}\big(\underline{\boldsymbol{C}}[\nabla\boldsymbol{u}]\big)\boldsymbol{k} \qquad \text{and} \qquad \nabla\boldsymbol{u} = \boldsymbol{u}_{,s}\otimes\boldsymbol{\tau} + \boldsymbol{u}_{,\nu}\otimes\boldsymbol{\nu}\,, \tag{4.76}$$

according to (4.56) and (4.60). Then, using (4.76) and denoting by $(\,\cdot\,)_{,s}$ the derivative $\frac{\mathrm{d}}{\mathrm{d}s}(\,\cdot\,)$ with respect to s, we can write

$$\begin{aligned} \boldsymbol{p}_a\cdot\boldsymbol{f}(\nabla\boldsymbol{u}) \; &= \; -\boldsymbol{p}_a\cdot\boldsymbol{A}^{-1}\big(\underline{\boldsymbol{C}}[\boldsymbol{u}_{,s}\otimes\boldsymbol{\tau}]\big)\boldsymbol{k} - \boldsymbol{p}_a\cdot\boldsymbol{A}^{-1}\big(\underline{\boldsymbol{C}}[\boldsymbol{u}_{,\nu}\otimes\boldsymbol{\nu}]\big)\boldsymbol{k} \\ &= \; -\big\{\boldsymbol{p}_a\cdot\boldsymbol{A}^{-1}\big(\underline{\boldsymbol{C}}[\boldsymbol{u}\otimes\boldsymbol{\tau}]\big)\boldsymbol{k}\big\}_{,s} + (\boldsymbol{p}_a)_{,s}\cdot\boldsymbol{A}^{-1}\big(\underline{\boldsymbol{C}}[\boldsymbol{u}\otimes\boldsymbol{\tau}]\big)\boldsymbol{k} \\ &\quad\; +\boldsymbol{p}_a\cdot\boldsymbol{A}^{-1}\big(\underline{\boldsymbol{C}}[\boldsymbol{u}\otimes\boldsymbol{\tau}_{,s}]\big)\boldsymbol{k} - \boldsymbol{p}_a\cdot\boldsymbol{A}^{-1}\big(\underline{\boldsymbol{C}}[\boldsymbol{u}_{,\nu}\otimes\boldsymbol{\nu}]\big)\boldsymbol{k}\,. \end{aligned} \tag{4.77}$$

Consider a family of virtual displacements $\boldsymbol{u}(\boldsymbol{r}\,;\varepsilon)$ which satisfies the prescribed boundary conditions on $\partial\Omega_u$. If we denote the derivative with respect to the parameter ε by a superposed dot, then we have

$$\dot{\boldsymbol{u}}(s) = \boldsymbol{0} \quad \text{on } \partial\Omega_u\,. \tag{4.78}$$

The *variation* (or virtual work) corresponding to edge loads is

$$\int_{\partial\Omega_t} (\boldsymbol{p}_u \cdot \dot{\boldsymbol{u}} + \boldsymbol{p}_a \cdot \dot{\boldsymbol{a}})\,\mathrm{d}s = \int_{\partial\Omega_t} \big(\boldsymbol{p}_u \cdot \dot{\boldsymbol{u}} + \boldsymbol{p}_a \cdot \boldsymbol{f}(\nabla\dot{\boldsymbol{u}})\big)\,\mathrm{d}s\,, \tag{4.79}$$

where the last term can be written by (4.77) as

$$\begin{aligned} \boldsymbol{p}_a \cdot \boldsymbol{f}(\nabla\dot{\boldsymbol{u}}) &= -\big\{\boldsymbol{p}_a \cdot \boldsymbol{A}^{-1}\big(\underline{\boldsymbol{C}}[\dot{\boldsymbol{u}} \otimes \boldsymbol{\tau}]\big)\boldsymbol{k}\big\}_{,s} + (\boldsymbol{p}_a)_{,s} \cdot \boldsymbol{A}^{-1}\big(\underline{\boldsymbol{C}}[\dot{\boldsymbol{u}} \otimes \boldsymbol{\tau}]\big)\boldsymbol{k} \\ &\quad + \boldsymbol{p}_a \cdot \boldsymbol{A}^{-1}\big(\underline{\boldsymbol{C}}[\dot{\boldsymbol{u}} \otimes \boldsymbol{\tau}_{,s}]\big)\boldsymbol{k} - \boldsymbol{p}_a \cdot \boldsymbol{A}^{-1}\big(\underline{\boldsymbol{C}}[\dot{\boldsymbol{u}}_{,\nu} \otimes \boldsymbol{\nu}]\big)\boldsymbol{k}\,. \end{aligned} \tag{4.80}$$

In view of (4.78), we can extend the integral $\int_{\partial\Omega_t}$ to the whole boundary $\int_{\partial\Omega}$ and compute the first term in the right-hand side of Eq. (4.80)

$$\int_{\partial\Omega_t} \big\{\boldsymbol{p}_a \cdot \boldsymbol{A}^{-1}\big(\underline{\boldsymbol{C}}[\dot{\boldsymbol{u}} \otimes \boldsymbol{\tau}]\big)\boldsymbol{k}\big\}_{,s}\,\mathrm{d}s = \int_{\partial\Omega} \big\{\boldsymbol{p}_a \cdot \boldsymbol{A}^{-1}\big(\underline{\boldsymbol{C}}[\dot{\boldsymbol{u}} \otimes \boldsymbol{\tau}]\big)\boldsymbol{k}\big\}_{,s}\,\mathrm{d}s = 0,$$

since it is an exact differential. Thus, its contribution vanishes (for smooth boundary $\partial\Omega$), implying that

$$\left(\int_{\partial\Omega_t} \big\{\boldsymbol{p}_a \cdot \boldsymbol{A}^{-1}\big(\underline{\boldsymbol{C}}[\boldsymbol{u} \otimes \boldsymbol{\tau}]\big)\boldsymbol{k}\big\}_{,s}\,\mathrm{d}s\right)^{\dot{}} = 0.$$

This means that the above integral contributes a fixed constant to the overall energy and can be suppressed without loss of generality. This leaves, in view of (4.77),

$$\int_{\partial\Omega_t} (\boldsymbol{p}_u \cdot \boldsymbol{u} + \boldsymbol{p}_a \cdot \boldsymbol{a})\,\mathrm{d}s = \int_{\partial\Omega_t} \big(\boldsymbol{\ell} \cdot \boldsymbol{u} + \boldsymbol{c} \cdot \boldsymbol{u}_{,\nu}\big)\,\mathrm{d}s\,, \tag{4.81}$$

where

$$\begin{aligned} &\boldsymbol{\ell} \cdot \boldsymbol{u} := \boldsymbol{p}_u \cdot \boldsymbol{u} + (\boldsymbol{p}_a)_{,s} \cdot \boldsymbol{A}^{-1}\big(\underline{\boldsymbol{C}}[\boldsymbol{u} \otimes \boldsymbol{\tau}]\big)\boldsymbol{k} + \boldsymbol{p}_a \cdot \boldsymbol{A}^{-1}\big(\underline{\boldsymbol{C}}[\boldsymbol{u} \otimes \boldsymbol{\tau}_{,s}]\big)\boldsymbol{k} \quad \text{and} \\ &\boldsymbol{c} \cdot \boldsymbol{u}_{,\nu} := -\boldsymbol{p}_a \cdot \boldsymbol{A}^{-1}\big(\underline{\boldsymbol{C}}[\boldsymbol{u}_{,\nu} \otimes \boldsymbol{\nu}]\big)\boldsymbol{k}\,. \end{aligned} \tag{4.82}$$

Notice that the right-hand sides in (4.82) are *linear* in $\boldsymbol{u}$ and $\boldsymbol{u}_{,\nu}$, respectively, i.e. the vectors $\boldsymbol{\ell}$ and $\boldsymbol{c}$ are independent of $\boldsymbol{u}$ and $\boldsymbol{u}_{,\nu}$.

4.2.4 Lateral Loads

Let us consider the lateral loads in more detail and write their contribution to the potential energy. Denote by $\partial\mathscr{K}^+$ and $\partial\mathscr{K}^-$ the two lateral faces, defined by $\zeta = \frac{h}{2}$ and $\zeta = -\frac{h}{2}$, respectively.

As mentioned at the beginning of Sect. 4.2.3, we designate by $\tilde{\boldsymbol{t}}^{\pm} = \hat{\boldsymbol{t}}(\boldsymbol{r}, \pm\frac{h}{2})$ the stress vector on the lateral faces, where we have $\tilde{\boldsymbol{t}}^{+} = \tilde{\boldsymbol{P}}^{+}\boldsymbol{k}$, $\tilde{\boldsymbol{t}}^{-} = -\tilde{\boldsymbol{P}}^{-}\boldsymbol{k}$, see also Fig. 4.5.

Recall that

$$\tilde{\boldsymbol{t}}^{+} + \tilde{\boldsymbol{t}}^{-} = h\,\boldsymbol{P}_0'\boldsymbol{k} + O(h^3) \quad \text{and} \quad \tilde{\boldsymbol{t}}^{+} - \tilde{\boldsymbol{t}}^{-} = 2\boldsymbol{P}_0\boldsymbol{k} + O(h^2),$$

cf. (4.36). If we assume that $\tilde{\boldsymbol{t}}^{\pm} = O(h^3)$, then the above relations imply that

$$\boldsymbol{P}_0\boldsymbol{k} = O(h^2) = \boldsymbol{P}_0'\boldsymbol{k}\,,$$

which have been used in the foregoing derivation in Sect. 4.2.3.

On the basis of $\tilde{\boldsymbol{t}}^{\pm} = O(h^3)$, we can write

$$\tilde{\boldsymbol{t}}^{\pm} = h^3\,\boldsymbol{t}^{\pm} + o(h^3), \tag{4.83}$$

where the fields $\boldsymbol{t}^{\pm}$ are functions which depend only on $\boldsymbol{r}$, with $|\boldsymbol{t}^{\pm}| = O(1)$.

The associated load potential is given by

$$-\left(\int_{\partial\mathscr{K}^+} \tilde{\boldsymbol{t}}^{+}\cdot\tilde{\boldsymbol{u}}^{+}\,\mathrm{d}a + \int_{\partial\mathscr{K}^-} \tilde{\boldsymbol{t}}^{-}\cdot\tilde{\boldsymbol{u}}^{-}\,\mathrm{d}a\right), \tag{4.84}$$

where we have denoted by

$$\tilde{\boldsymbol{u}}^{\pm} = \hat{\boldsymbol{u}}\left(\boldsymbol{r}, \pm\frac{h}{2}\right) = \boldsymbol{u} \pm \frac{h}{2}\boldsymbol{a} + \frac{h^2}{8}\boldsymbol{b} \pm \frac{h^3}{48}\boldsymbol{c} + o(h^3)\,, \tag{4.85}$$

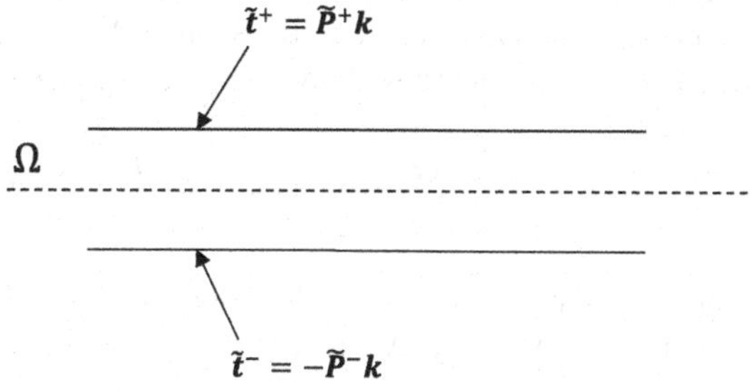

Fig. 4.5 The stress vector on the lateral faces

in view of (4.23). On the basis of (4.83) and (4.85), we get

$$\tilde{t}^+ \cdot \tilde{u}^+ = h^3\, t^+ \cdot u + o(h^3) \quad \text{and} \quad \tilde{t}^- \cdot \tilde{u}^- = h^3\, t^- \cdot u + o(h^3).$$

Thus, the lateral load potential (4.84) can be written as

$$-\left(\int_{\partial\mathscr{K}^+} \tilde{t}^+ \cdot \tilde{u}^+ \,\mathrm{d}a + \int_{\partial\mathscr{K}^-} \tilde{t}^- \cdot \tilde{u}^- \,\mathrm{d}a \right) = -\int_{\Omega} g \cdot u \,\mathrm{d}a + o(h^3), \tag{4.86}$$

where we have introduced the notation

$$g := h^3\left(t^+ + t^-\right) \quad \text{(net lateral traction)}. \tag{4.87}$$

Finally, in view of (4.74), (4.81) and (4.86), we obtain the potential energy

$$\begin{aligned} E[u] &= \int_{\Omega} \left[F(\nabla u, \nabla\nabla u) - g \cdot u\right] \mathrm{d}a - \int_{\partial\Omega_t} \left(\boldsymbol{\ell} \cdot u + c \cdot u_{,\nu}\right) \mathrm{d}s\,, \quad \text{where} \\ F(\nabla u, \nabla\nabla u) &:= W = \frac{1}{2}\, h\, \bar{H}_0 \cdot \underline{C}[\bar{H}_0] + \frac{1}{24}\, h^3\, \bar{H}_0' \cdot \underline{C}[\bar{H}_0']\,, \quad \text{in which} \\ &\bar{H}_0 = \nabla u + \bar{a} \otimes k\,, \qquad \bar{H}_0' = \nabla\bar{a} + \bar{b} \otimes k\,, \qquad \text{and} \\ &\bar{a} = -A^{-1}\left(\underline{C}[\nabla u]\right)k\,, \qquad \bar{b} = -A^{-1}\left(\underline{C}[\nabla\bar{a}]\right)k\,. \end{aligned} \tag{4.88}$$

4.3 Equilibrium Equations

Let us extract the equilibrium equations for plates from the condition that the potential energy E is stationary, as presented in the case of three-dimensional theory in Sect. 3.3. That is, we write for an equilibrium state the virtual-work equation

$$\frac{\mathrm{d}}{\mathrm{d}\varepsilon} E\left[u(r\,;\varepsilon)\right]_{|\varepsilon=0} = 0\,, \tag{4.89}$$

where $u(r\,;\varepsilon)$ is a family of virtual displacements with parameter ε, which satisfies the prescribed boundary conditions on $\partial\Omega_u$. If we designate the derivative with respect to ε by a superposed dot, then the Eq. (4.89) yields

$$\int_{\Omega} \left(\dot{F} - g \cdot \dot{u}\right) \mathrm{d}a = \int_{\partial\Omega_t} \left(\boldsymbol{\ell} \cdot \dot{u} + c \cdot \dot{u}_{,\nu}\right) \mathrm{d}s\,. \tag{4.90}$$

Note that, at any fixed $r \in \Omega$, the values of ∇u, $\nabla\nabla u$ and u are independent. Then, the derivative of $F(\nabla u, \nabla\nabla u)$ can be written in the form

$$\begin{aligned}\dot{F} &= \frac{\partial F}{\partial u_{i,\alpha}}\,\dot{u}_{i,\alpha} + \frac{\partial F}{\partial u_{i,\alpha\beta}}\,\dot{u}_{i,\alpha\beta} \\ &= \dot{u}_i\left[\left(\frac{\partial F}{\partial u_{i,\alpha\beta}}\right)_{,\alpha\beta} - \left(\frac{\partial F}{\partial u_{i,\alpha}}\right)_{,\alpha}\right] + \varphi_{\alpha,\alpha}\,,\end{aligned} \tag{4.91}$$

where we define φ_α by

$$\varphi_\alpha := \dot{u}_i\left[\frac{\partial F}{\partial u_{i,\alpha}} - \left(\frac{\partial F}{\partial u_{i,\alpha\beta}}\right)_{,\beta}\right] + \frac{\partial F}{\partial u_{i,\alpha\beta}}\,\dot{u}_{i,\beta}\,. \tag{4.92}$$

If

$$N_{i\alpha} := \frac{\partial F}{\partial u_{i,\alpha}}\,, \qquad M_{i\alpha\beta} := \frac{\partial F}{\partial u_{i,\alpha\beta}} \qquad \text{and} \qquad T_{i\alpha} := N_{i\alpha} - M_{i\alpha\beta,\beta}\,, \tag{4.93}$$

then Eqs. (4.91) and (4.92) become

$$\dot{F} = \dot{u}_i\left(M_{i\alpha\beta,\alpha\beta} - N_{i\alpha,\alpha}\right) + M_{i\alpha\beta}\,\dot{u}_{i,\beta} + \varphi_{\alpha,\alpha} = -T_{i\alpha,\alpha}\,\dot{u}_i + \varphi_{\alpha,\alpha} \tag{4.94}$$

and, respectively,

$$\varphi_\alpha = \dot{u}_i\left(N_{i\alpha} - M_{i\alpha\beta,\beta}\right) + M_{i\alpha\beta}\,\dot{u}_{i,\beta} = T_{i\alpha}\,\dot{u}_i + M_{i\alpha\beta}\,\dot{u}_{i,\beta}\,. \tag{4.95}$$

According to (2.143), we have

$$\int_\Omega \varphi_{\alpha,\alpha}\,\mathrm{d}a = \int_{\partial\Omega} \varphi_\alpha \nu_\alpha\,\mathrm{d}s\,, \tag{4.96}$$

where $\boldsymbol{\nu} = \nu_\alpha \boldsymbol{e}_\alpha$ is the unit outward normal to $\partial\Omega$. Thus, integrating the relation (4.94), we derive

$$\int_\Omega \dot{F}\,\mathrm{d}a = -\int_\Omega T_{i\alpha,\alpha}\,\dot{u}_i\,\mathrm{d}a + \int_{\partial\Omega} \varphi_\alpha \nu_\alpha\,\mathrm{d}s\,. \tag{4.97}$$

Inserting this in the Eq. (4.90), we obtain the form

$$\int_{\partial\Omega} \varphi_\alpha \nu_\alpha\,\mathrm{d}s - \int_\Omega \left(T_{i\alpha,\alpha} + g_i\right)\dot{u}_i\,\mathrm{d}a = \int_{\partial\Omega_t} \left(\ell_i\,\dot{u}_i + c_i\,\dot{u}_{i,\nu}\right)\mathrm{d}s\,. \tag{4.98}$$

The displacement gradient on the boundary curve can be decomposed in the form (4.60), i.e.

$$u_{i,\beta} = u_{i,s}\,\tau_\beta + u_{i,\nu}\,\nu_\beta\,, \qquad \text{with} \quad \tau_\beta = \boldsymbol{\tau}\cdot\boldsymbol{e}_\beta\,. \tag{4.99}$$

Using (4.95) and (4.99), we deduce

$$
\begin{aligned}
\varphi_\alpha \nu_\alpha &= T_{i\alpha}\nu_\alpha \dot{u}_i + M_{i\alpha\beta}\nu_\alpha \dot{u}_{i,\beta} \\
&= T_{i\alpha}\nu_\alpha \dot{u}_i + M_{i\alpha\beta}\nu_\alpha \nu_\beta \dot{u}_{i,\nu} + M_{i\alpha\beta}\nu_\alpha \tau_\beta \dot{u}_{i,s} \,.
\end{aligned}
\tag{4.100}
$$

If we integrate the last relation and take into account that

$$
M_{i\alpha\beta}\nu_\alpha \tau_\beta \dot{u}_{i,s} = (M_{i\alpha\beta}\nu_\alpha \tau_\beta \dot{u}_i)_{,s} - (M_{i\alpha\beta}\nu_\alpha \tau_\beta)_{,s}\, \dot{u}_i \,,
$$

then we obtain in the case of smooth boundary $\partial\Omega$ (where $\boldsymbol{\tau}$, $\boldsymbol{\nu}$ are continuous)

$$
\int_{\partial\Omega} \varphi_\alpha \nu_\alpha \,\mathrm{d}s = \int_{\partial\Omega} \big\{\big[T_{i\alpha}\nu_\alpha - (M_{i\alpha\beta}\nu_\alpha\tau_\beta)_{,s}\big]\dot{u}_i + M_{i\alpha\beta}\nu_\alpha\nu_\beta\, \dot{u}_{i,\nu}\big\}\,\mathrm{d}s. \tag{4.101}
$$

Substituting this relation into (4.98) and noting that $\dot{\boldsymbol{u}} = \boldsymbol{0}$, $\dot{\boldsymbol{u}}_{,\nu} = \boldsymbol{0}$ on $\partial\Omega_u$, we get finally the equation

$$
\begin{aligned}
&\int_{\partial\Omega_t} \big\{\big[T_{i\alpha}\nu_\alpha - (M_{i\alpha\beta}\nu_\alpha\tau_\beta)_{,s}\big]\dot{u}_i + M_{i\alpha\beta}\nu_\alpha\nu_\beta\, \dot{u}_{i,\nu}\big\}\,\mathrm{d}s \\
&\qquad - \int_{\Omega} \big(T_{i\alpha,\alpha} + g_i\big)\, \dot{u}_i \,\mathrm{d}a = \int_{\partial\Omega_t} \big(\ell_i\, \dot{u}_i + c_i\, \dot{u}_{i,\nu}\big)\,\mathrm{d}s \,,
\end{aligned}
\tag{4.102}
$$

which is the principle of virtual work for elastic plates.

Note that the fields $\dot{\boldsymbol{u}}$ and $\dot{\boldsymbol{u}}_{,\nu}$ are arbitrary and independent on $\partial\Omega_t$. Then, we can employ a similar procedure as in the three-dimensional theory (based on the fundamental lemma of the calculus of variations) to obtain from (4.102) the local equations of equilibrium

$$
T_{i\alpha,\alpha} + g_i = 0 \ \text{ in } \Omega \,, \tag{4.103}
$$

and the boundary conditions

$$
T_{i\alpha}\nu_\alpha - (M_{i\alpha\beta}\nu_\alpha\tau_\beta)_{,s} = \ell_i \,, \qquad M_{i\alpha\beta}\nu_\alpha\nu_\beta = c_i \ \text{ on } \partial\Omega_t \,, \tag{4.104}
$$

whereas $\boldsymbol{u}$ and $\boldsymbol{u}_{,\nu}$ are assigned on $\partial\Omega_u$ (for a *clamped* edge).

On a *pinned* edge, $\boldsymbol{u}_{,\nu}$ is not assigned and the condition $M_{i\alpha\beta}\nu_\alpha\nu_\beta = c_i$ holds on $\partial\Omega_u$ too. The relations (4.104) are also called the "Kirchhoff boundary conditions".

4.4 Isotropic Plates

Let us consider plates made of an isotropic (uniform) material and derive the relevant equations. In this case, the tensor of elastic moduli has the expression (see Sect. 3.2.1)

$$
\underline{C}[\boldsymbol{T}] = \lambda(\operatorname{tr}\boldsymbol{T})\,\boldsymbol{I} + 2\mu \operatorname{sym}\boldsymbol{T} \,. \tag{4.105}
$$

Also, the acoustic tensor $\boldsymbol{A}$ used in our derivation is (cf. (4.52))

$$\begin{aligned} \boldsymbol{A}\boldsymbol{a} &= \big(\underline{\boldsymbol{C}}[\boldsymbol{a} \otimes \boldsymbol{k}]\big)\boldsymbol{k} \;=\; \big[\lambda(\boldsymbol{a} \cdot \boldsymbol{k})\boldsymbol{I} + \mu(\boldsymbol{a} \otimes \boldsymbol{k} + \boldsymbol{k} \otimes \boldsymbol{a})\big]\boldsymbol{k} \\ &= \lambda(\boldsymbol{a} \cdot \boldsymbol{k})\boldsymbol{k} + \mu\big[\boldsymbol{a} + (\boldsymbol{a} \cdot \boldsymbol{k})\boldsymbol{k}\big] \;=\; (\lambda + \mu)(\boldsymbol{a} \cdot \boldsymbol{k})\boldsymbol{k} + \mu\, \boldsymbol{a} \\ &= \big[(\lambda + 2\mu)\boldsymbol{k} \otimes \boldsymbol{k} + \mu\mathbf{1}\big]\boldsymbol{a}\,, \end{aligned} \tag{4.106}$$

so we have

$$\boldsymbol{A} \;=\; \mu\mathbf{1} \;+\; (\lambda + 2\mu)\boldsymbol{k} \otimes \boldsymbol{k}\,. \tag{4.107}$$

Thus, the tensor $\boldsymbol{A}$ is positive definite if and only if

$$\mu > 0 \qquad \text{and} \qquad \lambda + 2\mu > 0\,. \tag{4.108}$$

Remark According to Theorem 3.1, the strain energy density U in the three-dimensional theory is positive definite if and only if $\mu > 0$ and $\kappa = \lambda + \frac{2}{3}\,\mu > 0$. Then,

$$\lambda + 2\mu \;=\; \lambda + \frac{2}{3}\,\mu + \frac{4}{3}\,\mu \;=\; \kappa + \frac{4}{3}\,\mu \;>\; 0.$$

Thus, the conditions (4.108) can be inferred from (3.28). □

By virtue of (4.107) and (4.108), we have the inverse tensor

$$\boldsymbol{A}^{-1} \;=\; \frac{1}{\mu}\,\mathbf{1} + \frac{1}{\lambda + 2\mu}\,\boldsymbol{k} \otimes \boldsymbol{k}\,. \tag{4.109}$$

4.4.1 Displacement Gradients

We decompose the displacement vector $\boldsymbol{u}$ as an in-plane part $\boldsymbol{v}$ and transverse component w as follows

$$\boldsymbol{u} = \boldsymbol{v} + w\,\boldsymbol{k}\,, \qquad \text{where} \quad \boldsymbol{v} := \mathbf{1}\boldsymbol{u}, \quad w := \boldsymbol{k} \cdot \boldsymbol{u}\,. \tag{4.110}$$

Also, we decompose the vectors $\boldsymbol{a}$ and $\boldsymbol{b}$ in the in-plane and transverse parts, introducing thereby the following notations

$$\begin{aligned} \boldsymbol{a} &= \boldsymbol{\alpha} + a\,\boldsymbol{k}\,, \qquad \text{with} \quad \boldsymbol{\alpha} := \mathbf{1}\boldsymbol{a}, \quad a := \boldsymbol{k} \cdot \boldsymbol{a}\,, \\ \boldsymbol{b} &= \boldsymbol{\beta} + b\,\boldsymbol{k}\,, \qquad \text{with} \quad \boldsymbol{\beta} := \mathbf{1}\boldsymbol{b}, \quad b := \boldsymbol{k} \cdot \boldsymbol{b}\,. \end{aligned} \tag{4.111}$$

Then, for the displacement gradient we obtain

$$\begin{aligned}\nabla \boldsymbol{u} &= \nabla \boldsymbol{v} + \nabla(w\,\boldsymbol{k}) = \nabla \boldsymbol{v} + (w\,\boldsymbol{k})_{,\alpha} \otimes \boldsymbol{e}_\alpha = \nabla \boldsymbol{v} + \boldsymbol{k} \otimes (w_{,\alpha}\,\boldsymbol{e}_\alpha) \\ &= \nabla \boldsymbol{v} + \boldsymbol{k} \otimes \nabla w\,,\end{aligned} \tag{4.112}$$

and similarly

$$\nabla \boldsymbol{a} = \nabla \boldsymbol{\alpha} + \boldsymbol{k} \otimes \nabla a\,. \tag{4.113}$$

Let us denote by

$$\theta := \mathrm{tr}\,(\nabla \boldsymbol{v}) = \mathrm{div}\,\boldsymbol{v} \tag{4.114}$$

the in-plane dilatation. In view of $\boldsymbol{A}\boldsymbol{a} = -\big(\underline{\boldsymbol{C}}[\nabla \boldsymbol{u}]\big)\boldsymbol{k}$, we compute

$$\underline{\boldsymbol{C}}[\nabla \boldsymbol{u}] = \lambda\,\theta\,\boldsymbol{I} + 2\mu\,\mathrm{sym}(\nabla \boldsymbol{v}) + \mu(\nabla w \otimes \boldsymbol{k} + \boldsymbol{k} \otimes \nabla w) \tag{4.115}$$

and

$$\big(\underline{\boldsymbol{C}}[\nabla \boldsymbol{u}]\big)\boldsymbol{k} = \lambda\,\theta\,\boldsymbol{k} + \mu \nabla w\,. \tag{4.116}$$

Thus, in view of (4.109) and (4.116), we obtain

$$\boldsymbol{a} = -\boldsymbol{A}^{-1}\big(\underline{\boldsymbol{C}}[\nabla \boldsymbol{u}]\big)\boldsymbol{k} = -\lambda\,\theta\,\boldsymbol{A}^{-1}\boldsymbol{k} - \mu\,\boldsymbol{A}^{-1}\nabla w = -\frac{\lambda}{\lambda + 2\mu}\,\theta\,\boldsymbol{k} - \nabla w \tag{4.117}$$

and so, comparing with $(4.111)_1$, we find

$$\boldsymbol{\alpha} = -\nabla w \qquad \text{and} \qquad a = -\frac{\lambda}{\lambda + 2\mu}\,\theta\,. \tag{4.118}$$

Similarly, starting from (4.113) we get

$$\big(\underline{\boldsymbol{C}}[\nabla \boldsymbol{a}]\big)\boldsymbol{k} = \lambda(\mathrm{div}\,\boldsymbol{\alpha})\boldsymbol{k} + \mu \nabla a \tag{4.119}$$

and

$$\begin{aligned}\boldsymbol{b} = -\boldsymbol{A}^{-1}\big(\underline{\boldsymbol{C}}[\nabla \boldsymbol{a}]\big)\boldsymbol{k} &= -\frac{\lambda}{\lambda + 2\mu}\,(\mathrm{div}\,\boldsymbol{\alpha})\boldsymbol{k} - \nabla a \\ &= \frac{\lambda}{\lambda + 2\mu}\,(\Delta w)\boldsymbol{k} + \frac{\lambda}{\lambda + 2\mu}\,\nabla\theta\,,\end{aligned} \tag{4.120}$$

where we have used (4.118) and $\mathrm{div}(\nabla w) = \Delta w$, the Laplace operator. Then, with the notations $(4.111)_2$ we have

$$\boldsymbol{\beta} = -\nabla a = \frac{\lambda}{\lambda + 2\mu}\,\nabla\theta \quad \text{and} \quad b = -\frac{\lambda}{\lambda + 2\mu}\,(\mathrm{div}\,\boldsymbol{\alpha}) = \frac{\lambda}{\lambda + 2\mu}\,\Delta w\,. \tag{4.121}$$

4.4.2 Strain Energy

Let us write the form of the strain energy density $W = F(\nabla \boldsymbol{u}, \nabla\nabla \boldsymbol{u})$ for isotropic plates. In view of $(4.88)_2$, we have

$$W = \frac{1}{2}\, h\, \boldsymbol{P}_0 \cdot \boldsymbol{H}_0 + \frac{1}{24}\, h^3 \boldsymbol{P}'_0 \cdot \boldsymbol{H}'_0\,, \tag{4.122}$$

where

$$\begin{aligned} \boldsymbol{P}_0 \cdot \boldsymbol{H}_0 &= (\boldsymbol{P}_0\mathbf{1} + \underbrace{\boldsymbol{P}_0\boldsymbol{k}}_{=0} \otimes \boldsymbol{k}) \cdot (\nabla \boldsymbol{u} + \boldsymbol{a} \otimes \boldsymbol{k}) = \boldsymbol{P}_0\mathbf{1} \cdot \nabla \boldsymbol{u} \qquad \text{and} \\ \boldsymbol{P}'_0 \cdot \boldsymbol{H}'_0 &= (\boldsymbol{P}'_0\mathbf{1} + \underbrace{\boldsymbol{P}'_0\boldsymbol{k}}_{=0} \otimes \boldsymbol{k}) \cdot (\nabla \boldsymbol{a} + \boldsymbol{b} \otimes \boldsymbol{k}) = \boldsymbol{P}'_0\mathbf{1} \cdot \nabla \boldsymbol{a}\,. \end{aligned} \tag{4.123}$$

Using the relations (4.111), (4.112) and (4.118), we have

$$\begin{aligned} \boldsymbol{H}_0 &= \nabla \boldsymbol{u} + \boldsymbol{a} \otimes \boldsymbol{k} = \nabla \boldsymbol{v} + \boldsymbol{k} \otimes \nabla w + \boldsymbol{\alpha} \otimes \boldsymbol{k} + a\, \boldsymbol{k} \otimes \boldsymbol{k} \\ &= \nabla \boldsymbol{v} + \boldsymbol{k} \otimes \nabla w - \nabla w \otimes \boldsymbol{k} + a\, \boldsymbol{k} \otimes \boldsymbol{k}\,, \end{aligned} \tag{4.124}$$

so

$$\begin{aligned} \boldsymbol{P}_0\mathbf{1} &= \big(\underline{\boldsymbol{C}}[\boldsymbol{H}_0]\big)\mathbf{1} = \big[\lambda(\mathrm{tr}\, \boldsymbol{H}_0)\boldsymbol{I} + 2\mu(\mathrm{sym}\, \boldsymbol{H}_0)\big]\mathbf{1} \\ &= \big[\lambda(\mathrm{div}\, \boldsymbol{v} + a)\boldsymbol{I} + 2\mu(\mathrm{sym}\nabla \boldsymbol{v} + a\, \boldsymbol{k} \otimes \boldsymbol{k})\big]\mathbf{1} \\ &= \lambda(\theta + a)\mathbf{1} + 2\mu(\mathrm{sym}\nabla \boldsymbol{v}) \;=\; \frac{2\lambda\mu}{\lambda + 2\mu}\, \theta\, \mathbf{1} + 2\mu(\mathrm{sym}\nabla \boldsymbol{v})\,. \end{aligned} \tag{4.125}$$

Analogously, starting from $\boldsymbol{H}'_0 = \nabla \boldsymbol{a} + \boldsymbol{b} \otimes \boldsymbol{k}$ and using the relations (4.113), (4.118) and (4.121), we deduce

$$\begin{aligned} \boldsymbol{P}'_0\mathbf{1} &= \big(\underline{\boldsymbol{C}}[\boldsymbol{H}'_0]\big)\mathbf{1} = \lambda(\mathrm{div}\, \boldsymbol{\alpha} + b)\mathbf{1} + 2\mu(\mathrm{sym}\nabla \boldsymbol{\alpha}) \\ &= -\frac{2\lambda\mu}{\lambda + 2\mu}\, (\Delta w)\mathbf{1} - 2\mu(\nabla\nabla w)\,, \end{aligned} \tag{4.126}$$

since $\nabla\nabla w$ is symmetric. Inserting (4.125) and (4.126) into (4.123), we obtain

$$\begin{aligned} \boldsymbol{P}_0 \cdot \boldsymbol{H}_0 &= \Big[\frac{2\lambda\mu}{\lambda + 2\mu}\, \theta\, \mathbf{1} + 2\mu(\mathrm{sym}\nabla \boldsymbol{v})\Big] \cdot (\nabla \boldsymbol{v} + \boldsymbol{k} \otimes \nabla w) \\ &= \frac{2\lambda\mu}{\lambda + 2\mu}\, \theta^2 + 2\mu\, \big|\mathrm{sym}\nabla \boldsymbol{v}\,\big|^2 \end{aligned} \tag{4.127}$$

and

$$\begin{aligned}
\boldsymbol{P}_0'\cdot\boldsymbol{H}_0' &= -\Big[\frac{2\lambda\mu}{\lambda+2\mu}\,(\Delta w)\mathbf{1} - 2\mu(\nabla\nabla w)\Big]\cdot(\nabla\boldsymbol{\alpha}+\boldsymbol{k}\otimes\nabla a)\\
&= \Big[\frac{2\lambda\mu}{\lambda+2\mu}\,(\Delta w)\mathbf{1} - 2\mu(\nabla\nabla w)\Big]\cdot\Big(\nabla\nabla w+\frac{\lambda}{\lambda+2\mu}\,\boldsymbol{k}\otimes\nabla\theta\Big)\\
&= \frac{2\lambda\mu}{\lambda+2\mu}\,(\Delta w)^2 + 2\mu\,\big|\nabla\nabla w\big|^2
\end{aligned}\tag{4.128}$$

Finally, by substituting (4.127) and (4.128) into (4.122), we can write the following decomposition of the strain energy density

$$\begin{aligned}
W &= W_s + W_b\,, \qquad \text{where}\\
W_s &= h\Big[\frac{\lambda\mu}{\lambda+2\mu}\,(\operatorname{div}\boldsymbol{v})^2 + \mu\,\big|\operatorname{sym}\nabla\boldsymbol{v}\,\big|^2\Big] \qquad \text{(stretching energy)},\\
W_b &= \frac{1}{12}\,h^3\Big[\frac{\lambda\mu}{\lambda+2\mu}\,(\Delta w)^2 + \mu\,\big|\nabla\nabla w\,\big|^2\Big] \qquad \text{(bending energy)}.
\end{aligned}\tag{4.129}$$

Remark We see that the strain energy "decouples", in the sense that the stretching energy W_s involves only the in-plane displacement $\boldsymbol{v}$, while the bending energy W_b involves only the transverse displacement w. This "decoupling" does not occur if Ω is not the midsurface of the plate, or for general material symmetry. □

An alternative expression of the bending energy W_b can be derived from (4.129) using the relation

$$\big|\nabla\nabla w\,\big|^2 = (\Delta w)^2 - 2\det(\nabla\nabla w). \tag{4.130}$$

To prove this equality, we employ the two-dimensional Cayley-Hamilton theorem

$$\boldsymbol{M}^2 - (\operatorname{tr}\boldsymbol{M})\boldsymbol{M} + (\det\boldsymbol{M})\mathbf{1} = \mathbf{0}, \qquad \text{for any } \boldsymbol{M} = M_{\alpha\beta}\boldsymbol{e}_\alpha\otimes\boldsymbol{e}_\beta\,. \tag{4.131}$$

Applying the trace operator, we deduce that for any symmetric tensor $\boldsymbol{M}$,

$$(\operatorname{tr}\boldsymbol{M})^2 = |\boldsymbol{M}|^2 + 2\det\boldsymbol{M}\,.$$

If we write this relation for $\boldsymbol{M} = \nabla\nabla w$ and use that $\operatorname{tr}(\nabla\nabla w) = \Delta w$, then we obtain the Eq. (4.130). Further, inserting the relation (4.130) into $(4.129)_3$, we get the following alternative expression of the bending energy

$$W_b = \frac{1}{2}\,D\big[\,(\Delta w)^2 - 2(1-\nu)\det(\nabla\nabla w)\big], \tag{4.132}$$

with

$$D := \frac{1}{12}\,h^3\,\frac{4\mu(\lambda+\mu)}{\lambda+2\mu} = \frac{E\,h^3}{12(1-\nu^2)} \qquad \text{(bending stiffness)}, \tag{4.133}$$

where E is the Young modulus and ν the Poisson ratio of the material, given by

$$E = \frac{\mu(3\lambda + 2\mu)}{\lambda + \mu}, \qquad \nu = \frac{\lambda}{2(\lambda + \mu)}. \tag{4.134}$$

One can also express the Lamé constants and bulk modulus in terms of E , ν as

$$\mu = \frac{E}{2(1+\nu)}, \qquad \lambda = \frac{E\nu}{(1-2\nu)(1+\nu)}, \qquad \kappa = \frac{E}{3(1-2\nu)}. \tag{4.135}$$

With the above notation, the bending energy $(4.129)_3$ can also be written in the form

$$W_b = \frac{1}{2} D\Big[\nu(\Delta w)^2 + (1-\nu)\big|\nabla\nabla w\,\big|^2\Big]. \tag{4.136}$$

4.4.3 Decoupled Equilibrium Equations

The local equilibrium equations have the general form (see Sect. 4.3)

$$T_{i\alpha,\alpha} + g_i = 0 \qquad (i = 1, 2, 3), \tag{4.137}$$

with

$$T_{i\alpha} = N_{i\alpha} - M_{i\alpha\beta,\beta} \quad \text{and} \quad N_{i\alpha} = \frac{\partial W}{\partial u_{i,\alpha}}, \qquad M_{i\alpha\beta} = \frac{\partial W}{\partial u_{i,\alpha\beta}}. \tag{4.138}$$

Let us write the specific form of these equations for the case of isotropic plates and show that the system of equilibrium equations decouples into two problems: one for the bending and one for the stretching of plates. Thus, we obtain the classical plate-bending equation and, respectively, the so-called "generalized plane stress" problem.

In view of the notation (4.110), we have $u_\alpha = v_\alpha$ and $u_3 = w$. Hence, for $i = 3$ the relations (4.138) and (4.129) yield

$$N_{3\alpha} = \frac{\partial W}{\partial u_{3,\alpha}} = \frac{\partial W_b}{\partial w_{,\alpha}} = 0, \tag{4.139}$$

since the bending energy W_b is a function of $\nabla\nabla w$ only. Further, using (4.136) and $(4.138)_3$ we have

$$\begin{aligned} M_{3\alpha\beta} = \frac{\partial W_b}{\partial w_{,\alpha\beta}} &= \frac{1}{2} D \frac{\partial}{\partial w_{,\alpha\beta}} \big[\nu(w_{,\gamma\gamma})^2 + (1-\nu) w_{,\gamma\delta}\, w_{,\gamma\delta}\big] \\ &= D\big[\nu(w_{,\gamma\gamma})\delta_{\alpha\beta} + (1-\nu) w_{,\alpha\beta}\big]. \end{aligned} \tag{4.140}$$

The last relation can also be written in the form

$$M_{3\alpha\beta} = D\big[(\Delta w)\delta_{\alpha\beta} - (1-\nu)\kappa_{\alpha\beta}\big], \qquad \text{where} \qquad \kappa_{\alpha\beta} := (\Delta w)\delta_{\alpha\beta} - w_{,\alpha\beta}\,. \tag{4.141}$$

Hence, we obtain

$$T_{3\alpha} = \underbrace{N_{3\alpha}}_{=0} - M_{3\alpha\beta,\beta} = -D\big[(\Delta w)_{,\beta}\delta_{\alpha\beta} - (1-\nu)\kappa_{\alpha\beta,\beta}\big],$$

i.e.

$$T_{3\alpha} = -D\,(\Delta w)_{,\alpha}\,, \tag{4.142}$$

since we have the relation[1]

$$\kappa_{\alpha\beta,\beta} = (\Delta w)_{,\beta}\delta_{\alpha\beta} - w_{,\alpha\beta\beta} = (\Delta w)_{,\alpha} - (w_{,\beta\beta})_{,\alpha} = 0. \tag{4.143}$$

From (4.142) we get

$$T_{3\alpha,\alpha} = -D\,(\Delta w)_{,\alpha\alpha} = -D\;\Delta(\Delta w)\,. \tag{4.144}$$

Thus, the equilibrium equation for $i = 3$ reads

$$T_{3\alpha,\alpha} + g_3 = 0 \quad \text{in } \Omega, \tag{4.145}$$

i.e.

$$D\;\Delta(\Delta w) = g_3 \quad \text{in } \Omega. \tag{4.146}$$

This is the *classical plate-bending equation* for the determination of the transverse displacement w. To this equation we adjoin the following boundary conditions:

$$w = w_0 \quad \text{and} \quad \boldsymbol{\nu}\cdot\nabla w = w_{\nu 0} \quad \text{are assigned on } \partial\Omega_u\,. \tag{4.147}$$

On the part $\partial\Omega_t$, the boundary conditions (4.104) read

$$T_{3\alpha}\nu_\alpha - (M_{3\alpha\beta}\nu_\alpha\tau_\beta)_{,s} = \ell_3\,, \qquad M_{3\alpha\beta}\nu_\alpha\nu_\beta = c_3\,. \tag{4.148}$$

In view of (4.140) and (4.142), we have

[1] One can show that this equality is equivalent to the (linearized) Mainardi-Codazzi equations on the deformed surface.

$$\begin{aligned} T_{3\alpha}\nu_\alpha &= -D\,(\Delta w)_{,\alpha}\,\nu_\alpha\,, \\ M_{3\alpha\beta}\nu_\alpha\tau_\beta &= D\,(1-\nu)w_{,\alpha\beta}\,\nu_\alpha\tau_\beta\,, \\ M_{3\alpha\beta}\nu_\alpha\nu_\beta &= D\big[\nu(\Delta w) + (1-\nu)w_{,\alpha\beta}\,\nu_\alpha\nu_\beta\big], \end{aligned} \tag{4.149}$$

since $\nu_\alpha\tau_\alpha = \boldsymbol{\nu}\cdot\boldsymbol{\tau} = 0$ and $\nu_\alpha\nu_\alpha = \boldsymbol{\nu}\cdot\boldsymbol{\nu} = 1$. Then, the boundary conditions (4.148) can be written in terms of w in the form

$$\begin{aligned} D\,(\Delta w)_{,\alpha}\,\nu_\alpha + D\,(1-\nu)(w_{,\alpha\beta}\,\nu_\alpha\tau_\beta)_{,s} &= -\ell_3 \quad \text{and} \\ D\big[\nu(\Delta w) + (1-\nu)w_{,\alpha\beta}\,\nu_\alpha\nu_\beta\big] &= c_3 \quad \text{on } \partial\Omega_t\,. \end{aligned} \tag{4.150}$$

On the other hand, if we write the Eqs. (4.138) for $i = 1, 2$, we get

$$\begin{aligned} N_{\alpha\beta} = \frac{\partial W_s}{\partial v_{\alpha,\beta}} &= h\,\frac{\partial}{\partial v_{\alpha,\beta}}\Big[\frac{\lambda\mu}{\lambda+2\mu}\,(v_{\gamma,\gamma})^2 + \frac{\mu}{4}\,(v_{\gamma,\delta}+v_{\delta,\gamma})(v_{\gamma,\delta}+v_{\delta,\gamma})\Big] \\ &= h\Big[\frac{2\lambda\mu}{\lambda+2\mu}\,(v_{\gamma,\gamma})\delta_{\alpha\beta} + \mu\,(v_{\alpha,\beta}+v_{\beta,\alpha})\Big], \end{aligned} \tag{4.151}$$

so

$$\begin{aligned} N_{\alpha\beta} = N_{\beta\alpha} &= 2h\Big(\frac{\lambda\mu}{\lambda+2\mu}\,\varepsilon_{\gamma\gamma}\,\delta_{\alpha\beta} + \mu\,\varepsilon_{\alpha\beta}\Big), \quad \text{where} \\ \varepsilon_{\alpha\beta} &:= \frac{1}{2}(v_{\alpha,\beta}+v_{\beta,\alpha}). \end{aligned} \tag{4.152}$$

Also, from (4.138) we deduce that the twisting moments are vanishing

$$M_{\gamma\alpha\beta} = \frac{\partial W}{\partial u_{\gamma,\alpha\beta}} = \frac{\partial W_s}{\partial v_{\gamma,\alpha\beta}} = 0, \tag{4.153}$$

since the stretching energy W_s is a function of $\nabla\boldsymbol{v}$ only. The last relation means that the twisting moments $M_{\gamma\alpha\beta}$ are of order $o(h^3)$ and they are not taken into account in this model. Thus, on the boundary part $\partial\Omega_t$ the edge loads satisfy $c_\gamma = 0$, see the *Remark* below.

By virtue of (4.153), we see that $T_{\alpha\beta} = N_{\alpha\beta}$ and the equilibrium equations (for $i = 1, 2$) reduce to

$$N_{\alpha\beta,\beta} + g_\alpha = 0 \quad \text{in } \Omega. \tag{4.154}$$

The boundary conditions are: $\boldsymbol{v}$ is assigned on $\partial\Omega_u$ and

$$N_{\alpha\beta}\,\nu_\beta = \ell_\alpha \quad \text{on } \partial\Omega_t\,. \tag{4.155}$$

This is the so-called *generalized plane stress problem*, which characterizes the stretching of isotropic elastic plates.

Using (4.152), we can write the equilibrium equations (4.154) in terms of the in-plane displacement $\boldsymbol{v}$ as follows

$$h\left[\mu\, \Delta v_\alpha + \frac{\mu(3\lambda+2\mu)}{\lambda+2\mu}\, v_{\beta,\beta\alpha}\right] + g_\alpha = 0 \quad \text{in } \Omega, \tag{4.156}$$

or, in invariant form,

$$\frac{1}{2}\, C\, [(1-\nu)\, \Delta \boldsymbol{v} + (1+\nu)\, \nabla(\operatorname{div} \boldsymbol{v})] + \mathbf{1}\boldsymbol{g} = \mathbf{0} \quad \text{in } \Omega, \tag{4.157}$$

where $C := h\,\dfrac{E}{1-\nu^2}$ is the so-called stretching stiffness of the plate.

In conclusion, we observe that the bending problem (4.145)–(4.147) and the stretching problem (4.154)–(4.156) decouple. The first problem can be solved to find the transverse displacement w, whereas the latter determines the in-plane displacements v_α. We mention that, in general, the stretching and bending problems do *not* decouple (e.g., for the case of curved shells, see Chap. 5).

Remark Consider the integral term $\displaystyle\int_{\partial\Omega_t} \boldsymbol{c}\cdot \boldsymbol{u}_{,\nu}\; \mathrm{d}s$, which appears in the edge load potential. We decompose the normal derivative

$$\boldsymbol{u}_{,\nu} = (\nabla \boldsymbol{u})\boldsymbol{\nu} = (\nabla \boldsymbol{v} + \boldsymbol{k}\otimes \nabla w)\boldsymbol{\nu} = \boldsymbol{v}_{,\nu} + (\boldsymbol{\nu}\cdot\nabla w)\boldsymbol{k} = \boldsymbol{v}_{,\nu} + w_{,\nu}\,\boldsymbol{k}\,. \tag{4.158}$$

In view of (4.104) and $M_{\gamma\alpha\beta} = 0$, we can write

$$\boldsymbol{c} = M_{i\alpha\beta}\nu_\alpha\nu_\beta\,\boldsymbol{e}_i = M_{3\alpha\beta}\nu_\alpha\nu_\beta\,\boldsymbol{k} = M\,\boldsymbol{k}\,, \qquad \text{with } M := M_{3\alpha\beta}\nu_\alpha\nu_\beta\,. \tag{4.159}$$

Then, from (4.158) and (4.159) we obtain the relation

$$\boldsymbol{c}\cdot\boldsymbol{u}_{,\nu} = M\, w_{,\nu}\,, \tag{4.160}$$

which can be applied in the expression of the potential energy (4.88). □

4.5 Energy Minimizers

Let us investigate the minimizers of the potential energy, in the context of the general plate theory presented in Sect. 4.2.

We know that equilibria minimize the energy in the three-dimensional theory, see Sect. 3.3. But, in the plate model the potential energy is the $O(h^3)$ truncation of the actual energy, which may not *itself* be minimized. Hence, the following question arises: Do plate equilibria minimize the plate energy?

Recall that the potential energy is given by

$$E[\boldsymbol{u}] = \int_{\Omega} W \,\mathrm{d}a - \int_{\Omega} \boldsymbol{g}\cdot\boldsymbol{u}\,\mathrm{d}a - \int_{\partial\Omega_t} \left(\boldsymbol{\ell}\cdot\boldsymbol{u} + \boldsymbol{c}\cdot\boldsymbol{u}_{,\nu}\right)\mathrm{d}s\,, \tag{4.161}$$

according to (4.88). If we denote by

$$\boldsymbol{E} := \operatorname{sym}\bar{\boldsymbol{H}}_0 = \operatorname{sym}(\nabla\boldsymbol{u} + \bar{\boldsymbol{a}}\otimes\boldsymbol{k}), \qquad \boldsymbol{E}' := \operatorname{sym}\bar{\boldsymbol{H}}_0' = \operatorname{sym}(\nabla\bar{\boldsymbol{a}} + \bar{\boldsymbol{b}}\otimes\boldsymbol{k}), \tag{4.162}$$

then the strain energy density can be written as

$$W = \frac{1}{2}\,h\,\boldsymbol{E}\cdot\underline{\boldsymbol{C}}[\boldsymbol{E}] + \frac{1}{24}\,h^3\,\boldsymbol{E}'\cdot\underline{\boldsymbol{C}}[\boldsymbol{E}']\,. \tag{4.163}$$

Note that $E[\boldsymbol{u}]$ is a quadratic functional of the displacement vector $\boldsymbol{u}$.

The answer to the above question is given by the following result.

Theorem 4.1 *Let $E[\boldsymbol{u}]$ be the potential energy of linearly elastic plates and let $\boldsymbol{u}$, $\boldsymbol{u}^*$ be two displacement fields which satisfy the same boundary conditions on $\partial\Omega_u$, namely*

$$\boldsymbol{u} = \boldsymbol{u}_0\,, \qquad \boldsymbol{u}_{,\nu} = \boldsymbol{u}_{\nu 0} \qquad on\ \ \partial\Omega_u\,, \tag{4.164}$$

and also

$$\boldsymbol{u}^* = \boldsymbol{u}_0\,, \qquad \boldsymbol{u}^*_{,\nu} = \boldsymbol{u}_{\nu 0} \qquad on\ \ \partial\Omega_u\,.$$

If $\boldsymbol{u}$ represents an equilibrium state of the plate, then $\boldsymbol{u}$ minimizes the potential energy, i.e.

$$E[\boldsymbol{u}^*] \ge E[\boldsymbol{u}] \qquad for\ all\ \ \boldsymbol{u}^* \neq \boldsymbol{u}\,. \tag{4.165}$$

The latter inequality is strict, provided that $\partial\Omega_u$ has a positive length measure.

Proof Denote by $\delta\boldsymbol{u}$ the difference of the two displacement fields. Then, we have $\boldsymbol{u}^* = \boldsymbol{u} + \delta\boldsymbol{u}$ and in view of (4.164)

$$\delta\boldsymbol{u} = \boldsymbol{0} \qquad \text{and} \qquad (\delta\boldsymbol{u})_{,\nu} = \delta(\boldsymbol{u}_{,\nu}) = \boldsymbol{0} \qquad \text{on}\ \ \partial\Omega_u\,. \tag{4.166}$$

Let us consider the family of displacement fields $\boldsymbol{u}(\varepsilon) := \boldsymbol{u} + \varepsilon(\delta\boldsymbol{u})$, where ε is a parameter. Then, the displacements $\boldsymbol{u}(\varepsilon)$ satisfy the boundary conditions (4.164) and the energy $\tilde{E}(\varepsilon) := E[\boldsymbol{u}(\varepsilon)]$ is a quadratic function of ε. Hence, denoting the derivative with respect to ε by a superposed dot, we have

$$\tilde{E}(\varepsilon) = \tilde{E}(0) + \varepsilon\,\dot{\tilde{E}}(0) + \frac{1}{2}\,\varepsilon^2\,\ddot{\tilde{E}}(0)\,, \tag{4.167}$$

where $\ddot{\tilde{E}}(0)$ is given by

$$\ddot{\tilde{E}}(0) = \int_{\Omega} \Psi\,\mathrm{d}a\,, \quad \text{with} \quad \Psi := h\,(\delta\boldsymbol{E})\cdot\underline{\boldsymbol{C}}[\delta\boldsymbol{E}] + \frac{1}{12}\,h^3\,(\delta\boldsymbol{E}')\cdot\underline{\boldsymbol{C}}[\delta\boldsymbol{E}'] \tag{4.168}$$

and $\delta \boldsymbol{E} = \boldsymbol{E}(\delta \boldsymbol{u})$, $\delta \boldsymbol{E}' = \boldsymbol{E}'(\delta \boldsymbol{u})$. Putting $\varepsilon = 1$ in (4.167) we get

$$E[\boldsymbol{u}^*] - E[\boldsymbol{u}] = \dot{\bar{E}}(0) + \frac{1}{2}\int_{\Omega} \Psi \, \mathrm{d}a\,. \tag{4.169}$$

Now, if $\boldsymbol{u}$ is an equilibrium state, then we have

$$\dot{\bar{E}}(0) = \frac{\mathrm{d}}{\mathrm{d}\varepsilon} E\big[\boldsymbol{u}(\varepsilon)\big]\Big|_{\varepsilon=0} = 0\,,$$

according to (4.89). Thus, the relation (4.169) becomes

$$E[\boldsymbol{u}^*] - E[\boldsymbol{u}] = \frac{1}{2}\int_{\Omega} \Psi \, \mathrm{d}a\,. \tag{4.170}$$

Note that $\Psi \geq 0$, in view of the definition (4.168) and the fact that $\underline{\boldsymbol{C}}$ is positive definite on the set of symmetric tensors. Therefore, the right-hand side of (4.170) is non-negative and the inequality (4.165) is proved, i.e. $\boldsymbol{u}$ is a minimizer of the energy.

Let us investigate the case when the relation (4.165) holds with equality: by virtue of $\Psi \geq 0$, we see that the integral in (4.170) can vanish only if $\Psi = 0$ pointwise in Ω. But since Ψ is the sum of two positive definite forms (cf. (4.168)), we then infer

$$(\delta \boldsymbol{E}) \cdot \underline{\boldsymbol{C}}[\delta \boldsymbol{E}] = 0 \quad \text{and} \quad (\delta \boldsymbol{E}') \cdot \underline{\boldsymbol{C}}[\delta \boldsymbol{E}'] = 0,$$

which shows that

$$\delta \boldsymbol{E} = \boldsymbol{E}(\delta \boldsymbol{u}) = \boldsymbol{0} \quad \text{and} \quad \delta \boldsymbol{E}' = \boldsymbol{E}'(\delta \boldsymbol{u}) = \boldsymbol{0}. \tag{4.171}$$

But, in view of the definition (4.162) and the relations (4.111)–(4.113), we have

$$\begin{aligned} \boldsymbol{E}(\delta \boldsymbol{u}) &= \operatorname{sym}\big[\nabla(\delta \boldsymbol{u}) + (\delta \boldsymbol{a}) \otimes \boldsymbol{k}\big] \\ &= \operatorname{sym}\big[\nabla(\delta \boldsymbol{v}) + \boldsymbol{k} \otimes \big(\delta \boldsymbol{\alpha} + \nabla(\delta w)\big)\big] + (\delta a)\, \boldsymbol{k} \otimes \boldsymbol{k} \qquad \text{and} \\ \boldsymbol{E}'(\delta \boldsymbol{u}) &= \operatorname{sym}\big[\nabla(\delta \boldsymbol{a}) + (\delta \boldsymbol{b}) \otimes \boldsymbol{k}\big] \\ &= \operatorname{sym}\big[\nabla(\delta \boldsymbol{\alpha}) + \boldsymbol{k} \otimes \big(\delta \boldsymbol{\beta} + \nabla(\delta a)\big)\big] + (\delta b)\, \boldsymbol{k} \otimes \boldsymbol{k}\,. \end{aligned} \tag{4.172}$$

Hence, the relations (4.171) and (4.172) yield

$$\begin{aligned} &\operatorname{sym}\big[\nabla(\delta \boldsymbol{v})\big] = \boldsymbol{0}\,, \quad \delta \boldsymbol{\alpha} = -\nabla(\delta w), \quad \delta a = 0\,, \quad \text{and} \\ &\operatorname{sym}\big[\nabla(\delta \boldsymbol{\alpha})\big] = \boldsymbol{0}\,, \quad \delta \boldsymbol{\beta} = -\nabla(\delta a), \quad \delta b = 0\,, \end{aligned}$$

which reduce to

$$\begin{aligned} \operatorname{sym}\big[\nabla(\delta \boldsymbol{v})\big] = \boldsymbol{0}\,, \quad \delta \boldsymbol{\alpha} = -\nabla(\delta w), \quad \nabla\nabla(\delta w) = \boldsymbol{0}\,, \\ \delta a = 0\,, \quad \delta \boldsymbol{\beta} = \boldsymbol{0}, \quad \delta b = 0 \quad \text{in } \Omega\,. \end{aligned} \tag{4.173}$$

Assume now that the boundary $\partial\Omega_u$ is non-empty and has a positive length measure. In view of (4.166) and $\boldsymbol{u} = \boldsymbol{v} + w\,\boldsymbol{k}$, $\boldsymbol{u}_{,\nu} = \boldsymbol{v}_{,\nu} + w_{,\nu}\,\boldsymbol{k}$, it follows that

$$\delta\boldsymbol{v} = \boldsymbol{0}, \quad \delta\boldsymbol{v}_{,\nu} = \boldsymbol{0} \qquad \text{and} \qquad \delta w = 0, \quad \delta w_{,\nu} = 0 \qquad \text{on} \quad \partial\Omega_u\,. \tag{4.174}$$

Hence, $\delta w_{,s} = 0$ on $\partial\Omega_u$ too and

$$\nabla(\delta w) = \delta(\nabla w) = \delta(w_{,s}\boldsymbol{\tau} + w_{,\nu}\boldsymbol{\nu}) = \boldsymbol{0} \qquad \text{on} \quad \partial\Omega_u\,. \tag{4.175}$$

But from $(4.173)_3$ we have $\nabla(\delta w) = \text{const.}$, so from (4.175) we deduce

$$\nabla(\delta w) = \boldsymbol{0} \qquad \text{in} \quad \overline{\Omega}\,. \tag{4.176}$$

Thus, δw is constant in Ω and by $(4.174)_3$ we get

$$\delta w = 0 \qquad \text{in} \quad \Omega\,. \tag{4.177}$$

Also, from $(4.173)_2$ it follows that

$$\delta\boldsymbol{\alpha} = \boldsymbol{0} \qquad \text{in} \quad \Omega\,. \tag{4.178}$$

Next, the condition $\text{sym}\big[\nabla(\delta\boldsymbol{v})\big] = \boldsymbol{0}$ in $(4.173)_1$ shows that $\nabla(\delta\boldsymbol{v})$ is skew-symmetric, i.e. there exists a scalar field $\omega(\boldsymbol{r})$ such that

$$\nabla(\delta\boldsymbol{v}) \;=\; \omega(\boldsymbol{e}_1\otimes\boldsymbol{e}_2 - \boldsymbol{e}_2\otimes\boldsymbol{e}_1) = \omega\,\boldsymbol{\varepsilon} \qquad \text{with} \quad \boldsymbol{\varepsilon} = \varepsilon_{\alpha\beta}\boldsymbol{e}_\alpha\otimes\boldsymbol{e}_\beta\,, \tag{4.179}$$

where $\varepsilon_{\alpha\beta}$ is the two-dimensional alternator ($\varepsilon_{12} = -\varepsilon_{21} = 1$, $\varepsilon_{11} = \varepsilon_{22} = 0$). From (4.179) we get

$$\delta v_{\alpha,\beta} \;=\; \omega\,\varepsilon_{\alpha\beta}\,, \qquad \text{so} \quad (\delta v_\alpha)_{,\beta\gamma} \;=\; \omega_{,\gamma}\,\varepsilon_{\alpha\beta}\,,$$

which implies

$$0 \;=\; \varepsilon_{\beta\gamma}\,(\delta v_\alpha)_{,\beta\gamma} \;=\; \varepsilon_{\beta\gamma}\,\varepsilon_{\alpha\beta}\,\omega_{,\gamma} \;=\; -\delta_{\alpha\gamma}\,\omega_{,\gamma} \;=\; -\omega_{,\alpha}\,.$$

Thus, $\nabla\omega = \boldsymbol{0}$ in Ω, so ω is constant, and by integrating (4.179) we find

$$\delta\boldsymbol{v} \;=\; (\omega\,\boldsymbol{\varepsilon})\boldsymbol{r} \;+\; \tilde{\boldsymbol{c}}\,, \qquad \text{with} \quad \omega = \text{const.}\,, \; \tilde{\boldsymbol{c}} = \text{const.} \tag{4.180}$$

But since $\delta\boldsymbol{v} = \boldsymbol{0}$ on $\partial\Omega_u$ (cf. (4.174)), we also have $(\delta\boldsymbol{v})_{,s} = \boldsymbol{0}$, and so

$$\boldsymbol{0} = \big[\nabla(\delta\boldsymbol{v})\big]\boldsymbol{\tau} \;=\; (\omega\,\boldsymbol{\varepsilon})\boldsymbol{\tau} \;=\; \omega\,\boldsymbol{\nu}\,,$$

which means that $\omega = 0$. Hence, the relation (4.180) reduces to $\delta\boldsymbol{v} = \tilde{\boldsymbol{c}}$ (constant), so $\delta\boldsymbol{v} = \boldsymbol{0}$ in Ω.

In conclusion, we have proved that

$$\delta \boldsymbol{u} = \delta \boldsymbol{v} + (\delta w)\boldsymbol{k} = \boldsymbol{0}, \qquad \text{provided} \quad \Psi = 0 \quad \text{pointwise in } \Omega .$$

This means that $\boldsymbol{u}^* = \boldsymbol{u}$, so the inequality (4.165) is strict for any $\boldsymbol{u}^* \neq \boldsymbol{u}$. The proof is complete. □

As a consequence of this result, we can prove the uniqueness of solution to the linear plate equations.

Theorem 4.2 *Assume that the part of the boundary $\partial\Omega_u$ is non-empty and has a positive length measure. Then, the equilibrium equations (4.103) with boundary conditions (4.104) and (4.164) has at most one solution $\boldsymbol{u}$.*

Proof Let $\boldsymbol{u}$ and $\boldsymbol{u}^*$ be two solutions of the boundary-value problem for linearly elastic plates. In view of (4.165), we have $E[\boldsymbol{u}^*] \geq E[\boldsymbol{u}]$, as well as $E[\boldsymbol{u}] \geq E[\boldsymbol{u}^*]$. Thus, we obtain $E[\boldsymbol{u}] = E[\boldsymbol{u}^*]$. By virtue of Theorem 4.1, this is possible only if $\boldsymbol{u}^* = \boldsymbol{u}$, i.e. the two solutions coincide. □

Remark In the case when $\partial\Omega_u = \emptyset$, the inequality (4.165) is not necessarily strict. More precisely, we have equality in (4.165) if and only if $\boldsymbol{u}$ and $\boldsymbol{u}^*$ differ by a rigid body displacement field of three-dimensional plate. □

To prove this, let us consider two displacement fields $\boldsymbol{u}$ and $\boldsymbol{u}^* = \boldsymbol{u} + \delta\boldsymbol{u}$ such that $E[\boldsymbol{u}] = E[\boldsymbol{u}^*]$. Then, according to the proof of Theorem 4.1, we derive that the relations (4.173) hold. Further, the first Eq. $(4.173)_1$ can be exploited to obtain the relation (4.180), which can be written in the form

$$\delta \boldsymbol{v} = -\omega\, \boldsymbol{k} \times \boldsymbol{r} + \tilde{\boldsymbol{c}}\,, \qquad \omega = \text{const.}\,, \ \tilde{\boldsymbol{c}} = \text{const.} \tag{4.181}$$

Also, Eq. (4.173) imply that $\nabla(\delta w) = \boldsymbol{d}$ (constant) in Ω, so

$$\delta w = \boldsymbol{d} \cdot \boldsymbol{r} + q \quad \text{and} \quad \delta \boldsymbol{\alpha} = -\boldsymbol{d} \qquad (\text{with } q = \text{const.}). \tag{4.182}$$

Altogether, from (4.173), (4.181) and (4.182) we obtain that

$$\begin{aligned} \delta \boldsymbol{u} &= \delta \boldsymbol{v} + (\delta w)\boldsymbol{k} \;=\; -\omega\, \boldsymbol{k} \times \boldsymbol{r} + (\boldsymbol{d} \cdot \boldsymbol{r})\boldsymbol{k} + \boldsymbol{c}\,, \\ \delta \boldsymbol{a} &= \delta \boldsymbol{\alpha} + (\delta a)\boldsymbol{k} = -\boldsymbol{d} \quad \text{and} \quad \delta \boldsymbol{b} = \delta \boldsymbol{\beta} + (\delta b)\boldsymbol{k} = \boldsymbol{0}, \end{aligned} \tag{4.183}$$

where $\boldsymbol{d} = d_\alpha \boldsymbol{e}_\alpha$ and $\boldsymbol{c} := \tilde{\boldsymbol{c}} + q\, \boldsymbol{k}$ are arbitrary constant vectors.

Thus, if we recompose the three-dimensional displacement in the form

$$\tilde{\boldsymbol{u}}(\hat{\boldsymbol{r}}) = \boldsymbol{u}(\boldsymbol{r}) + \zeta\, \boldsymbol{a}(\boldsymbol{r}) + \frac{1}{2}\,\zeta^2 \boldsymbol{b}(\boldsymbol{r}) + \dots\,, \qquad \text{with } \hat{\boldsymbol{r}} = \boldsymbol{r} + \zeta \boldsymbol{k}\,,$$

then we find using (4.183)

$$\begin{aligned} \delta \tilde{\boldsymbol{u}} &= \delta \boldsymbol{u} + \zeta\, (\delta \boldsymbol{a}) + \frac{1}{2}\,\zeta^2 (\delta \boldsymbol{b}) \\ &= -\omega\, \boldsymbol{k} \times \hat{\boldsymbol{r}} + (\boldsymbol{d} \cdot \hat{\boldsymbol{r}})\boldsymbol{k} - (\hat{\boldsymbol{r}} \cdot \boldsymbol{k})\boldsymbol{d} + \boldsymbol{c} \;=\; (\boldsymbol{d} \times \boldsymbol{k} - \omega\, \boldsymbol{k}) \times \hat{\boldsymbol{r}} + \boldsymbol{c}\,, \end{aligned} \tag{4.184}$$

which represents a rigid body displacement field in three-dimensional linear elasticity (with infinitesimal rotation vector $\boldsymbol{d} \times \boldsymbol{k} - \omega \boldsymbol{k}$ and translation $\boldsymbol{c}$). The statement is proved.

Concerning the uniqueness of solutions in the case $\partial \Omega_u = \emptyset$, we obtain that any two solutions of the boundary-value problem (4.103), (4.104) differ by a rigid body displacement field of the form (4.184) for the three-dimensional plate.

4.6 Exercises

4.1 For uniform isotropic plates we have the classical plate-bending equation

$$D \Delta (\Delta w) = g ,$$

where g is the transverse distributed loading.

(a) The lateral loading is uniform and the plate has an elliptical boundary described by $f(r_1, r_2) = 0$, where

$$f(r_1, r_2) = \left(\frac{r_1}{a_1}\right)^2 + \left(\frac{r_2}{a_2}\right)^2 - 1.$$

The plate is clamped along its entire edge ($w = 0 = w_{,\nu}$ on $\partial\Omega$). Show that this problem is solved by taking $w = Cf^2$, where C is a suitable constant.

(b) Consider a circular plate under a distributed loading that depends only on radius r in a polar coordinate system: $g = g(r)$. Assume a solution of the form $w(r)$ depending only on r, and show that

$$\Delta w = \frac{1}{r}\left(r w'\right)' \qquad \text{where} \qquad (\,)' = \frac{d(\,)}{dr}.$$

Find the general solution to the plate bending equation in terms of four integration constants, assuming that $g(r) = \text{const.}$

(c) A solid circular plate is clamped at $r = R$ and subjected to a concentrated point load f at $r = 0$. Show that the bounded solution is

$$w(r) = \frac{f}{8\pi D}\left[\frac{1}{2}\left(R^2 - r^2\right) + r^2 \ln\left(\frac{r}{R}\right)\right].$$

(*Hint*: assume $g = 0$ and require f to be in equilibrium with the net transverse shear force at the external boundary).

(d) An annular plate with internal and external radii a and b is clamped at its inner and outer edges. There is no distributed load. The outer edge does not move, but the inner edge is fixed to a rigid cylinder that is displaced vertically by amount W. Obtain the displacement field $w(r)$ and find the total pull-out force on the cylinder required to produce the deformation.

4.2 Derive the boundary conditions for a plate with piecewise smooth boundary ($\boldsymbol{\tau}$ discontinuous at a finite number of points on $\partial\Omega$), associated with the load potential discussed in this chapter, but modified by the subtraction of

$$\sum_{i=1}^{n} \boldsymbol{f}_i \cdot \boldsymbol{u}_i \,,$$

where $\boldsymbol{f}_i$ are discrete forces applied at the corners of the plate and $\boldsymbol{u}_i$ are the displacements at the corners.

4.3 In a transversely isotropic material, the elastic moduli are given (in Cartesian components) by

$$\begin{aligned} C_{ijkl} &= \lambda\delta_{ij}\delta_{kl} + \mu_T\left(\delta_{ik}\delta_{jl} + \delta_{il}\delta_{jk}\right) + \alpha\left(\delta_{ij}m_k m_l + \delta_{kl}m_i m_j\right) \\ &\quad + (\mu_L - \mu_T)\left(\delta_{jl}m_i m_k + \delta_{jk}m_i m_l + \delta_{il}m_j m_k + \delta_{ik}m_j m_l\right) + \beta m_i m_j m_k m_l \,, \end{aligned}$$

where δ_{ij} is the Kronecker delta and α, β, λ, μ_T and μ_L are uniform material constants; and the *unit* vector $\boldsymbol{m}$ with components m_i, is the axis of transverse isotropy (unit normal to the isotropic plane), assumed to be uniform. It may be shown that μ_T is the shear modulus for shearing in planes transverse to $\boldsymbol{m}$, whereas μ_L is the shear modulus for shearing parallel to $\boldsymbol{m}$. The other moduli may be interpreted similarly.

(a) Derive the equations and boundary conditions of the order-h^3 plate model assuming that $\boldsymbol{m}$ is
  (i) orthogonal to the midplane Ω (i.e., $\boldsymbol{m} = \boldsymbol{k}$);
  (ii) tangential to the midplane Ω (i.e., $\boldsymbol{m} = m_\alpha \boldsymbol{e}_\alpha$).
  State the relevant strong-ellipticity conditions in both cases and show that the in-plane and transverse displacements decouple, as they do in the case of isotropy.

(b) Derive the order-h^3 set of equations and boundary conditions in the case when the axis $\boldsymbol{m}$ is oblique to Ω; i.e., $\boldsymbol{m} = m_\alpha \boldsymbol{e}_\alpha + m\,\boldsymbol{k}$, with $m_\alpha m_\alpha + m^2 = 1$. Show that the equations for in-plane and transverse displacements are coupled in this case.

References

1. Ciarlet, P.G.: Mathematical Elasticity, vol. II. Theory of Plates, North-Holland, Amsterdam (1997)
2. Naghdi, P.M.: The theory of shells and plates. In: Flügge, W. (ed.) Handbuch der Physik, vol. VIa/2, pp. 425–640. Springer, Berlin (1972)

Chapter 5
Linear Shell Theory

Abstract In this chapter the dimension reduction procedure is extended to curved shells, again in the context of linear elasticity. The resulting model is used to obtain some simple solutions to problems of practical interest.

5.1 Geometry of the Curved Shell

Let us consider a shell-like body having a curved reference configuration $\mathscr{K}$ with midsurface Ω and thickness h. Denote by (θ^1, θ^2) the convected curvilinear coordinates on the surface Ω and by ζ the thickness coordinate. That is, the position vector of a generic point can be written as

$$\boldsymbol{x}(\theta^\alpha, \zeta) = \boldsymbol{r}(\theta^\alpha) + \zeta\, \boldsymbol{n}(\theta^\alpha), \qquad -\frac{h}{2} \leq \zeta \leq \frac{h}{2}\,, \tag{5.1}$$

where $\boldsymbol{r}(\theta^\alpha)$ is a parametrization of the midsurface Ω (see Fig. 5.1), $\boldsymbol{a}_\alpha = \boldsymbol{r}_{,\alpha}$ are the covariant base vectors, and $\boldsymbol{n}(\theta^\alpha)$ is the unit normal vector given by

$$\boldsymbol{n} = \frac{1}{\sqrt{a}}\boldsymbol{a}_1 \times \boldsymbol{a}_2, \qquad \text{with} \quad a = \det(a_{\alpha\beta})_{2\times 2}\,, \quad a_{\alpha\beta} = \boldsymbol{a}_\alpha \cdot \boldsymbol{a}_\beta\,. \tag{5.2}$$

We assume that $|\boldsymbol{a}_1 \times \boldsymbol{a}_2| > 0$, so we have $a > 0$.

Let $\boldsymbol{\kappa} = \kappa_{\alpha\beta}\boldsymbol{a}^\alpha \otimes \boldsymbol{a}^\beta = \kappa^\alpha_\beta \boldsymbol{a}_\alpha \otimes \boldsymbol{a}^\beta$ be the curvature tensor of the surface Ω, i.e. (cf. (2.69)–(2.73))

$$\boldsymbol{n}_{,\alpha} = -\kappa_{\alpha\beta}\,\boldsymbol{a}^\beta = -\kappa^\beta_\alpha\,\boldsymbol{a}_\beta \qquad \text{and} \qquad \kappa_{\alpha\beta} = \kappa_{\beta\alpha}\,. \tag{5.3}$$

Then, we obtain

$$\begin{aligned} \mathrm{d}\boldsymbol{n} &= \boldsymbol{n}_{,\alpha}\,\mathrm{d}\theta^\alpha = \boldsymbol{n}_{,\alpha}(\boldsymbol{a}^\alpha \cdot \mathrm{d}\boldsymbol{r}) = (\boldsymbol{n}_{,\alpha} \otimes \boldsymbol{a}^\alpha)\,\mathrm{d}\boldsymbol{r} \\ &= -(\kappa_{\alpha\beta}\boldsymbol{a}^\alpha \otimes \boldsymbol{a}^\beta)\,\mathrm{d}\boldsymbol{r} = -\boldsymbol{\kappa}\,\mathrm{d}\boldsymbol{r}\,. \end{aligned} \tag{5.4}$$

D. J. Steigmann et al., *Lecture Notes on the Theory of Plates and Shells*,
Solid Mechanics and Its Applications 274,
https://doi.org/10.1007/978-3-031-25674-5_5

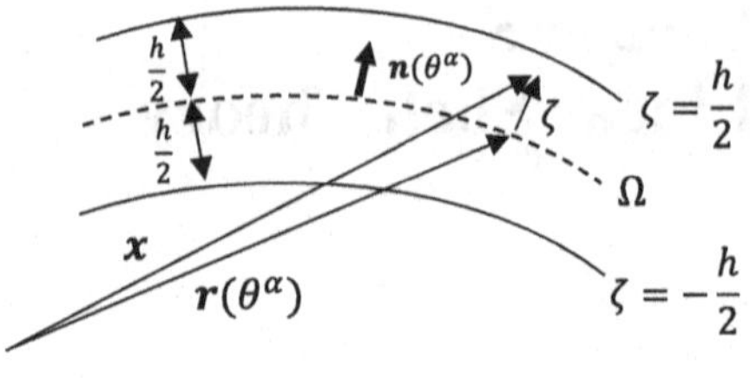

Fig. 5.1 Reference configuration of the curved shell

Thus, using (5.1) and (5.4) we can write

$$\mathrm{d}\boldsymbol{x} = \mathrm{d}\boldsymbol{r} + \zeta\,\mathrm{d}\boldsymbol{n} + \boldsymbol{n}\,\mathrm{d}\zeta = (\mathbf{1} - \zeta\boldsymbol{\kappa})\,\mathrm{d}\boldsymbol{r} + \boldsymbol{n}\,\mathrm{d}\zeta, \tag{5.5}$$

where $\mathbf{1} = \boldsymbol{I} - \boldsymbol{n} \otimes \boldsymbol{n}$ is the projection tensor on the tangent plane T_Ω (and the identity tensor in T_Ω). If we denote by

$$\boldsymbol{\mu} := \mathbf{1} - \zeta\boldsymbol{\kappa} \qquad \text{and} \qquad \mu := \det \boldsymbol{\mu}, \tag{5.6}$$

then the relation (5.5) can be put in the form

$$\mathrm{d}\boldsymbol{x} = \boldsymbol{\mu}\,\mathrm{d}\boldsymbol{r} + \boldsymbol{n}\,\mathrm{d}\zeta = \boldsymbol{G}(\mathrm{d}\boldsymbol{r} + \boldsymbol{n}\,\mathrm{d}\zeta), \qquad \text{where} \quad \boldsymbol{G} := \boldsymbol{\mu} + \boldsymbol{n} \otimes \boldsymbol{n}\,. \tag{5.7}$$

Also, from (2.50) we have

$$\mathrm{d}\boldsymbol{r} = \boldsymbol{a}_\alpha\,\mathrm{d}\theta^\alpha = \mathrm{d}\boldsymbol{r}_1 + \mathrm{d}\boldsymbol{r}_2, \qquad \text{with} \ \ \mathrm{d}\boldsymbol{r}_1 := \boldsymbol{a}_1\,\mathrm{d}\theta^1, \ \ \mathrm{d}\boldsymbol{r}_2 := \boldsymbol{a}_2\,\mathrm{d}\theta^2\,. \tag{5.8}$$

Substituting (5.8) into (5.7) we find

$$\begin{aligned} \mathrm{d}\boldsymbol{x} &= \boldsymbol{G}(\mathrm{d}\boldsymbol{r}_1 + \mathrm{d}\boldsymbol{r}_2 + \boldsymbol{n}\,\mathrm{d}\zeta) = \mathrm{d}\boldsymbol{x}_1 + \mathrm{d}\boldsymbol{x}_2 + \mathrm{d}\boldsymbol{x}_3, \qquad \text{where} \\ \mathrm{d}\boldsymbol{x}_\alpha &:= \boldsymbol{G}\,\mathrm{d}\boldsymbol{r}_\alpha \qquad \text{and} \qquad \mathrm{d}\boldsymbol{x}_3 := \boldsymbol{G}\,\boldsymbol{n}\,\mathrm{d}\zeta = \boldsymbol{n}\,\mathrm{d}\zeta\,. \end{aligned} \tag{5.9}$$

Then, the elemental volume is given by

$$\begin{aligned} \mathrm{d}v &= \big[\mathrm{d}\boldsymbol{x}_1,\, \mathrm{d}\boldsymbol{x}_2,\, \mathrm{d}\boldsymbol{x}_3\big] = (\det \boldsymbol{G})\big[\mathrm{d}\boldsymbol{r}_1,\, \mathrm{d}\boldsymbol{r}_2,\, \boldsymbol{n}\,\mathrm{d}\zeta\big] \\ &= (\det \boldsymbol{G})(\mathrm{d}\boldsymbol{r}_1 \times \mathrm{d}\boldsymbol{r}_2) \cdot \boldsymbol{n}\,\mathrm{d}\zeta = (\det \boldsymbol{G})(\sqrt{a}\,\mathrm{d}\theta^1\mathrm{d}\theta^2)\,\mathrm{d}\zeta, \end{aligned}$$

i.e.

$$\mathrm{d}v = (\det \boldsymbol{G})\,\mathrm{d}a\,\mathrm{d}\zeta, \tag{5.10}$$

since the elemental area of Ω is $\mathrm{d}a = \sqrt{a}\,\mathrm{d}\theta^1\mathrm{d}\theta^2$. In view of (5.6) and $(5.7)_2$ we have

$$\det \boldsymbol{G} = \det \boldsymbol{\mu} = \det(\mathbf{1} - \zeta\boldsymbol{\kappa})$$

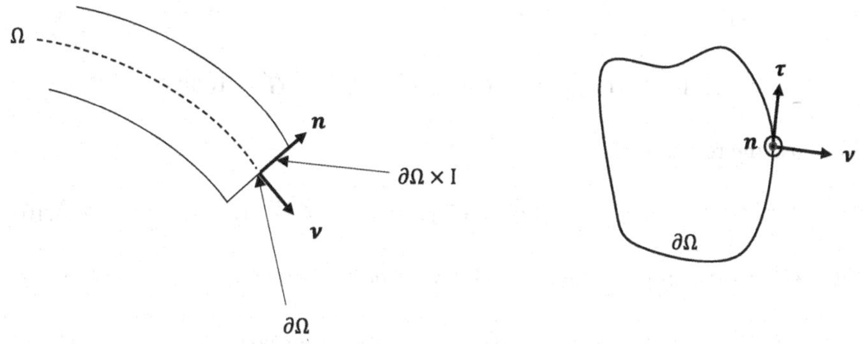

Fig. 5.2 The edge boundary of the reference configuration

and taking into account that the mean curvature H and the Gaussian curvature K satisfy

$$2H = \operatorname{tr} \boldsymbol{\kappa} \qquad \text{and} \qquad K = \det \boldsymbol{\kappa},$$

we get

$$\det \boldsymbol{G} = \mu = 1 - 2H\zeta + K\zeta^2 . \tag{5.11}$$

Inserting (5.11) into (5.10) we find

$$\mathrm{d}v = \mu \, \mathrm{d}a \, \mathrm{d}\zeta = (1 - 2H\zeta + K\zeta^2) \, \mathrm{d}a \, \mathrm{d}\zeta . \tag{5.12}$$

By virtue of (2.108), the relation (5.11) can be written as

$$\mu = (1 - \zeta\kappa_1)(1 - \zeta\kappa_2), \tag{5.13}$$

where κ_α are the principal curvatures. Thus, we have

$$\mu > 0 \quad \text{if} \quad |\zeta| < \min\{r_1, r_2\}, \quad \text{with} \quad r_\alpha := \frac{1}{|\kappa_\alpha|} . \tag{5.14}$$

In this case, the transformation $(\theta^\alpha, \zeta) \mapsto \boldsymbol{x}$ is one-one and orientation preserving.

Consider now the edge boundary surface $\partial\Omega \times I$, where $I = \left[-\frac{h}{2}, \frac{h}{2} \right]$. Let s be the arclength parameter along the boundary curve $\partial\Omega$, while $\boldsymbol{\tau} = \frac{\mathrm{d}\boldsymbol{r}}{\mathrm{d}s}$ is the unit tangent vector and $\boldsymbol{\nu} = \boldsymbol{\tau} \times \boldsymbol{n}$ the normal vector (see Fig. 5.2)

Using (s, ζ) as coordinates on the surface $\partial\Omega \times I$, the (oriented) area measure is

$$\boldsymbol{N} \, \mathrm{d}a = \mathrm{d}\boldsymbol{x}_\tau \times \mathrm{d}\boldsymbol{x}_n ,$$

where, in view of (5.7), we have

$$\mathrm{d}\boldsymbol{x}_n = \boldsymbol{G}(\boldsymbol{n} \, \mathrm{d}\zeta) = \boldsymbol{n} \, \mathrm{d}\zeta \qquad \text{and} \qquad \mathrm{d}\boldsymbol{x}_\tau = \boldsymbol{G}(\mathrm{d}\boldsymbol{r}) = \boldsymbol{\mu}(\mathrm{d}\boldsymbol{r}) \quad \text{with} \quad \mathrm{d}\boldsymbol{r} = \boldsymbol{\tau} \, \mathrm{d}s .$$

Thus, we obtain

$$N\,\mathrm{d}a = (\boldsymbol{G}\,\boldsymbol{\tau}\,\mathrm{d}s) \times (\boldsymbol{G}\,\boldsymbol{n}\,\mathrm{d}\zeta) = (\boldsymbol{G}\,\boldsymbol{\tau} \times \boldsymbol{G}\,\boldsymbol{n})\,\mathrm{d}s\,\mathrm{d}\zeta\,. \tag{5.15}$$

Let us use the relation (see, e.g., [1])

$$\boldsymbol{G}\,\boldsymbol{\tau} \times \boldsymbol{G}\,\boldsymbol{n} = \boldsymbol{G}^*(\boldsymbol{\tau} \times \boldsymbol{n}) = \boldsymbol{G}^*\boldsymbol{\nu}, \tag{5.16}$$

where $\boldsymbol{G}^*$ designates the cofactor of $\boldsymbol{G}$. By virtue of (5.6) and (5.7), we deduce

$$\boldsymbol{G}^* = (\det \boldsymbol{G})\boldsymbol{G}^{-1} = \mu(\boldsymbol{\mu}^{-1} + \boldsymbol{n} \otimes \boldsymbol{n}), \tag{5.17}$$

where $\boldsymbol{\mu}^{-1}$ is the tensor in the basis $\{\boldsymbol{a}_\alpha \otimes \boldsymbol{a}_\beta\}$ such that $\boldsymbol{\mu}\boldsymbol{\mu}^{-1} = \boldsymbol{\mu}^{-1}\boldsymbol{\mu} = \mathbf{1}$ (the two-dimensional inverse of $\boldsymbol{\mu}$ in the tangent plane T_Ω). Using (5.16) and (5.17), the relation (5.15) becomes

$$N\,\mathrm{d}a = \boldsymbol{G}^*\boldsymbol{\nu}\,\mathrm{d}\zeta\,\mathrm{d}s = \boldsymbol{\mu}^*\boldsymbol{\nu}\,\mathrm{d}\zeta\,\mathrm{d}s, \tag{5.18}$$

where

$$\boldsymbol{\mu}^* := \mu\,\boldsymbol{\mu}^{-1} \tag{5.19}$$

is the two-dimensional cofactor of $\boldsymbol{\mu}$ in the tangent plane, with $\boldsymbol{\mu}\boldsymbol{\mu}^* = \boldsymbol{\mu}^*\boldsymbol{\mu} = \mu\,\mathbf{1}$. Since N is a unit vector, from Eq. (5.18) we get

$$N := \frac{\boldsymbol{\mu}^*\boldsymbol{\nu}}{|\boldsymbol{\mu}^*\boldsymbol{\nu}|} \quad \text{and} \quad \mathrm{d}a = |\boldsymbol{\mu}^*\boldsymbol{\nu}|\,\mathrm{d}\zeta\,\mathrm{d}s\,. \tag{5.20}$$

Further, if we use the two-dimensional Cayley-Hamilton formula

$$\boldsymbol{\kappa}^2 - 2H\boldsymbol{\kappa} + K\,\mathbf{1} = \mathbf{0}, \tag{5.21}$$

we derive that the two-dimensional cofactor of $\boldsymbol{\kappa}$ is given by

$$\boldsymbol{\kappa}^* = -\boldsymbol{\kappa} + 2H\mathbf{1}\,. \tag{5.22}$$

Then, we obtain for the two-dimensional cofactor of $\boldsymbol{\mu}$ the formula

$$\boldsymbol{\mu}^* = \mathbf{1} - \zeta\boldsymbol{\kappa}^* = \mathbf{1} + \zeta(\boldsymbol{\kappa} - 2H\mathbf{1}) \tag{5.23}$$

and for the inverse we have

$$\boldsymbol{\mu}^{-1} = \frac{1}{\mu}\,\boldsymbol{\mu}^* = \frac{1}{1 - 2H\zeta + K\zeta^2}\left[\mathbf{1} + \zeta(\boldsymbol{\kappa} - 2H\mathbf{1})\right]. \tag{5.24}$$

Let us decompose the symmetric tensor $\boldsymbol{\kappa}$ in the tensor basis generated by $\boldsymbol{\nu}$ and $\boldsymbol{\tau}$ and denote its components as follows

$$\boldsymbol{\kappa} = \kappa_\nu \boldsymbol{\nu} \otimes \boldsymbol{\nu} + \kappa_\tau \boldsymbol{\tau} \otimes \boldsymbol{\tau} + \tau(\boldsymbol{\tau} \otimes \boldsymbol{\nu} + \boldsymbol{\nu} \otimes \boldsymbol{\tau}). \tag{5.25}$$

Then, we have $2H = \mathrm{tr}\,\boldsymbol{\kappa} = \kappa_\nu + \kappa_\tau$ and using (5.23) we obtain

$$\begin{aligned} \boldsymbol{\mu}^* \boldsymbol{\nu} &= \big[\mathbf{1} + \zeta(\boldsymbol{\kappa} - 2H\mathbf{1})\big]\boldsymbol{\nu} = \boldsymbol{\nu} + \zeta\boldsymbol{\kappa}\boldsymbol{\nu} - 2H\zeta\boldsymbol{\nu} \\ &= \boldsymbol{\nu} + \zeta(\kappa_\nu \boldsymbol{\nu} + \tau\boldsymbol{\tau}) - \zeta(\kappa_\nu + \kappa_\tau)\boldsymbol{\nu} = (1 - \zeta\kappa_\tau)\boldsymbol{\nu} + (\zeta\tau)\boldsymbol{\tau}\,. \end{aligned} \tag{5.26}$$

Hence, we get

$$|\boldsymbol{\mu}^* \boldsymbol{\nu}|^2 = (1 - \zeta\kappa_\tau)^2 + (\zeta\tau)^2 \tag{5.27}$$

and the relation $(5.20)_2$ becomes

$$\mathrm{d}a = \sqrt{(1 - \zeta\kappa_\tau)^2 + \zeta^2\tau^2}\; \mathrm{d}\zeta\, \mathrm{d}s\,. \tag{5.28}$$

5.2 Kinematics

Let us denote by $\tilde{\boldsymbol{u}}(\boldsymbol{x})$ the three-dimensional displacement field and by $\tilde{\boldsymbol{H}}(\boldsymbol{x})$ its gradient, so we have

$$\mathrm{d}\tilde{\boldsymbol{u}} = \tilde{\boldsymbol{H}}\mathrm{d}\boldsymbol{x}\,. \tag{5.29}$$

We write these fields as functions of the coordinates (θ^α, ζ) using

$$\hat{\boldsymbol{u}}(\theta^\alpha, \zeta) := \tilde{\boldsymbol{u}}\big(\boldsymbol{r}(\theta^\alpha) + \zeta\boldsymbol{n}(\theta^\alpha)\big), \qquad \hat{\boldsymbol{H}}(\theta^\alpha, \zeta) := \tilde{\boldsymbol{H}}\big(\boldsymbol{r}(\theta^\alpha) + \zeta\boldsymbol{n}(\theta^\alpha)\big). \tag{5.30}$$

Then, we get

$$\mathrm{d}\hat{\boldsymbol{u}} = \hat{\boldsymbol{u}}_{,\alpha}\mathrm{d}\theta^\alpha + \hat{\boldsymbol{u}}'\mathrm{d}\zeta, \tag{5.31}$$

where we denote the derivatives $\boldsymbol{f}_{,\alpha} := \frac{\partial \boldsymbol{f}}{\partial \theta^\alpha}$ and $\boldsymbol{f}' := \frac{\partial \boldsymbol{f}}{\partial \zeta}$ for any field $\boldsymbol{f}(\theta^\alpha, \zeta)$. Taking into account that $\mathrm{d}\theta^\alpha = \boldsymbol{a}^\alpha \cdot \mathrm{d}\boldsymbol{r}$, the last relation can be written as

$$\mathrm{d}\hat{\boldsymbol{u}} = \big(\hat{\boldsymbol{u}}_{,\alpha} \otimes \boldsymbol{a}^\alpha\big)\mathrm{d}\boldsymbol{r} + \hat{\boldsymbol{u}}'\mathrm{d}\zeta = (\nabla\hat{\boldsymbol{u}})\mathrm{d}\boldsymbol{r} + \hat{\boldsymbol{u}}'\mathrm{d}\zeta, \tag{5.32}$$

where we designate by

$$\nabla\boldsymbol{f} = \boldsymbol{f}_{,\alpha} \otimes \boldsymbol{a}^\alpha \tag{5.33}$$

the surface gradient of any field $\boldsymbol{f}$ (see Sect. 2.7).

On the other hand, in view of Eqs. (5.7) and (5.29) we have

$$\mathrm{d}\hat{\boldsymbol{u}} = \hat{\boldsymbol{H}}\mathrm{d}\boldsymbol{x} = \hat{\boldsymbol{H}}(\boldsymbol{\mu}\,\mathrm{d}\boldsymbol{r} + \boldsymbol{n}\,\mathrm{d}\zeta) = \hat{\boldsymbol{H}}\boldsymbol{\mu}\,\mathrm{d}\boldsymbol{r} + \hat{\boldsymbol{H}}\boldsymbol{n}\,\mathrm{d}\zeta\,. \tag{5.34}$$

By comparison of (5.32) and (5.34) it follows that

$$\hat{\boldsymbol{H}}\boldsymbol{\mu} = \nabla\hat{\boldsymbol{u}} \quad \text{and} \quad \hat{\boldsymbol{H}}\boldsymbol{n} = \hat{\boldsymbol{u}}' . \tag{5.35}$$

Thus, we have

$$\hat{\boldsymbol{H}}\mathbf{1} = \hat{\boldsymbol{H}}\boldsymbol{\mu}\boldsymbol{\mu}^{-1} = (\nabla\hat{\boldsymbol{u}})\boldsymbol{\mu}^{-1}, \tag{5.36}$$

and using the decomposition

$$\hat{\boldsymbol{H}} = \hat{\boldsymbol{H}}\boldsymbol{I} = \hat{\boldsymbol{H}}(\mathbf{1}+\boldsymbol{n}\otimes\boldsymbol{n}) = \hat{\boldsymbol{H}}\mathbf{1}+\hat{\boldsymbol{H}}\boldsymbol{n}\otimes\boldsymbol{n},$$

we obtain from (5.35) and (5.36) the representation

$$\hat{\boldsymbol{H}} = (\nabla\hat{\boldsymbol{u}})\boldsymbol{\mu}^{-1} + \hat{\boldsymbol{u}}'\otimes\boldsymbol{n} . \tag{5.37}$$

This relation allows us to express the derivatives of $\hat{\boldsymbol{H}}$ with respect to ζ in the form

$$\begin{aligned}\hat{\boldsymbol{H}}' &= (\nabla\hat{\boldsymbol{u}}')\boldsymbol{\mu}^{-1} + (\nabla\hat{\boldsymbol{u}})(\boldsymbol{\mu}^{-1})' + \hat{\boldsymbol{u}}''\otimes\boldsymbol{n} \quad \text{and}\\ \hat{\boldsymbol{H}}'' &= (\nabla\hat{\boldsymbol{u}}'')\boldsymbol{\mu}^{-1} + 2(\nabla\hat{\boldsymbol{u}}')(\boldsymbol{\mu}^{-1})' + (\nabla\hat{\boldsymbol{u}})(\boldsymbol{\mu}^{-1})'' + \hat{\boldsymbol{u}}'''\otimes\boldsymbol{n} .\end{aligned} \tag{5.38}$$

In order to compute the derivatives of $\boldsymbol{\mu}^{-1}$ appearing in the above relations, we differentiate the equation $\boldsymbol{\mu}\boldsymbol{\mu}^{-1} = \mathbf{1}$ with respect to ζ and deduce

$$\boldsymbol{\mu}'\boldsymbol{\mu}^{-1} + \boldsymbol{\mu}(\boldsymbol{\mu}^{-1})' = \mathbf{0}, \quad \text{so} \quad (\boldsymbol{\mu}^{-1})' = -\boldsymbol{\mu}^{-1}\boldsymbol{\mu}'\boldsymbol{\mu}^{-1} . \tag{5.39}$$

The second derivative is

$$\boldsymbol{\mu}''\boldsymbol{\mu}^{-1} + 2\boldsymbol{\mu}'(\boldsymbol{\mu}^{-1})' + \boldsymbol{\mu}(\boldsymbol{\mu}^{-1})'' = \mathbf{0}, \quad \text{so} \quad (\boldsymbol{\mu}^{-1})'' = \boldsymbol{\mu}^{-1}(2\boldsymbol{\mu}'\boldsymbol{\mu}^{-1}\boldsymbol{\mu}' - \boldsymbol{\mu}'')\boldsymbol{\mu}^{-1}. \tag{5.40}$$

But since $\boldsymbol{\mu} = \mathbf{1} - \zeta\boldsymbol{\kappa}$, we have $\boldsymbol{\mu}' = -\boldsymbol{\kappa}$ and $\boldsymbol{\mu}'' = \mathbf{0}$, so the relations (5.39) and (5.40) reduce to

$$(\boldsymbol{\mu}^{-1})' = \boldsymbol{\mu}^{-1}\boldsymbol{\kappa}\,\boldsymbol{\mu}^{-1} \quad \text{and} \quad (\boldsymbol{\mu}^{-1})'' = 2\boldsymbol{\mu}^{-1}\boldsymbol{\kappa}\,\boldsymbol{\mu}^{-1}\boldsymbol{\kappa}\,\boldsymbol{\mu}^{-1}. \tag{5.41}$$

As in the case of plates, we expand the displacement field with respect to ζ as

$$\hat{\boldsymbol{u}}(\theta^\alpha,\zeta) = \boldsymbol{u}(\theta^\alpha) + \zeta\,\boldsymbol{a}(\theta^\alpha) + \frac{1}{2}\zeta^2\,\boldsymbol{b}(\theta^\alpha) + \frac{1}{6}\zeta^3\,\boldsymbol{c}(\theta^\alpha) + \cdots, \tag{5.42}$$

with

$$\boldsymbol{u} = \hat{\boldsymbol{u}}_0, \qquad \boldsymbol{a} = \hat{\boldsymbol{u}}_0', \qquad \boldsymbol{b} = \hat{\boldsymbol{u}}_0'', \qquad \boldsymbol{c} = \hat{\boldsymbol{u}}_0''', \tag{5.43}$$

where $\boldsymbol{f}_0 := \boldsymbol{f}_{|\zeta=0} = \boldsymbol{f}(\theta^\alpha,0)$ for any field $\boldsymbol{f}(\theta^\alpha,\zeta)$.

Thus, if we put $\zeta = 0$ in the relations (5.6), (5.24) and (5.41), we obtain

$$\boldsymbol{\mu}_0 = \mathbf{1}, \qquad (\boldsymbol{\mu}^{-1})_0 = \mathbf{1}, \qquad (\boldsymbol{\mu}^{-1})'_0 = \boldsymbol{\kappa}, \qquad (\boldsymbol{\mu}^{-1})''_0 = 2\boldsymbol{\kappa}^2. \tag{5.44}$$

Further, taking $\zeta = 0$ in the Eqs. (5.37), (5.38) and using the relations (5.43), (5.44), we derive the following expressions for the displacement gradient and its derivatives on the midsurface

$$\begin{aligned}\hat{\boldsymbol{H}}_0 &= \nabla\boldsymbol{u} + \boldsymbol{a}\otimes\boldsymbol{n},\\ \hat{\boldsymbol{H}}'_0 &= \nabla\boldsymbol{a} + (\nabla\boldsymbol{u})\boldsymbol{\kappa} + \boldsymbol{b}\otimes\boldsymbol{n},\\ \hat{\boldsymbol{H}}''_0 &= \nabla\boldsymbol{b} + 2(\nabla\boldsymbol{a})\boldsymbol{\kappa} + 2(\nabla\boldsymbol{u})\boldsymbol{\kappa}^2 + \boldsymbol{c}\otimes\boldsymbol{n}\,.\end{aligned} \tag{5.45}$$

5.3 Derivation of the Shell Model

To perform the integration over the thickness, we will require integrals such as

$$\int_{\mathscr{K}} U\,\mathrm{d}v = \int_{\Omega}\int_I \mu\,U\,\mathrm{d}\zeta\,\mathrm{d}a, \tag{5.46}$$

where $I = \left[-\frac{h}{2}, \frac{h}{2}\right]$ and $\mu = 1 - 2H\zeta + K\zeta^2$, cf. (5.11). Let us denote the areal strain energy density by

$$W = \int_I G\,\mathrm{d}\zeta, \qquad \text{where} \quad G := \mu\,U\,. \tag{5.47}$$

As in the case of plates (see Sect. 4.2.1), we define $f_0 = f\big|_{\zeta=0}$ for any field $f(\boldsymbol{r},\zeta)$, and we obtain

$$\int_I G\,\mathrm{d}\zeta = h\,G_0 + \frac{1}{24}\,h^3 G''_0 + o(h^3). \tag{5.48}$$

Now, in view of (5.11) and (5.47), we have

$$\begin{aligned}G_0 &= \mu_0 U_0 = U_0,\\ G'_0 &= \mu'_0 U_0 + \mu_0 U'_0 = U'_0 - 2H\,U_0,\\ G''_0 &= \mu''_0 U_0 + 2\mu'_0 U'_0 + \mu_0 U''_0 = U''_0 - 4H\,U'_0 + 2K\,U_0\,.\end{aligned} \tag{5.49}$$

Hence, inserting (5.49) into (5.48) we get

$$W = \int_I G\,\mathrm{d}\zeta = h\big(1 + \frac{1}{12}\,h^2 K\big)U_0 + \frac{1}{24}h^3\big(U''_0 - 4HU'_0\big) + o(h^3). \tag{5.50}$$

We also need integrals of the type

$$\int_{\partial \mathscr{K}_I} J \, da = \int_{\partial \Omega} \int_I \mathscr{I} \, d\zeta \, ds, \qquad \text{with} \quad \mathscr{I} = |\boldsymbol{\mu}^* \boldsymbol{\nu}| \, J, \tag{5.51}$$

for some function J, where $\partial \mathscr{K}_I = \partial \Omega \times I$ is the edge boundary, cf. (5.20). Similar to (5.48) and (5.49), we can write

$$\int_I \mathscr{I} \, d\zeta = h \, \mathscr{I}_0 + \frac{1}{24} h^3 \mathscr{I}_0'' + o(h^3), \tag{5.52}$$

where

$$\begin{aligned} \mathscr{I}_0 &= |\boldsymbol{\mu}^* \boldsymbol{\nu}|_0 \, J_0, \\ \mathscr{I}_0' &= |\boldsymbol{\mu}^* \boldsymbol{\nu}|_0' \, J_0 + |\boldsymbol{\mu}^* \boldsymbol{\nu}|_0 \, J_0', \\ \mathscr{I}_0'' &= |\boldsymbol{\mu}^* \boldsymbol{\nu}|_0'' \, J_0 + 2|\boldsymbol{\mu}^* \boldsymbol{\nu}|_0' \, J_0' + |\boldsymbol{\mu}^* \boldsymbol{\nu}|_0 \, J_0'' . \end{aligned} \tag{5.53}$$

In view of (5.27), we have

$$|\boldsymbol{\mu}^* \boldsymbol{\nu}| = \sqrt{(1 - \zeta \kappa_\tau)^2 + (\zeta \tau)^2}$$

and, hence, we get

$$|\boldsymbol{\mu}^* \boldsymbol{\nu}|_0 = 1, \qquad |\boldsymbol{\mu}^* \boldsymbol{\nu}|_0' = -\kappa_\tau, \qquad |\boldsymbol{\mu}^* \boldsymbol{\nu}|_0'' = \tau^2. \tag{5.54}$$

Substituting (5.53) and (5.54) in (5.52), we obtain

$$\int_I \mathscr{I} \, d\zeta \; = \; h \, J_0 + \frac{1}{24} h^3 \big(J_0'' - 2\kappa_\tau \, J_0' + \tau^2 J_0 \big) + o(h^3). \tag{5.55}$$

Thus, by virtue of (5.46) and (5.50), we can put the strain energy in the form

$$\int_{\mathscr{K}} U \, dv = \int_\Omega W \, da, \qquad \text{where} \quad W = \int_I \mu \, U \, d\zeta \tag{5.56}$$

and

$$W = h \big(1 + \frac{1}{12} h^2 K \big) U_0 + \frac{1}{24} h^3 \big(U_0'' - 4 H U_0' \big) + o(h^3). \tag{5.57}$$

Here, we can write

$$\begin{aligned} U_0 \; &= \; U(\boldsymbol{H}_0) \; = \; \tfrac{1}{2} \, \boldsymbol{P}_0 \cdot \boldsymbol{H}_0 \; = \; \tfrac{1}{2} \, \boldsymbol{H}_0 \cdot \underline{\boldsymbol{C}}[\boldsymbol{H}_0], \\ U_0' \; &= \; \boldsymbol{P}_0 \cdot \boldsymbol{H}_0' \qquad \text{and} \qquad U_0'' \; = \; \boldsymbol{P}_0' \cdot \boldsymbol{H}_0' + \boldsymbol{P}_0 \cdot \boldsymbol{H}_0'', \end{aligned} \tag{5.58}$$

with

$$\boldsymbol{P}_0 \; = \; \underline{\boldsymbol{C}}[\boldsymbol{H}_0], \qquad \boldsymbol{P}_0' \; = \; \underline{\boldsymbol{C}}[\boldsymbol{H}_0'] \, . \tag{5.59}$$

In the last relation we have assumed the homogeneity condition $\underline{\boldsymbol{C}}' = \boldsymbol{0}$, i.e. the elastical properties do not depend on ζ.

Using the expressions for $\boldsymbol{H}_0$, $\boldsymbol{H}'_0$, $\boldsymbol{H}''_0$ given by (5.45), the above relations yield

$$\begin{aligned}W = {} & \tfrac{1}{2}h\big(1+\tfrac{1}{12}h^2K\big)\boldsymbol{P}_0\cdot(\nabla\boldsymbol{u}+\boldsymbol{a}\otimes\boldsymbol{n})\\ & +\tfrac{1}{24}h^3\Big\{\boldsymbol{P}'_0\cdot\big[\nabla\boldsymbol{a}+(\nabla\boldsymbol{u})\boldsymbol{\kappa}+\boldsymbol{b}\otimes\boldsymbol{n}\big]+\boldsymbol{P}_0\cdot\big[\nabla\boldsymbol{b}+2(\nabla\boldsymbol{a})\boldsymbol{\kappa}+2(\nabla\boldsymbol{u})\boldsymbol{\kappa}^2+\boldsymbol{c}\otimes\boldsymbol{n}\big]\\ & -4H\boldsymbol{P}_0\cdot\big[\nabla\boldsymbol{a}+(\nabla\boldsymbol{u})\boldsymbol{\kappa}+\boldsymbol{b}\otimes\boldsymbol{n}\big]\Big\}+o(h^3).\end{aligned}$$

Further, we use here the decomposition $\boldsymbol{P}_0=\boldsymbol{P}_0\boldsymbol{1}+\boldsymbol{P}_0\boldsymbol{n}\otimes\boldsymbol{n}$ and the relation $\boldsymbol{\kappa}^2=2H\boldsymbol{\kappa}-K\boldsymbol{1}$ and we get (after some algebraic calculations)

$$\begin{aligned}W = {} & \tfrac{1}{2}h\big(1-\tfrac{1}{12}h^2K\big)\boldsymbol{P}_0\boldsymbol{1}\cdot\nabla\boldsymbol{u}+\tfrac{1}{2}h\big(1+\tfrac{1}{12}h^2K\big)\boldsymbol{P}_0\boldsymbol{n}\cdot\boldsymbol{a}\\ & +\tfrac{1}{24}h^3\{\boldsymbol{P}'_0\boldsymbol{1}\cdot\big[\nabla\boldsymbol{a}+(\nabla\boldsymbol{u})\boldsymbol{\kappa}\big]+\boldsymbol{P}'_0\boldsymbol{n}\cdot\boldsymbol{b}\}\\ & +\tfrac{1}{24}h^3\big[\boldsymbol{P}_0\boldsymbol{1}\cdot\nabla\boldsymbol{b}-2\boldsymbol{P}_0\boldsymbol{1}\cdot(\nabla\boldsymbol{a})\boldsymbol{\kappa}^*+\boldsymbol{P}_0\boldsymbol{n}\cdot(\boldsymbol{c}-4H\boldsymbol{b})\big]+o(h^3),\end{aligned}\tag{5.60}$$

where $\boldsymbol{\kappa}^*=-\boldsymbol{\kappa}+2H\boldsymbol{1}$ is the cofactor introduced in (5.22).

Similarly, let us denote by $\partial\mathcal{K}_t$ the subset of the boundary surface $\partial\mathcal{K}$ where the tractions $\tilde{\boldsymbol{t}}$ are assigned. Note that the boundary $\partial\mathcal{K}_t$ consists of the lateral faces ($\zeta=\pm\frac{h}{2}$) and the edge boundary subset $\partial\Omega_t\times I$. In the case of *pure edge loads* $\tilde{\boldsymbol{t}}$ (no lateral loads), we can use the relation (5.51) for the function $J=\tilde{\boldsymbol{t}}\cdot\tilde{\boldsymbol{u}}$ and write

$$\int_{\partial\mathcal{K}_t}\tilde{\boldsymbol{t}}\cdot\tilde{\boldsymbol{u}}\,\mathrm{d}a=\int_{\partial\Omega_t}\Big(\int_I|\boldsymbol{\mu}^*\boldsymbol{\nu}|\,\tilde{\boldsymbol{t}}\cdot\tilde{\boldsymbol{u}}\,\mathrm{d}\zeta\Big)\,\mathrm{d}s\,.\tag{5.61}$$

Taking into account the expansions (5.42) and

$$\tilde{\boldsymbol{t}}\;=\;\boldsymbol{t}_0+\zeta\,\boldsymbol{t}'_0+\frac{1}{2}\zeta^2\boldsymbol{t}''_0+O(\zeta^3)\tag{5.62}$$

we deduce

$$J\;=\;\tilde{\boldsymbol{t}}\cdot\tilde{\boldsymbol{u}}\;=\;\boldsymbol{t}_0\cdot\boldsymbol{u}+\zeta(\boldsymbol{t}_0\cdot\boldsymbol{a}+\boldsymbol{t}'_0\cdot\boldsymbol{u})+\frac{1}{2}\zeta^2\big(\boldsymbol{t}_0\cdot\boldsymbol{b}+2\boldsymbol{t}'_0\cdot\boldsymbol{a}+\boldsymbol{t}''_0\cdot\boldsymbol{u}\big)+O(\zeta^3).\tag{5.63}$$

Hence, we have

$$J_0=\boldsymbol{t}_0\cdot\boldsymbol{u},\qquad J'_0=\boldsymbol{t}_0\cdot\boldsymbol{a}+\boldsymbol{t}'_0\cdot\boldsymbol{u},\qquad J''_0=\boldsymbol{t}_0\cdot\boldsymbol{b}+2\boldsymbol{t}'_0\cdot\boldsymbol{a}+\boldsymbol{t}''_0\cdot\boldsymbol{u}$$

and the relation (5.55) yields

$$\begin{aligned}\int_I|\boldsymbol{\mu}^*\boldsymbol{\nu}|\,\tilde{\boldsymbol{t}}\cdot\tilde{\boldsymbol{u}}\,\mathrm{d}\zeta = {} & h\big(\boldsymbol{t}_0\cdot\boldsymbol{u}\big)+\frac{h^3}{24}\big[\big(\boldsymbol{t}_0\cdot\boldsymbol{b}+2\boldsymbol{t}'_0\cdot\boldsymbol{a}+\boldsymbol{t}''_0\cdot\boldsymbol{u}\big)\\ & -2\kappa_\tau(\boldsymbol{t}_0\cdot\boldsymbol{a}+\boldsymbol{t}'_0\cdot\boldsymbol{u})+\tau^2(\boldsymbol{t}_0\cdot\boldsymbol{u})\big]+o(h^3).\end{aligned}\tag{5.64}$$

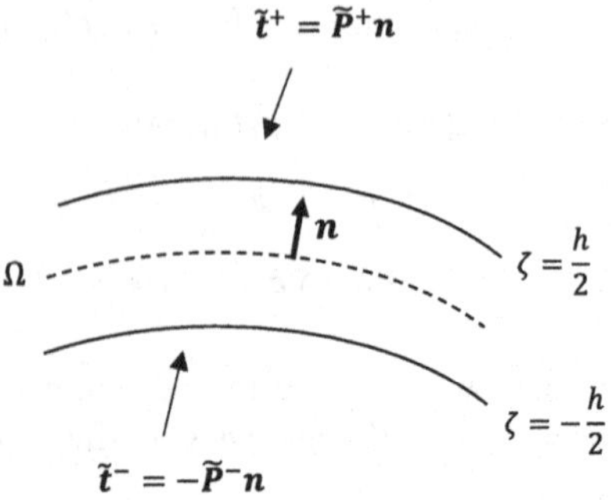

Fig. 5.3 The assigned tractions on the lateral surfaces

In view of (5.61) and (5.64), we obtain

$$\int_{\partial\mathscr{K}_t} \tilde{\boldsymbol{t}}\cdot\tilde{\boldsymbol{u}}\,\mathrm{d}a = \int_{\partial\Omega_t}\big(\boldsymbol{p}_u\cdot\boldsymbol{u}+\boldsymbol{p}_a\cdot\boldsymbol{a}+\boldsymbol{p}_b\cdot\boldsymbol{b}\big)\,\mathrm{d}s + o(h^3), \tag{5.65}$$

where

$$\begin{aligned}\boldsymbol{p}_u &= h\big(1+\frac{1}{24}h^2\tau^2\big)\boldsymbol{t}_0+\frac{1}{24}h^3\big(\boldsymbol{t}_0''-2\kappa_\tau\,\boldsymbol{t}_0'\big),\\ \boldsymbol{p}_a &= \frac{1}{12}h^3\big(\boldsymbol{t}_0'-\kappa_\tau\,\boldsymbol{t}_0\big),\qquad \boldsymbol{p}_b=\frac{1}{24}h^3\,\boldsymbol{t}_0\,.\end{aligned} \tag{5.66}$$

Let us consider now the contribution of *lateral loads* to the potential energy. Denote by $\tilde{\boldsymbol{t}}^{+}$ and $\tilde{\boldsymbol{t}}^{-}$ the tractions assigned on the lateral surfaces $\zeta=\frac{h}{2}$ and $\zeta=-\frac{h}{2}$, respectively, see Fig. 5.3.

Remark The unit normal $\boldsymbol{n}$ to the midsurface Ω is also orthogonal to the lateral surfaces $\zeta=\pm\frac{h}{2}$. Indeed, according to (5.1), the parametric representations of the lateral surfaces are $\boldsymbol{x}^{\pm}(\theta^\alpha)=\boldsymbol{r}(\theta^\alpha)\pm\frac{h}{2}\boldsymbol{n}(\theta^\alpha)$. Then, the tangent base vectors to the lateral surfaces are $\boldsymbol{a}^{\pm}_{\alpha}=\boldsymbol{x}^{\pm}_{,\alpha}=\boldsymbol{a}_\alpha\pm\frac{h}{2}\boldsymbol{n}_{,\alpha}$, so we have $\boldsymbol{n}\cdot\boldsymbol{a}^{\pm}_{\alpha}=\boldsymbol{n}\cdot\boldsymbol{a}_\alpha\pm\frac{h}{2}\boldsymbol{n}\cdot\boldsymbol{n}_{,\alpha}=0$, i.e. $\boldsymbol{n}$ is orthogonal to the lateral surfaces $\zeta=\pm\frac{h}{2}$. □

Let us suppose that the lateral loads admit the representation

$$\tilde{\boldsymbol{t}}^{\pm} = h^3\boldsymbol{t}^{\pm} + o(h^3),\qquad \text{with } |\boldsymbol{t}^{\pm}| = O(1). \tag{5.67}$$

Then, using the expansion (5.42), we deduce

$$\tilde{\boldsymbol{t}}^{\pm}\cdot\tilde{\boldsymbol{u}} = h^3\boldsymbol{t}^{\pm}\cdot\boldsymbol{u} + o(h^3)\qquad \text{on the lateral faces } \zeta=\pm\frac{h}{2}. \tag{5.68}$$

Taking into account the contribution (5.68) of lateral loads in the integral (5.65), we obtain the final form

$$\int_{\partial \mathscr{K}_t} \tilde{\boldsymbol{t}} \cdot \tilde{\boldsymbol{u}} \,\mathrm{da} = \int_{\partial \Omega_t} \left(\boldsymbol{p}_u \cdot \boldsymbol{u} + \boldsymbol{p}_a \cdot \boldsymbol{a} + \boldsymbol{p}_b \cdot \boldsymbol{b}\right) \mathrm{ds} + \int_{\Omega} \boldsymbol{g} \cdot \boldsymbol{u} \,\mathrm{da} + o(h^3), \tag{5.69}$$

where the net lateral traction is

$$\boldsymbol{g} \;=\; h^3\left(\boldsymbol{t}^+ + \boldsymbol{t}^-\right). \tag{5.70}$$

5.4 Optimal Expression for Potential Energy

To derive an optimal expression of the potential energy, we proceed in the same manner as in the case of plates, see Sect. 4.2.3. For the stress vectors on the lateral faces $\tilde{\boldsymbol{t}}^{\pm}$ we have (cf. (4.36))

$$\begin{aligned} \tilde{\boldsymbol{t}}^{+} + \tilde{\boldsymbol{t}}^{-} &= h\, \boldsymbol{P}_0' \boldsymbol{n} + O(h^3), \\ \tilde{\boldsymbol{t}}^{+} - \tilde{\boldsymbol{t}}^{-} &= 2\boldsymbol{P}_0 \boldsymbol{n} + O(h^2). \end{aligned} \tag{5.71}$$

Hence, if assume that $\tilde{\boldsymbol{t}}^{\pm} = O(h^3)$, we deduce

$$\boldsymbol{P}_0 \boldsymbol{n} \;=\; O(h^2), \qquad \boldsymbol{P}_0' \boldsymbol{n} \;=\; O(h^2). \tag{5.72}$$

Thus, we can put $\boldsymbol{P}_0 \boldsymbol{n} = \boldsymbol{0}$ and $\boldsymbol{P}_0' \boldsymbol{n} = \boldsymbol{0}$ in the coefficients of h^3 in the potential energy. Also, as in the case of plates, one can show that the states with $\boldsymbol{P}_0 \boldsymbol{n} = \boldsymbol{0}$ minimize U_0 in the coefficient of h. So, as before, we put

$$\boldsymbol{P}_0 \boldsymbol{n} \;=\; \boldsymbol{0} \qquad \text{and} \qquad \boldsymbol{P}_0' \boldsymbol{n} \;=\; \boldsymbol{0} \tag{5.73}$$

in *all* terms of the potential energy. Also, in view of (5.59) and (5.45), we can write

$$\begin{aligned} \boldsymbol{P}_0 \boldsymbol{n} \;&=\; \left(\underline{\boldsymbol{C}}[\boldsymbol{H}_0]\right)\boldsymbol{n} \;=\; \left(\underline{\boldsymbol{C}}[\nabla \boldsymbol{u} + \boldsymbol{a} \otimes \boldsymbol{n}]\right)\boldsymbol{n}\,, \\ \boldsymbol{P}_0' \boldsymbol{n} \;&=\; \left(\underline{\boldsymbol{C}}[\boldsymbol{H}_0']\right)\boldsymbol{n} \;=\; \left(\underline{\boldsymbol{C}}[\nabla \boldsymbol{a} + (\nabla \boldsymbol{u})\boldsymbol{\kappa} + \boldsymbol{b} \otimes \boldsymbol{n}]\right)\boldsymbol{n}\,. \end{aligned} \tag{5.74}$$

Thus, on the basis of (5.73) and (5.74), we put $\boldsymbol{a} = \bar{\boldsymbol{a}}$ and $\boldsymbol{b} = \bar{\boldsymbol{b}}$ in *all* terms, where the vectors $\bar{\boldsymbol{a}}$ and $\bar{\boldsymbol{b}}$ are determined by

$$\boldsymbol{A}_n \bar{\boldsymbol{a}} \;=\; -\left(\underline{\boldsymbol{C}}[\nabla \boldsymbol{u}]\right)\boldsymbol{n} \qquad \text{and} \qquad \boldsymbol{A}_n \bar{\boldsymbol{b}} \;=\; -\left(\underline{\boldsymbol{C}}[\nabla \bar{\boldsymbol{a}} + (\nabla \boldsymbol{u})\boldsymbol{\kappa}]\right)\boldsymbol{n}, \tag{5.75}$$

where the tensor $\boldsymbol{A}_n$ (*acoustic tensor* based on $\boldsymbol{n}$) is defined by

$$\boldsymbol{A}_n \boldsymbol{w} \;=\; \left(\underline{\boldsymbol{C}}[\boldsymbol{w} \otimes \boldsymbol{n}]\right)\boldsymbol{n} \qquad \text{for any vector } \boldsymbol{w}. \tag{5.76}$$

The above relations yield

$$\boldsymbol{P}_0 = \boldsymbol{P}_0\mathbf{1} + \boldsymbol{P}_0\boldsymbol{n}\otimes\boldsymbol{n} = \bar{\boldsymbol{P}}_0\mathbf{1}, \qquad \boldsymbol{P}'_0 = \boldsymbol{P}'_0\mathbf{1} + \boldsymbol{P}'_0\boldsymbol{n}\otimes\boldsymbol{n} = \bar{\boldsymbol{P}}'_0\mathbf{1}, \tag{5.77}$$

with

$$\bar{\boldsymbol{P}}_0\mathbf{1} = \big(\underline{\boldsymbol{C}}[\nabla\boldsymbol{u} + \bar{\boldsymbol{a}}\otimes\boldsymbol{n}]\big)\mathbf{1}, \qquad \bar{\boldsymbol{P}}'_0\mathbf{1} = \big(\underline{\boldsymbol{C}}[\nabla\bar{\boldsymbol{a}} + (\nabla\boldsymbol{u})\boldsymbol{\kappa} + \bar{\boldsymbol{b}}\otimes\boldsymbol{n}]\big)\mathbf{1}\,. \tag{5.78}$$

Then, the strain energy density (5.60) can be written in the form

$$\begin{aligned} W = {} & \tfrac{1}{2}\,h\big(1 - \tfrac{1}{12}\,h^2K\big)\bar{\boldsymbol{P}}_0\mathbf{1}\cdot\nabla\boldsymbol{u} \\ & + \tfrac{1}{24}\,h^3\,\bar{\boldsymbol{P}}'_0\mathbf{1}\cdot\big[\nabla\bar{\boldsymbol{a}} + (\nabla\boldsymbol{u})\boldsymbol{\kappa}\big] \\ & + \tfrac{1}{24}\,h^3\big[\bar{\boldsymbol{P}}_0\mathbf{1}\cdot\nabla\bar{\boldsymbol{b}} - 2\bar{\boldsymbol{P}}_0\mathbf{1}\cdot(\nabla\bar{\boldsymbol{a}})\boldsymbol{\kappa}^*\big] + o(h^3). \end{aligned} \tag{5.79}$$

Note that the first two lines in (5.79) involve the quadratic forms $\bar{\boldsymbol{H}}_0\cdot\underline{\boldsymbol{C}}[\bar{\boldsymbol{H}}_0]$ and $\bar{\boldsymbol{H}}'_0\cdot\underline{\boldsymbol{C}}[\bar{\boldsymbol{H}}'_0]$, which are positive definite forms. Indeed, in view of (5.45), (5.59) and (5.77), we have

$$\begin{aligned} &\bar{\boldsymbol{P}}_0\mathbf{1}\cdot\nabla\boldsymbol{u} = \bar{\boldsymbol{P}}_0\mathbf{1}\cdot(\nabla\boldsymbol{u} + \bar{\boldsymbol{a}}\otimes\boldsymbol{n}) = \bar{\boldsymbol{P}}_0\cdot\bar{\boldsymbol{H}}_0 = \bar{\boldsymbol{H}}_0\cdot\underline{\boldsymbol{C}}[\bar{\boldsymbol{H}}_0] \quad \text{and} \\ &\bar{\boldsymbol{P}}'_0\mathbf{1}\cdot\big[\nabla\bar{\boldsymbol{a}} + (\nabla\boldsymbol{u})\boldsymbol{\kappa}\big] = \bar{\boldsymbol{P}}'_0\mathbf{1}\cdot\big[\nabla\bar{\boldsymbol{a}} + (\nabla\boldsymbol{u})\boldsymbol{\kappa} + \bar{\boldsymbol{b}}\otimes\boldsymbol{n}\big] = \bar{\boldsymbol{P}}'_0\cdot\bar{\boldsymbol{H}}'_0 = \bar{\boldsymbol{H}}'_0\cdot\underline{\boldsymbol{C}}[\bar{\boldsymbol{H}}'_0]. \end{aligned} \tag{5.80}$$

Next, we want to estimate the third line in (5.79). Let us estimate first the term $\bar{\boldsymbol{P}}_0\mathbf{1}\cdot\nabla\bar{\boldsymbol{b}}$. To this aim, recall that the three-dimensional equilibrium equations (for vanishing body loads) are

$$\operatorname{Div}\tilde{\boldsymbol{P}} = \mathbf{0},$$

or equivalently (cf. (1.49)),

$$(\tilde{\boldsymbol{P}}_{,i})\,\boldsymbol{g}^i = \mathbf{0}\,. \tag{5.81}$$

Here, in view of (5.1), (5.6) and (5.24), we have

$$\boldsymbol{g}_\alpha = \mu\boldsymbol{a}_\alpha, \qquad \boldsymbol{g}^\alpha = \mu^{-1}\boldsymbol{a}^\alpha, \qquad \boldsymbol{g}_3 = \boldsymbol{g}^3 = \boldsymbol{n}\,. \tag{5.82}$$

Hence, if we evaluate the relation (5.81) on the midsurface Ω, we get

$$(\boldsymbol{P}_{0,\alpha})\boldsymbol{a}^\alpha + \boldsymbol{P}'_0\,\boldsymbol{n} = \mathbf{0}\,. \tag{5.83}$$

Denoting by div the surface divergence operator (see Sect. 2.7), we have

$$\operatorname{div}(\boldsymbol{P}_0\mathbf{1}) = \big[(\boldsymbol{P}_0\mathbf{1})_{,\alpha}\big]\boldsymbol{a}^\alpha\,. \tag{5.84}$$

We can write

$$\boldsymbol{P}_{0,\alpha} = (\boldsymbol{P}_0\mathbf{1} + \boldsymbol{P}_0\boldsymbol{n}\otimes\boldsymbol{n})_{,\alpha} = (\boldsymbol{P}_0\mathbf{1})_{,\alpha} + (\boldsymbol{P}_0\boldsymbol{n})_{,\alpha}\otimes\boldsymbol{n} + \boldsymbol{P}_0\boldsymbol{n}\otimes\boldsymbol{n}_{,\alpha}\,.$$

If we multiply this relation with $\boldsymbol{a}^{\alpha}$ and use (5.84), we get

$$(\boldsymbol{P}_{0,\alpha})\boldsymbol{a}^{\alpha} = \mathrm{div}(\boldsymbol{P}_0\mathbf{1}) - 2H\ \boldsymbol{P}_0\boldsymbol{n}, \tag{5.85}$$

since $\boldsymbol{n}_{,\alpha}\cdot\boldsymbol{a}^{\alpha} = -\kappa^{\alpha}_{\alpha} = -2H$. Substituting (5.85) in (5.83), we obtain the following exact equation on Ω

$$\mathrm{div}(\boldsymbol{P}_0\mathbf{1}) + \boldsymbol{P}_0'\,\boldsymbol{n} - 2H\ \boldsymbol{P}_0\boldsymbol{n} = \mathbf{0}. \tag{5.86}$$

On the other hand, note that

$$\mathrm{div}\big[(\boldsymbol{P}_0\mathbf{1})^T\boldsymbol{b}\big] = \boldsymbol{P}_0\mathbf{1}\cdot\nabla\boldsymbol{b} + \boldsymbol{b}\cdot\mathrm{div}(\boldsymbol{P}_0\mathbf{1}). \tag{5.87}$$

If we integrate this relation over Ω and use the divergence formula for surfaces (2.155), we deduce (alternatively, one can use Stokes' Theorem, see Sect. 2.6)

$$\int_{\partial\Omega}(\boldsymbol{P}_0\mathbf{1})^T\boldsymbol{b}\cdot\boldsymbol{\nu}\,\mathrm{d}s = \int_{\Omega}\big[\boldsymbol{P}_0\mathbf{1}\cdot\nabla\boldsymbol{b} + \boldsymbol{b}\cdot\mathrm{div}(\boldsymbol{P}_0\mathbf{1})\big]\,\mathrm{d}a$$

and inserting here (5.86) we deduce

$$\int_{\Omega}\boldsymbol{P}_0\mathbf{1}\cdot\nabla\boldsymbol{b}\ \mathrm{d}a = \int_{\partial\Omega}\boldsymbol{b}\cdot(\boldsymbol{P}_0\mathbf{1}\boldsymbol{\nu})\,\mathrm{d}s + \int_{\Omega}\boldsymbol{b}\cdot\big(\underbrace{\boldsymbol{P}_0'\boldsymbol{n}}_{=0} - 2H\ \underbrace{\boldsymbol{P}_0\boldsymbol{n}}_{=0}\big)\,\mathrm{d}a\,. \tag{5.88}$$

Further, we can also estimate the last term in (5.79) by using the relation

$$(\nabla\bar{\boldsymbol{a}})\boldsymbol{\kappa}^* = \bar{\boldsymbol{H}}_0'\,\boldsymbol{\kappa}^* - K\,\nabla\boldsymbol{u}\,. \tag{5.89}$$

Indeed, in view of (5.45) and $\boldsymbol{\kappa}\,\boldsymbol{\kappa}^* = K\mathbf{1}$, it holds

$$\bar{\boldsymbol{H}}_0'\,\boldsymbol{\kappa}^* = \big[\nabla\bar{\boldsymbol{a}} + (\nabla\boldsymbol{u})\boldsymbol{\kappa} + \bar{\boldsymbol{b}}\otimes\boldsymbol{n}\big]\boldsymbol{\kappa}^* = (\nabla\bar{\boldsymbol{a}})\boldsymbol{\kappa}^* + K\,\nabla\boldsymbol{u}$$

and the Eq. (5.89) is proved.

Taking into account relations (5.80), (5.88) and (5.89), we obtain from (5.79) the following form of the strain energy

$$\int_{\Omega}W\ \mathrm{d}a = \int_{\Omega}\bar{W}\ \mathrm{d}a + \frac{1}{24}h^3\int_{\partial\Omega}\bar{\boldsymbol{P}}_0\mathbf{1}\boldsymbol{\nu}\cdot\bar{\boldsymbol{b}}\ \mathrm{d}s + o(h^3), \tag{5.90}$$

where

$$\bar{W} = \big(1+\tfrac{1}{12}h^2K\big)\,W_1 + W_2 + W_3\,, \tag{5.91}$$

with

$$W_1 = \tfrac{1}{2} h\, \bar{\boldsymbol{P}}_0 \mathbf{1} \cdot \nabla \boldsymbol{u} \;=\; \tfrac{1}{2} h\, \bar{\boldsymbol{H}}_0 \cdot \underline{\boldsymbol{C}}[\bar{\boldsymbol{H}}_0],$$
$$W_2 = \tfrac{1}{24} h^3\, \bar{\boldsymbol{P}}_0' \mathbf{1} \cdot \big[\nabla \bar{\boldsymbol{a}} + (\nabla \boldsymbol{u})\boldsymbol{\kappa}\big] \;=\; \tfrac{1}{24} h^3\, \bar{\boldsymbol{H}}_0' \cdot \underline{\boldsymbol{C}}[\bar{\boldsymbol{H}}_0'], \tag{5.92}$$
$$W_3 = -\tfrac{1}{12} h^3\, \bar{\boldsymbol{P}}_0 \mathbf{1} \cdot \bar{\boldsymbol{H}}_0' \boldsymbol{\kappa}^* \;=\; -\tfrac{1}{12} h^3\, \underline{\boldsymbol{C}}[\bar{\boldsymbol{H}}_0] \cdot \bar{\boldsymbol{H}}_0' \boldsymbol{\kappa}^*.$$

For the edge loads we can write $\boldsymbol{t}_0 = \boldsymbol{P}_0 \boldsymbol{\nu} = \boldsymbol{P}_0 \mathbf{1} \boldsymbol{\nu}$ on $\partial\Omega_t$. So, in view of $(5.66)_3$, we have

$$\int_{\partial\Omega_t} \boldsymbol{p}_b \cdot \boldsymbol{b}\, \mathrm{d}s \;=\; \frac{1}{24} h^3 \int_{\partial\Omega_t} \boldsymbol{P}_0 \mathbf{1} \boldsymbol{\nu} \cdot \boldsymbol{b}\, \mathrm{d}s\,. \tag{5.93}$$

As in the plate theory (see relations (4.58)–(4.60)), we may show that the integral

$$\int_{\partial\Omega_u} \boldsymbol{P}_0 \mathbf{1} \boldsymbol{\nu} \cdot \boldsymbol{b}\, \mathrm{d}s \quad \text{is } \textit{fixed} \text{ by the boundary data,}$$

and may therefore be supressed in the overall energy, since it contributes a constant and does not affect the minimizers. Also, $\boldsymbol{u}$ and $\boldsymbol{a}$ are specified on $\partial\Omega_u$, where

$$\boldsymbol{a} = \bar{\boldsymbol{a}} \;=\; -\boldsymbol{A}_n^{-1}\big(\underline{\boldsymbol{C}}[\nabla \boldsymbol{u}]\big)\boldsymbol{n} \qquad \text{and} \qquad \nabla \boldsymbol{u} \;=\; \boldsymbol{u}_{,s} \otimes \boldsymbol{\tau} + \boldsymbol{u}_{,\nu} \otimes \boldsymbol{\nu}\,. \tag{5.94}$$

Thus, $\boldsymbol{u}(s)$ and $\boldsymbol{u}_{,\nu}(s)$ are specified on $\partial\Omega_u$.

Altogether, accounting for the contribution of edge loads and lateral loads, we obtain from (5.69), (5.90) and (5.93) that the potential energy has the following form

$$E = \bar{E} + o(h^3), \qquad \text{where}$$
$$\bar{E} = \int_{\Omega} \big(\bar{W} - \boldsymbol{g} \cdot \boldsymbol{u}\big)\, \mathrm{d}a \;-\; \int_{\partial\Omega_t} \big(\boldsymbol{p}_u \cdot \boldsymbol{u} + \boldsymbol{p}_a \cdot \bar{\boldsymbol{a}}\big)\, \mathrm{d}s, \tag{5.95}$$

and $\bar{W}$ is expressed by (5.91), while $\boldsymbol{p}_u$, $\boldsymbol{p}_a$ and $\boldsymbol{g}$ are given by (5.66) and (5.70).

The potential energy (5.95) agrees with the expression of the plate energy (4.74) (or (4.88)), when the shell is flat ($\boldsymbol{\kappa} = \mathbf{0}$ and $K = 0$).

5.4.1 Membrane Energy

Consider now the membrane energy W_1 in the shell model. In view of $(5.92)_1$ and the symmetries of $\underline{\boldsymbol{C}}$, we have

$$W_1 \;=\; \frac{1}{2} h\, \bar{\boldsymbol{P}}_0 \mathbf{1} \cdot \nabla \boldsymbol{u} \;=\; \frac{1}{2} h\, \bar{\boldsymbol{H}}_0 \cdot \underline{\boldsymbol{C}}[\bar{\boldsymbol{H}}_0] \;=\; \frac{1}{2} h\, \mathrm{sym}\, \bar{\boldsymbol{H}}_0 \cdot \underline{\boldsymbol{C}}[\mathrm{sym}\, \bar{\boldsymbol{H}}_0]. \tag{5.96}$$

If we decompose the displacement vector as

$$\boldsymbol{u} = u_\alpha \boldsymbol{a}^\alpha + w\, \boldsymbol{n},$$

then the surface gradient of the displacement is

$$\nabla \boldsymbol{u} \;=\; \boldsymbol{u}_{,\beta} \otimes \boldsymbol{a}^{\beta} \;=\; (u_{\alpha}\boldsymbol{a}^{\alpha} + w\,\boldsymbol{n})_{,\beta} \otimes \boldsymbol{a}^{\beta}$$

and using (2.82) we get

$$\nabla \boldsymbol{u} \;=\; \big[(u_{\alpha;\beta} - w\kappa_{\alpha\beta})\boldsymbol{a}^{\alpha} + (w_{,\beta} + u_{\alpha}\kappa^{\alpha}_{\beta})\boldsymbol{n}\big] \otimes \boldsymbol{a}^{\beta}, \tag{5.97}$$

where $u_{\alpha;\beta} = u_{\alpha,\beta} - u_{\gamma}\,\Gamma^{\gamma}_{\alpha\beta}$ is the covariant derivative.

If we denote by $\boldsymbol{\alpha}$ the vector

$$\boldsymbol{\alpha} \;:=\; -(\nabla w + \boldsymbol{\kappa}\,\boldsymbol{u}) \;=\; -(w_{,\beta} + u_{\alpha}\kappa^{\alpha}_{\beta})\boldsymbol{a}^{\beta}, \tag{5.98}$$

then the relation (5.97) can be written in the form

$$\nabla \boldsymbol{u} = (u_{\alpha;\beta} - w\kappa_{\alpha\beta})\boldsymbol{a}^{\alpha} \otimes \boldsymbol{a}^{\beta} - \boldsymbol{n} \otimes \boldsymbol{\alpha}\,. \tag{5.99}$$

Hence, we have

$$\mathrm{sym}(\nabla \boldsymbol{u}) = \mathrm{sym}\big[(u_{\alpha;\beta} - w\kappa_{\alpha\beta})\boldsymbol{a}^{\alpha} \otimes \boldsymbol{a}^{\beta}\big] - \mathrm{sym}(\boldsymbol{n} \otimes \boldsymbol{\alpha})\,. \tag{5.100}$$

If we introduce the *infinitesimal surface strain* $\boldsymbol{\varepsilon}$ given by

$$\boldsymbol{\varepsilon} \;=\; \varepsilon_{\alpha\beta}\boldsymbol{a}^{\alpha} \otimes \boldsymbol{a}^{\beta}, \qquad \text{with} \qquad \varepsilon_{\alpha\beta} \;:=\; \frac{1}{2}\,(u_{\alpha;\beta} + u_{\beta;\alpha}) - w\kappa_{\alpha\beta}\;, \tag{5.101}$$

then the Eq. (5.100) becomes

$$\mathrm{sym}(\nabla \boldsymbol{u}) = \;\boldsymbol{\varepsilon} - \mathrm{sym}(\boldsymbol{\alpha} \otimes \boldsymbol{n})\,. \tag{5.102}$$

Thus, in view of $(5.45)_1$ and (5.102) we deduce

$$\mathrm{sym}\,\bar{\boldsymbol{H}}_0 \;=\; \mathrm{sym}(\nabla \boldsymbol{u} + \bar{\boldsymbol{a}} \otimes \boldsymbol{n}) \;=\; \boldsymbol{\varepsilon} + \mathrm{sym}\big[(\bar{\boldsymbol{a}} - \boldsymbol{\alpha}) \otimes \boldsymbol{n}\big]. \tag{5.103}$$

Also, the vector $\bar{\boldsymbol{a}}$ satisfies (cf. $(5.75)_1$)

$$\begin{aligned}\bar{\boldsymbol{a}} \;&=\; -\boldsymbol{A}_n^{-1}\big(\underline{\boldsymbol{C}}[\nabla \boldsymbol{u}]\big)\boldsymbol{n} \;=\; -\boldsymbol{A}_n^{-1}\big(\underline{\boldsymbol{C}}[\mathrm{sym}\nabla \boldsymbol{u}]\big)\boldsymbol{n}\\ &=\; -\boldsymbol{A}_n^{-1}\big(\underline{\boldsymbol{C}}[\boldsymbol{\varepsilon}]\big)\boldsymbol{n} \;+\; \boldsymbol{A}_n^{-1}\big(\underline{\boldsymbol{C}}[\boldsymbol{\alpha} \otimes \boldsymbol{n}]\big)\boldsymbol{n},\end{aligned}$$

where we have used (5.102). Taking into account (5.76), the last relation yields

$$\bar{\boldsymbol{a}} \;=\; -\boldsymbol{A}_n^{-1}\big(\underline{\boldsymbol{C}}[\boldsymbol{\varepsilon}]\big)\boldsymbol{n} \;+\; \boldsymbol{\alpha}\,. \tag{5.104}$$

Substituting (5.104) in (5.103), we obtain

$$\operatorname{sym} \bar{\boldsymbol{H}}_0 = \boldsymbol{\varepsilon} - \operatorname{sym}\{\boldsymbol{A}_n^{-1}(\underline{\boldsymbol{C}}[\boldsymbol{\varepsilon}])\boldsymbol{n} \otimes \boldsymbol{n}\}. \tag{5.105}$$

On the basis of (5.96) and (5.105), we see that the energy density W_1 is a positive definite function of $\boldsymbol{\varepsilon}$.

Thus, in view of (5.66), (5.70) and (5.95), the membrane energy can be represented as

$$\begin{aligned} E &= h\, E_M + o(h), \qquad \text{where} \\ h\, E_M &= \int_{\Omega} W_1 \,\mathrm{d}a - \int_{\partial\Omega_t} h\, \boldsymbol{t}_0 \cdot \boldsymbol{u} \,\mathrm{d}s \end{aligned} \tag{5.106}$$

and

$$W_1 = \frac{1}{2} h \operatorname{sym} \bar{\boldsymbol{H}}_0 \cdot \underline{\boldsymbol{C}}[\operatorname{sym} \bar{\boldsymbol{H}}_0], \tag{5.107}$$

with

$$\boldsymbol{t}_0 = \bar{\boldsymbol{P}}_0 \mathbf{1} \boldsymbol{\nu}, \qquad \bar{\boldsymbol{P}}_0 = \underline{\boldsymbol{C}}[\operatorname{sym} \bar{\boldsymbol{H}}_0]. \tag{5.108}$$

5.4.2 Pure Bending

Suppose that the membrane energy is vanishing, $E_M \equiv 0$. Then, from (5.107) and (5.108) we get

$$\operatorname{sym} \bar{\boldsymbol{H}}_0 \equiv \mathbf{0} \quad \text{and} \quad \bar{\boldsymbol{P}}_0 = \mathbf{0}, \quad \boldsymbol{t}_0 = \mathbf{0}, \tag{5.109}$$

and the relation (5.105) yields

$$\boldsymbol{\varepsilon} \equiv \mathbf{0}. \tag{5.110}$$

Thus, taking into account (5.92) and (5.66), the energy (5.95) reduces to

$$E = h^3 E_B + o(h^3), \tag{5.111}$$

where the bending energy E_B is given by

$$\begin{aligned} E_B = &\int_{\Omega} \Big\{ \frac{1}{24} \bar{\boldsymbol{P}}_0' \mathbf{1} \cdot \big[\nabla \bar{\boldsymbol{a}} + (\nabla \boldsymbol{u})\boldsymbol{\kappa}\big] - \boldsymbol{g} \cdot \boldsymbol{u} \Big\} \,\mathrm{d}a \\ &- \int_{\partial\Omega_t} \Big[\frac{1}{24} \big(\boldsymbol{t}_0'' - 2\kappa_\tau \boldsymbol{t}_0'\big) \cdot \boldsymbol{u} + \frac{1}{12} \boldsymbol{t}_0' \cdot \bar{\boldsymbol{a}} \Big] \mathrm{d}s. \end{aligned} \tag{5.112}$$

According to (5.101) and (5.110) we have

$$0 = \varepsilon_{\alpha\beta} = \frac{1}{2}(u_{\alpha;\beta} + u_{\beta;\alpha}) - w\kappa_{\alpha\beta},$$

so the symmetric part of $u_{\alpha;\beta}$ is equal to

$$\frac{1}{2}\left(u_{\alpha;\beta}+u_{\beta;\alpha}\right) = w\kappa_{\alpha\beta}\,. \tag{5.113}$$

Hence, we can represent $u_{\alpha;\beta}$ in the form

$$u_{\alpha;\beta} = w\,\kappa_{\alpha\beta}+\omega\,\tilde{\varepsilon}_{\alpha\beta}\,, \tag{5.114}$$

with

$$\omega = \frac{1}{2}\,\tilde{\varepsilon}^{\alpha\beta}\,u_{\alpha;\beta}\,, \tag{5.115}$$

where the skew-symmetric tensor $\tilde{\boldsymbol{\varepsilon}} = \tilde{\varepsilon}_{\alpha\beta}\boldsymbol{a}^{\alpha}\otimes\boldsymbol{a}^{\beta} = \tilde{\varepsilon}^{\alpha\beta}\boldsymbol{a}_{\alpha}\otimes\boldsymbol{a}_{\beta}$ is the alternator tensor defined by (2.54). Indeed, to prove (5.115), we multiply the relation (5.114) by $\tilde{\varepsilon}^{\alpha\beta}$ and derive

$$\tilde{\varepsilon}^{\alpha\beta}\,u_{\alpha;\beta} = w\,\tilde{\varepsilon}^{\alpha\beta}\,\kappa_{\alpha\beta}+\omega\,\tilde{\varepsilon}^{\alpha\beta}\,\tilde{\varepsilon}_{\alpha\beta} = \omega\,e_{\alpha\beta}\,e_{\alpha\beta} = 2\omega,$$

since $\kappa_{\alpha\beta}$ is symmetric and $\tilde{\varepsilon}_{\alpha\beta} = \sqrt{a}\,e_{\alpha\beta}\,,\ \tilde{\varepsilon}^{\alpha\beta} = \dfrac{1}{\sqrt{a}}\,e_{\alpha\beta}$.

Remark The alternator tensor describes a rotation about $\boldsymbol{n}$, since

$$\tilde{\boldsymbol{\varepsilon}}\,\boldsymbol{v} = -\boldsymbol{n}\times\boldsymbol{v}, \qquad \text{for any vector} \quad \boldsymbol{v} = v^{\alpha}\boldsymbol{a}_{\alpha}+v_3\boldsymbol{n}\,. \tag{5.116}$$

Indeed, using (2.61) we get

$$\begin{aligned}\tilde{\boldsymbol{\varepsilon}}\,\boldsymbol{v} &= (\tilde{\varepsilon}_{\alpha\beta}\boldsymbol{a}^{\alpha}\otimes\boldsymbol{a}^{\beta})(v^{\gamma}\boldsymbol{a}_{\gamma}+v_3\boldsymbol{n}) = \tilde{\varepsilon}_{\alpha\beta}\,v^{\beta}\,\boldsymbol{a}^{\alpha} = -\tilde{\varepsilon}_{\alpha\beta}\,v^{\alpha}\,\boldsymbol{a}^{\beta}\\ &= -\boldsymbol{n}\times(v^{\alpha}\boldsymbol{a}_{\alpha}) = -\boldsymbol{n}\times\boldsymbol{v}\,.\end{aligned}$$

This shows that $-\boldsymbol{n}$ is the axial vector of the skew-symmetric tensor $\tilde{\boldsymbol{\varepsilon}}$. □

By virtue of (5.99) and (5.114), we obtain

$$\nabla\boldsymbol{u} = \omega\,\tilde{\boldsymbol{\varepsilon}}-\boldsymbol{n}\otimes\boldsymbol{\alpha}, \tag{5.117}$$

which implies

$$(\nabla\boldsymbol{u})\boldsymbol{\kappa} = \omega\,\tilde{\boldsymbol{\varepsilon}}\,\boldsymbol{\kappa}-\boldsymbol{n}\otimes\boldsymbol{\kappa}\,\boldsymbol{\alpha}, \tag{5.118}$$

since $\boldsymbol{\kappa}$ is symmetric. On the other hand, using the relation (5.104) with $\boldsymbol{\varepsilon} = \boldsymbol{0}$ we deduce that $\bar{\boldsymbol{a}} = \boldsymbol{\alpha}$, so we get

$$\nabla\bar{\boldsymbol{a}} = \nabla\boldsymbol{\alpha} = \boldsymbol{\alpha}_{,\gamma}\otimes\boldsymbol{a}^{\gamma} = \left(\alpha_{\beta;\gamma}\,\boldsymbol{a}^{\beta}+\alpha_{\beta}\,\kappa^{\beta}_{\gamma}\,\boldsymbol{n}\right)\otimes\boldsymbol{a}^{\gamma},$$

where we have employed the formula (2.82) for the vector $\boldsymbol{\alpha} = \alpha_{\beta}\,\boldsymbol{a}^{\beta}$. Hence, the last equation yields

$$\nabla\bar{\boldsymbol{a}} = \nabla\boldsymbol{\alpha} = \alpha_{\beta;\gamma}\,\boldsymbol{a}^\beta\otimes\boldsymbol{a}^\gamma + \boldsymbol{n}\otimes\boldsymbol{\kappa}\,\boldsymbol{\alpha}\,. \tag{5.119}$$

From (5.118) and (5.119) we obtain the relation

$$\nabla\bar{\boldsymbol{a}} + (\nabla\boldsymbol{u})\boldsymbol{\kappa} = \boldsymbol{B}, \tag{5.120}$$

where the tensor $\boldsymbol{B}$ is the *bending strain* defined by

$$\boldsymbol{B} = B_{\beta\gamma}\,\boldsymbol{a}^\beta\otimes\boldsymbol{a}^\gamma, \qquad \text{with} \qquad B_{\beta\gamma} := \alpha_{\beta;\gamma} + \omega\,\tilde{\varepsilon}_{\beta\lambda}\,\kappa^\lambda_\gamma\,. \tag{5.121}$$

If we insert here the relation $\alpha_\beta = -(w_{,\beta} + \kappa^\lambda_\beta\,u_\lambda)$ (cf. (5.98)) and use (5.114), then we derive

$$B_{\alpha\beta} = \omega(\tilde{\varepsilon}_{\alpha\lambda}\,\kappa^\lambda_\beta + \tilde{\varepsilon}_{\beta\lambda}\,\kappa^\lambda_\alpha) - (w_{;\alpha\beta} + \kappa^\lambda_{\alpha;\beta}\,u_\lambda + w\,\kappa^\lambda_\alpha\,\kappa_{\beta\lambda}). \tag{5.122}$$

By virtue of the Mainardi-Codazzi equations (2.119) we have here $\kappa^\lambda_{\alpha;\beta} = \kappa^\lambda_{\beta;\alpha}$. Hence, the bending tensor (5.122) is symmetric, i.e.

$$B_{\alpha\beta} = B_{\beta\alpha}\,. \tag{5.123}$$

Now, using Eq. $(5.75)_2$ we write

$$\bar{\boldsymbol{b}} = -\boldsymbol{A}_n^{-1}\big(\underline{\boldsymbol{C}}[\nabla\bar{\boldsymbol{a}} + (\nabla\boldsymbol{u})\boldsymbol{\kappa}]\big)\boldsymbol{n}$$

and the relation (5.120) yields

$$\bar{\boldsymbol{b}} = -\boldsymbol{A}_n^{-1}\big(\underline{\boldsymbol{C}}[\boldsymbol{B}]\big)\boldsymbol{n}\,. \tag{5.124}$$

Thus, from $(5.45)_2$, (5.120) and (5.124) we deduce

$$\operatorname{sym}\bar{\boldsymbol{H}}_0' = \operatorname{sym}\big[\nabla\bar{\boldsymbol{a}} + (\nabla\boldsymbol{u})\boldsymbol{\kappa} + \bar{\boldsymbol{b}}\otimes\boldsymbol{n}\big] = \operatorname{sym}\boldsymbol{B} - \operatorname{sym}\big\{\boldsymbol{A}_n^{-1}\big(\underline{\boldsymbol{C}}[\boldsymbol{B}]\big)\boldsymbol{n}\otimes\boldsymbol{n}\big\}.$$

Since $\boldsymbol{B}$ is symmetric, we obtain

$$\operatorname{sym}\bar{\boldsymbol{H}}_0' = \boldsymbol{B} - \operatorname{sym}\big\{\boldsymbol{A}_n^{-1}\big(\underline{\boldsymbol{C}}[\boldsymbol{B}]\big)\boldsymbol{n}\otimes\boldsymbol{n}\big\}. \tag{5.125}$$

Note the similarity of the last relation to the equation for surface strains (5.105).

According to $(5.92)_2$, the bending energy term W_2 can be written as

$$\begin{aligned} W_2 &= \tfrac{1}{24}\,h^3\,\bar{\boldsymbol{P}}_0'\boldsymbol{1}\cdot\big[\nabla\bar{\boldsymbol{a}} + (\nabla\boldsymbol{u})\boldsymbol{\kappa}\big] = \tfrac{1}{24}\,h^3\,\bar{\boldsymbol{H}}_0'\cdot\underline{\boldsymbol{C}}[\bar{\boldsymbol{H}}_0'] \\ &= \tfrac{1}{24}\,h^3\operatorname{sym}\bar{\boldsymbol{H}}_0'\cdot\underline{\boldsymbol{C}}[\operatorname{sym}\bar{\boldsymbol{H}}_0'] \end{aligned} \tag{5.126}$$

and the relation (5.125) shows that W_2 is a positive definite function of the bending strain $\boldsymbol{B}$, in the case of pure bending.

Remark Note that the bending tensor (5.122) reduces to

$$B_{\alpha\beta} = -w_{;\alpha\beta} \tag{5.127}$$

in the case of plates ($\boldsymbol{\kappa} = \mathbf{0}$). □

5.4.3 Estimate of the Coupling Term in the Energy

In order to estimate the coupling term W_3 which appears in the expression of the energy (5.91), let us introduce the inner product

$$\langle \boldsymbol{A}, \boldsymbol{B} \rangle = \boldsymbol{A} \cdot \underline{\boldsymbol{C}}[\boldsymbol{B}], \tag{5.128}$$

for any symmetric tensors $\boldsymbol{A}$, $\boldsymbol{B}$, as well as the induced norm

$$\|\boldsymbol{A}\| = \sqrt{\boldsymbol{A} \cdot \underline{\boldsymbol{C}}[\boldsymbol{A}]}\,. \tag{5.129}$$

Then, the energy term W_3 can be written as

$$W_3 = -\frac{1}{12} h^3 \underbrace{\bar{\boldsymbol{P}}_0 \mathbf{1}}_{=\underline{\boldsymbol{C}}[\bar{\boldsymbol{H}}_0]} \cdot \bar{\boldsymbol{H}}_0' \boldsymbol{\kappa}^* = -\frac{1}{12} h^3 \big\langle \operatorname{sym} \bar{\boldsymbol{H}}_0, \operatorname{sym}\big(\bar{\boldsymbol{H}}_0' \boldsymbol{\kappa}^*\big) \big\rangle . \tag{5.130}$$

Further, the Cauchy-Schwarz inequality yields

$$\Big| \big\langle \operatorname{sym} \bar{\boldsymbol{H}}_0, \operatorname{sym}\big(\bar{\boldsymbol{H}}_0' \boldsymbol{\kappa}^*\big) \big\rangle \Big| \le \big\| \operatorname{sym} \bar{\boldsymbol{H}}_0 \big\| \cdot \big\| \operatorname{sym}\big(\bar{\boldsymbol{H}}_0' \boldsymbol{\kappa}^*\big) \big\| . \tag{5.131}$$

To estimate the last norm, we use the spectral representation of the curvature tensor $\boldsymbol{\kappa}$ in the form

$$\boldsymbol{\kappa} = \kappa_1 \boldsymbol{u}_1 \otimes \boldsymbol{u}_1 + \kappa_2 \boldsymbol{u}_2 \otimes \boldsymbol{u}_2, \tag{5.132}$$

where κ_1 and κ_2 are the principal curvatures, and $\boldsymbol{u}_1$ and $\boldsymbol{u}_2$ the orthonormal principal vectors, see (2.106). In view of $2H = \kappa_1 + \kappa_2$ and $\mathbf{1} = \boldsymbol{u}_1 \otimes \boldsymbol{u}_1 + \boldsymbol{u}_2 \otimes \boldsymbol{u}_2$, we obtain for the cofactor $\boldsymbol{\kappa}^* = 2H\mathbf{1} - \boldsymbol{\kappa}$ the expression

$$\begin{aligned} \boldsymbol{\kappa}^* &= (\kappa_1 + \kappa_2)(\boldsymbol{u}_1 \otimes \boldsymbol{u}_1 + \boldsymbol{u}_2 \otimes \boldsymbol{u}_2) - (\kappa_1 \boldsymbol{u}_1 \otimes \boldsymbol{u}_1 + \kappa_2 \boldsymbol{u}_2 \otimes \boldsymbol{u}_2) \\ &= \kappa_2 \boldsymbol{u}_1 \otimes \boldsymbol{u}_1 + \kappa_1 \boldsymbol{u}_2 \otimes \boldsymbol{u}_2 . \end{aligned} \tag{5.133}$$

Let us assume without loss of generality that $\kappa_1 \ge \kappa_2$. Then, using (2.108) we get

$$\sqrt{H^2 - K} = \sqrt{\tfrac{1}{4}(\kappa_1 + \kappa_2)^2 - \kappa_1\kappa_2} = \sqrt{\tfrac{1}{4}(\kappa_1 - \kappa_2)^2} = \tfrac{1}{2}(\kappa_1 - \kappa_2)$$

and the relation (5.133) can be written equivalently

$$\boldsymbol{\kappa}^* = \frac{1}{2}(\kappa_1+\kappa_2)(\boldsymbol{u}_1\otimes\boldsymbol{u}_1+\boldsymbol{u}_2\otimes\boldsymbol{u}_2) + \frac{1}{2}(\kappa_1-\kappa_2)(\boldsymbol{u}_2\otimes\boldsymbol{u}_2-\boldsymbol{u}_1\otimes\boldsymbol{u}_1),$$

i.e.

$$\boldsymbol{\kappa}^* = H\,\mathbf{1}+\sqrt{H^2-K}\;\boldsymbol{\Delta}, \qquad \text{where} \quad \boldsymbol{\Delta} := \boldsymbol{u}_2\otimes\boldsymbol{u}_2-\boldsymbol{u}_1\otimes\boldsymbol{u}_1\,. \tag{5.134}$$

Using this relation, we can decompose

$$\operatorname{sym}\big(\bar{\boldsymbol{H}}_0'\boldsymbol{\kappa}^*\big) = H\operatorname{sym}\big(\bar{\boldsymbol{H}}_0'\mathbf{1}\big) + \sqrt{H^2-K}\operatorname{sym}\big(\bar{\boldsymbol{H}}_0'\boldsymbol{\Delta}\big)$$

and the triangle inequality implies

$$\big\|\operatorname{sym}\big(\bar{\boldsymbol{H}}_0'\boldsymbol{\kappa}^*\big)\big\| \le |H|\,\big\|\operatorname{sym}\big(\bar{\boldsymbol{H}}_0'\mathbf{1}\big)\big\| + \sqrt{H^2-K}\,\big\|\operatorname{sym}\big(\bar{\boldsymbol{H}}_0'\boldsymbol{\Delta}\big)\big\|\,. \tag{5.135}$$

If we designate by $\kappa = \max\limits_{\Omega}\{|\kappa_1|,|\kappa_2|\}$, then we have

$$|H| = \Big|\frac{1}{2}(\kappa_1+\kappa_2)\Big| \le \kappa, \qquad \sqrt{H^2-K} = \frac{1}{2}(\kappa_1-\kappa_2) \le \kappa$$

and the inequality (5.135) implies

$$\big\|\operatorname{sym}\big(\bar{\boldsymbol{H}}_0'\boldsymbol{\kappa}^*\big)\big\| \le \kappa\Big(\big\|\operatorname{sym}\big(\bar{\boldsymbol{H}}_0'\mathbf{1}\big)\big\| + \big\|\operatorname{sym}\big(\bar{\boldsymbol{H}}_0'\boldsymbol{\Delta}\big)\big\|\Big)\,. \tag{5.136}$$

In order to estimate the two norms in the right-hand side, let us denote by $\boldsymbol{A}_1$, $\boldsymbol{A}_2$ the tensors defined by

$$\boldsymbol{A}_1\,\boldsymbol{v} := \big(\underline{\boldsymbol{C}}[\boldsymbol{v}\otimes\boldsymbol{u}_1]\big)\boldsymbol{u}_1, \qquad \boldsymbol{A}_2\,\boldsymbol{v} := \big(\underline{\boldsymbol{C}}[\boldsymbol{v}\otimes\boldsymbol{u}_2]\big)\boldsymbol{u}_2, \quad \text{for any vector } \boldsymbol{v}. \tag{5.137}$$

Then, we have

$$\begin{aligned}\big\|\operatorname{sym}\,(\bar{\boldsymbol{H}}_0'\mathbf{1})\big\|^2 &= \operatorname{sym}\big(\bar{\boldsymbol{H}}_0'\mathbf{1}\big)\cdot\underline{\boldsymbol{C}}\big[\operatorname{sym}(\bar{\boldsymbol{H}}_0'\mathbf{1})\big] = \big(\bar{\boldsymbol{H}}_0'\mathbf{1}\big)\cdot\underline{\boldsymbol{C}}\big[\bar{\boldsymbol{H}}_0'\mathbf{1}\big]\\ &= \big(\bar{\boldsymbol{H}}_0'\boldsymbol{u}_1\otimes\boldsymbol{u}_1+\bar{\boldsymbol{H}}_0'\boldsymbol{u}_2\otimes\boldsymbol{u}_2\big)\cdot\underline{\boldsymbol{C}}\big[\bar{\boldsymbol{H}}_0'\boldsymbol{u}_1\otimes\boldsymbol{u}_1+\bar{\boldsymbol{H}}_0'\boldsymbol{u}_2\otimes\boldsymbol{u}_2\big]\\ &= \bar{\boldsymbol{H}}_0'\boldsymbol{u}_1\cdot\boldsymbol{A}_1(\bar{\boldsymbol{H}}_0'\boldsymbol{u}_1)+\bar{\boldsymbol{H}}_0'\boldsymbol{u}_2\cdot\boldsymbol{A}_2(\bar{\boldsymbol{H}}_0'\boldsymbol{u}_2)+2(\bar{\boldsymbol{H}}_0'\boldsymbol{u}_1\otimes\boldsymbol{u}_1)\cdot\underline{\boldsymbol{C}}\big[\bar{\boldsymbol{H}}_0'\boldsymbol{u}_2\otimes\boldsymbol{u}_2\big],\end{aligned} \tag{5.138}$$

since

$$(\bar{\boldsymbol{H}}_0'\boldsymbol{u}_1\otimes\boldsymbol{u}_1)\cdot\underline{\boldsymbol{C}}\big[\bar{\boldsymbol{H}}_0'\boldsymbol{u}_1\otimes\boldsymbol{u}_1\big] = \bar{\boldsymbol{H}}_0'\boldsymbol{u}_1\cdot\big(\underline{\boldsymbol{C}}[\bar{\boldsymbol{H}}_0'\boldsymbol{u}_1\otimes\boldsymbol{u}_1]\big)\boldsymbol{u}_1 = \bar{\boldsymbol{H}}_0'\boldsymbol{u}_1\cdot\boldsymbol{A}_1(\bar{\boldsymbol{H}}_0'\boldsymbol{u}_1).$$

Similarly, using (5.134) we get

$$\begin{aligned}\|\mathrm{sym}\,(\bar{\boldsymbol{H}}_0'\,\boldsymbol{\Delta})\|^2 &= \mathrm{sym}(\bar{\boldsymbol{H}}_0'\,\boldsymbol{\Delta})\cdot\underline{\boldsymbol{C}}[\mathrm{sym}(\bar{\boldsymbol{H}}_0'\,\boldsymbol{\Delta})] = (\bar{\boldsymbol{H}}_0'\,\boldsymbol{\Delta})\cdot\underline{\boldsymbol{C}}[\bar{\boldsymbol{H}}_0'\,\boldsymbol{\Delta}]\\ &= (\bar{\boldsymbol{H}}_0'\,\boldsymbol{u}_2\otimes\boldsymbol{u}_2-\bar{\boldsymbol{H}}_0'\,\boldsymbol{u}_1\otimes\boldsymbol{u}_1)\cdot\underline{\boldsymbol{C}}[\bar{\boldsymbol{H}}_0'\,\boldsymbol{u}_2\otimes\boldsymbol{u}_2-\bar{\boldsymbol{H}}_0'\,\boldsymbol{u}_1\otimes\boldsymbol{u}_1]\\ &= \bar{\boldsymbol{H}}_0'\,\boldsymbol{u}_2\cdot\boldsymbol{A}_2(\bar{\boldsymbol{H}}_0'\,\boldsymbol{u}_2)+\bar{\boldsymbol{H}}_0'\,\boldsymbol{u}_1\cdot\boldsymbol{A}_1(\bar{\boldsymbol{H}}_0'\,\boldsymbol{u}_1)-2(\bar{\boldsymbol{H}}_0'\,\boldsymbol{u}_1\otimes\boldsymbol{u}_1)\cdot\underline{\boldsymbol{C}}[\bar{\boldsymbol{H}}_0'\,\boldsymbol{u}_2\otimes\boldsymbol{u}_2].\end{aligned}\tag{5.139}$$

Hence, by summation of relations (5.138) and (5.139) we obtain

$$\|\mathrm{sym}(\bar{\boldsymbol{H}}_0'\,\boldsymbol{1})\|^2+\|\mathrm{sym}(\bar{\boldsymbol{H}}_0'\,\boldsymbol{\Delta})\|^2 = 2(\bar{\boldsymbol{H}}_0'\,\boldsymbol{u}_1\cdot\boldsymbol{A}_1(\bar{\boldsymbol{H}}_0'\,\boldsymbol{u}_1)+\bar{\boldsymbol{H}}_0'\,\boldsymbol{u}_2\cdot\boldsymbol{A}_2(\bar{\boldsymbol{H}}_0'\,\boldsymbol{u}_2)).\tag{5.140}$$

Further, using an inequality of the type $|x|+|y|\leq\sqrt{2(x^2+y^2)}$, we derive from (5.136) and (5.140) that

$$\|\mathrm{sym}(\bar{\boldsymbol{H}}_0'\,\boldsymbol{\kappa}^*)\| \;\leq\; 2\kappa\big[\bar{\boldsymbol{H}}_0'\,\boldsymbol{u}_1\cdot\boldsymbol{A}_1(\bar{\boldsymbol{H}}_0'\,\boldsymbol{u}_1)+\bar{\boldsymbol{H}}_0'\,\boldsymbol{u}_2\cdot\boldsymbol{A}_2(\bar{\boldsymbol{H}}_0'\,\boldsymbol{u}_2)\big]^{\frac{1}{2}}.\tag{5.141}$$

Finally, if we substitute this back into (5.131) we obtain the estimate

$$|W_3| \;\leq\; \frac{1}{6}\,h^3\kappa\;\|\mathrm{sym}\,\bar{\boldsymbol{H}}_0\|\;\big[\bar{\boldsymbol{H}}_0'\,\boldsymbol{u}_1\cdot\boldsymbol{A}_1(\bar{\boldsymbol{H}}_0'\,\boldsymbol{u}_1)+\bar{\boldsymbol{H}}_0'\,\boldsymbol{u}_2\cdot\boldsymbol{A}_2(\bar{\boldsymbol{H}}_0'\,\boldsymbol{u}_2)\big]^{\frac{1}{2}}.\tag{5.142}$$

Let us consider the strain energy density $\hat{W}$ given by

$$\hat{W} \;=\; W_1+W_2\,.\tag{5.143}$$

Then, by virtue of (5.91) and (5.142) we have

$$\begin{aligned}|\bar{W}-\hat{W}| &= \big|\,W_3+\big(\tfrac{1}{12}\,h^2K\big)\,W_1\big|\\ &\leq \frac{h^2}{6}\,(h\kappa)\;\|\mathrm{sym}\,\bar{\boldsymbol{H}}_0\|\;\big[\bar{\boldsymbol{H}}_0'\,\boldsymbol{u}_1\cdot\boldsymbol{A}_1(\bar{\boldsymbol{H}}_0'\,\boldsymbol{u}_1)+\bar{\boldsymbol{H}}_0'\,\boldsymbol{u}_2\cdot\boldsymbol{A}_2(\bar{\boldsymbol{H}}_0'\,\boldsymbol{u}_2)\big]^{\frac{1}{2}}\\ &\quad+\frac{1}{12}\,(h\kappa)^2\,W_1\,.\end{aligned}\tag{5.144}$$

Thus, we deduce the limit

$$\bar{W}\;\rightarrow\;\hat{W}\quad\text{as}\quad h\kappa\;\rightarrow\;0,\tag{5.145}$$

i.e., for a sufficiently thin shell. So, in this case the strain energy density is approximated by

$$W \;=\; W_1+W_2 \;=\; \frac{1}{2}\,h\,\bar{\boldsymbol{H}}_0\cdot\underline{\boldsymbol{C}}[\bar{\boldsymbol{H}}_0]\;+\;\frac{1}{24}\,h^3\,\bar{\boldsymbol{H}}_0'\cdot\underline{\boldsymbol{C}}[\bar{\boldsymbol{H}}_0'],\tag{5.146}$$

with potential energy

$$E = \int_{\Omega} \big(W - \boldsymbol{g}\cdot\boldsymbol{u}\big)\,\mathrm{d}a - \int_{\partial\Omega_t} \big(\boldsymbol{p}_u\cdot\boldsymbol{u} + \boldsymbol{p}_a\cdot\bar{\boldsymbol{a}}\big)\,\mathrm{d}s\,. \tag{5.147}$$

Here, the load vectors $\boldsymbol{p}_u\ \boldsymbol{p}_a$ are given by (5.66) and the net lateral traction $\boldsymbol{g}$ by (5.70).

Remark In the potential energy (5.147) we have $W = W_1 + W_2$ with

$$W_1 = \frac{1}{2}\,h\,\bar{\boldsymbol{H}}_0\cdot\underline{\boldsymbol{C}}[\bar{\boldsymbol{H}}_0], \qquad W_2 = \frac{1}{24}\,h^3\,\bar{\boldsymbol{H}}_0'\cdot\underline{\boldsymbol{C}}[\bar{\boldsymbol{H}}_0']\,. \tag{5.148}$$

□

For the membrane energy we have already obtained in Sect. 5.4.1 that

$$\begin{aligned} W_1 &= \tfrac{1}{2}\,h\,\mathrm{sym}\,\bar{\boldsymbol{H}}_0\cdot\underline{\boldsymbol{C}}[\mathrm{sym}\,\bar{\boldsymbol{H}}_0], \qquad \text{with} \\ \mathrm{sym}\,\bar{\boldsymbol{H}}_0 &= \boldsymbol{\varepsilon} - \mathrm{sym}\big\{\boldsymbol{A}_n^{-1}\big(\underline{\boldsymbol{C}}[\boldsymbol{\varepsilon}]\big)\boldsymbol{n}\otimes\boldsymbol{n}\big\}\,. \end{aligned} \tag{5.149}$$

In Sect. 5.4.2 we have obtained the energy term W_2 in case of *pure bending*. Let us reduce W_2 here in the general case.

Recall that (cf. (5.45))

$$\bar{\boldsymbol{H}}_0' = \nabla\bar{\boldsymbol{a}} + (\nabla\boldsymbol{u})\boldsymbol{\kappa} + \bar{\boldsymbol{b}}\otimes\boldsymbol{n}, \qquad \text{where} \quad \nabla\bar{\boldsymbol{a}} = \bar{\boldsymbol{a}}_{,\beta}\otimes\boldsymbol{a}^\beta\,. \tag{5.150}$$

In view of (5.75), the vector $\bar{\boldsymbol{a}}$ is determined by the equation

$$\boldsymbol{A}_n\bar{\boldsymbol{a}} = -\big(\underline{\boldsymbol{C}}[\nabla\boldsymbol{u}]\big)\boldsymbol{n} \tag{5.151}$$

where the tensor $\boldsymbol{A}_n$ satisfies the relation

$$\boldsymbol{A}_n\bar{\boldsymbol{a}} = \big(\underline{\boldsymbol{C}}[\bar{\boldsymbol{a}}\otimes\boldsymbol{n}]\big)\boldsymbol{n}, \tag{5.152}$$

according to (5.76). If we differentiate the last relation with respect to θ^β we get

$$(\boldsymbol{A}_n\bar{\boldsymbol{a}})_{,\beta} = \boldsymbol{A}_n\bar{\boldsymbol{a}}_{,\beta} + \big(\underline{\boldsymbol{C}}[\bar{\boldsymbol{a}}\otimes\boldsymbol{n}_{,\beta}]\big)\boldsymbol{n} + \big(\underline{\boldsymbol{C}}[\bar{\boldsymbol{a}}\otimes\boldsymbol{n}]\big)\boldsymbol{n}_{,\beta} + \big(\underline{\boldsymbol{C}}_{,\beta}[\bar{\boldsymbol{a}}\otimes\boldsymbol{n}]\big)\boldsymbol{n}\,. \tag{5.153}$$

On the other hand, the derivative of (5.151) yields

$$(\boldsymbol{A}_n\bar{\boldsymbol{a}})_{,\beta} = -\big(\underline{\boldsymbol{C}}\big[(\nabla\boldsymbol{u})_{,\beta}\big]\big)\boldsymbol{n} - \big(\underline{\boldsymbol{C}}[\nabla\boldsymbol{u}]\big)\boldsymbol{n}_{,\beta} - \big(\underline{\boldsymbol{C}}_{,\beta}[\nabla\boldsymbol{u}]\big)\boldsymbol{n}\,. \tag{5.154}$$

If we equate the relations (5.153) and (5.154), we find

$$\begin{aligned} \bar{\boldsymbol{a}}_{,\beta} = -\boldsymbol{A}_n^{-1}\big\{&\big(\underline{\boldsymbol{C}}[(\nabla\boldsymbol{u})_{,\beta} + \bar{\boldsymbol{a}}\otimes\boldsymbol{n}_{,\beta}]\big)\boldsymbol{n} + \big(\underline{\boldsymbol{C}}[\nabla\boldsymbol{u} + \bar{\boldsymbol{a}}\otimes\boldsymbol{n}]\big)\boldsymbol{n}_{,\beta} \\ &+ \big(\underline{\boldsymbol{C}}_{,\beta}[\nabla\boldsymbol{u} + \bar{\boldsymbol{a}}\otimes\boldsymbol{n}]\big)\boldsymbol{n}\,\big\}\,. \end{aligned} \tag{5.155}$$

Here, we have

$$(\nabla \boldsymbol{u})_{,\beta} = \left(\boldsymbol{u}_{,\alpha} \otimes \boldsymbol{a}^{\alpha}\right)_{,\beta} = \boldsymbol{u}_{;\alpha\beta} \otimes \boldsymbol{a}^{\alpha} + \boldsymbol{u}_{,\alpha} \otimes \boldsymbol{a}^{\alpha}{}_{;\beta}\,, \tag{5.156}$$

where

$$\boldsymbol{u}_{;\alpha\beta} = \boldsymbol{u}_{,\alpha\beta} - \boldsymbol{u}_{,\lambda}\,\Gamma^{\lambda}_{\alpha\beta}\,. \tag{5.157}$$

Further, in view of (2.81) and (2.79) we can write

$$\boldsymbol{a}^{\alpha}{}_{;\beta} = \boldsymbol{a}^{\alpha}{}_{,\beta} + \boldsymbol{a}^{\lambda}\,\Gamma^{\alpha}_{\lambda\beta} = \left(-\,\boldsymbol{a}^{\lambda}\,\Gamma^{\alpha}_{\lambda\beta} + \kappa^{\alpha}_{\beta}\,\boldsymbol{n}\right) + \boldsymbol{a}^{\lambda}\,\Gamma^{\alpha}_{\lambda\beta} = \kappa^{\alpha}_{\beta}\,\boldsymbol{n}, \tag{5.158}$$

so we have

$$\boldsymbol{u}_{,\alpha} \otimes \boldsymbol{a}^{\alpha}{}_{;\beta} = \boldsymbol{u}_{,\alpha} \otimes \kappa^{\alpha}_{\beta}\,\boldsymbol{n} = (\boldsymbol{u}_{,\alpha} \otimes \boldsymbol{a}^{\alpha})(\boldsymbol{\kappa}\,\boldsymbol{a}_{\beta} \otimes \boldsymbol{n}) = \left[(\nabla \boldsymbol{u})\boldsymbol{\kappa}\right]\boldsymbol{a}_{\beta} \otimes \boldsymbol{n}\,. \tag{5.159}$$

Using (5.159), the relation (5.156) becomes

$$(\nabla \boldsymbol{u})_{,\beta} = \boldsymbol{u}_{;\alpha\beta} \otimes \boldsymbol{a}^{\alpha} + \left[(\nabla \boldsymbol{u})\boldsymbol{\kappa}\right]\boldsymbol{a}_{\beta} \otimes \boldsymbol{n}\,. \tag{5.160}$$

Finally, if we insert this equation into (5.155) and (5.150), we obtain the expression of $\bar{\boldsymbol{H}}_0'$. Then, we are able to write the bending energy W_2 according to $(5.148)_2$.

5.5 Measures of Distortion

In the preceding sections, we have denoted by $a_{\alpha\beta}$ the metric tensor, $\kappa_{\alpha\beta}$ the curvature tensor, and $\Gamma^{\lambda}_{\alpha\beta}$ the Christoffel symbols of the reference midsurface. Let us consider now the tensors

$$\varepsilon_{\alpha\beta} = \frac{1}{2}\left(a'_{\alpha\beta} - a_{\alpha\beta}\right), \qquad \rho_{\alpha\beta} = b_{\alpha\beta} - \kappa_{\alpha\beta}, \qquad S^{\lambda}_{\alpha\beta} = \Gamma'^{\lambda}_{\alpha\beta} - \Gamma^{\lambda}_{\alpha\beta}, \tag{5.161}$$

where $a'_{\alpha\beta}$ designates the metric, $b_{\alpha\beta}$ the curvature tensor, and $\Gamma'^{\lambda}_{\alpha\beta}$ the Christoffel symbols in the deformed configuration. We want to characterize the tensors (5.161) in terms of the midsurface displacement field

$$\boldsymbol{u} = \boldsymbol{r}' - \boldsymbol{r}\,. \tag{5.162}$$

Here, $\boldsymbol{r}(\theta^{\alpha})$ is the position vector of the reference midsurface and $\boldsymbol{r}'$ describes the deformed midsurface. Let $\boldsymbol{r}'(\theta^{\alpha};\varepsilon)$ be a one-parameter family of positions with $\boldsymbol{r}'(\theta^{\alpha};0) = \boldsymbol{r}(\theta^{\alpha})$. Then, in the deformed configuration we have $\boldsymbol{a}'_{\alpha} = \boldsymbol{r}'_{,\alpha}$ and (cf. (2.51), (2.70) and (2.75))

$$a'_{\alpha\beta} = \boldsymbol{r}'_{,\alpha} \cdot \boldsymbol{r}'_{,\beta}, \qquad b_{\alpha\beta} = \boldsymbol{n}' \cdot \boldsymbol{r}'_{,\alpha\beta}, \qquad \Gamma'^{\lambda}_{\alpha\beta} = \boldsymbol{a}'^{\lambda} \cdot \boldsymbol{a}'_{\alpha,\beta}\,. \tag{5.163}$$

Also, we can expand $\boldsymbol{r}'$ about $\varepsilon = 0$ in the form

$$\boldsymbol{r}'(\theta^\alpha;\varepsilon) = \boldsymbol{r}(\theta^\alpha) + \varepsilon\,\dot{\boldsymbol{r}}'(\theta^\alpha;0) + \frac{1}{2}\,\varepsilon^2\,\ddot{\boldsymbol{r}}'(\theta^\alpha;0) + \cdots\,, \tag{5.164}$$

where a superposed dot stands for the derivative with respect to the parameter ε. Hence, in view of (5.162) and (5.164), we have

$$\dot{\boldsymbol{u}} = \dot{\boldsymbol{r}}' \qquad \text{and} \qquad \dot{\boldsymbol{a}}'_\alpha = \dot{\boldsymbol{r}}'_{,\alpha} = \dot{\boldsymbol{u}}_{,\alpha} \tag{5.165}$$

and from (5.161) we get

$$\dot{\varepsilon}_{\alpha\beta} = \frac{1}{2}\,\dot{a}'_{\alpha\beta}, \qquad \dot{\rho}_{\alpha\beta} = \dot{b}_{\alpha\beta}, \qquad \dot{S}^\lambda_{\alpha\beta} = \dot{\Gamma}'^\lambda_{\alpha\beta}\,. \tag{5.166}$$

We can choose $\varepsilon = 0$ in the above relations and denote for brevity $\dot{f}(0) := \dot{f}(\theta^\alpha;0)$ for any field f. Then, we get

$$\begin{aligned}\dot{\varepsilon}_{\alpha\beta}(0) &= \tfrac{1}{2}\,\dot{a}'_{\alpha\beta}(0) = \tfrac{1}{2}\,\big(\boldsymbol{a}'_\alpha\cdot\boldsymbol{a}'_\beta\big)^{\cdot}(0) = \tfrac{1}{2}\,\big(\dot{\boldsymbol{a}}'_\alpha\cdot\boldsymbol{a}'_\beta + \boldsymbol{a}'_\alpha\cdot\dot{\boldsymbol{a}}'_\beta\big)(0)\\ &= \tfrac{1}{2}\,\big(\dot{\boldsymbol{u}}_{,\alpha}(0)\cdot\boldsymbol{a}_\beta + \boldsymbol{a}_\alpha\cdot\dot{\boldsymbol{u}}_{,\beta}(0)\big) = \tfrac{1}{2}\,\big(\boldsymbol{u}_{,\alpha}\cdot\boldsymbol{a}_\beta + \boldsymbol{a}_\alpha\cdot\boldsymbol{u}_{,\beta}\big)^{\cdot}(0).\end{aligned} \tag{5.167}$$

Since the fields $\varepsilon_{\alpha\beta}$ and $\boldsymbol{u}$ vanish for $\varepsilon = 0$ and in the linear theory we neglect the terms of order $O(\varepsilon^2)$, from (5.167) we get

$$\varepsilon_{\alpha\beta} = \frac{1}{2}\,\big(\boldsymbol{u}_{,\alpha}\cdot\boldsymbol{a}_\beta + \boldsymbol{u}_{,\beta}\cdot\boldsymbol{a}_\alpha\big). \tag{5.168}$$

Using the components of $\boldsymbol{u} = u_\alpha\boldsymbol{a}^\alpha + w\,\boldsymbol{n}$ and the relation (2.82), we obtain the infinitesimal surface strain in the form

$$\varepsilon_{\alpha\beta} \;=\; \frac{1}{2}\,(u_{\alpha;\beta} + u_{\beta;\alpha}) - w\,\kappa_{\alpha\beta}\,, \tag{5.169}$$

which has already been written in (5.101).

For the bending measure $\rho_{\alpha\beta}$ let us prove the equation

$$\rho_{\alpha\beta} \;=\; \boldsymbol{n}\cdot\boldsymbol{u}_{;\alpha\beta}\,, \tag{5.170}$$

where $\boldsymbol{u}_{;\alpha\beta}$ is given by (5.157). Indeed, from $(5.166)_2$ and $(5.165)_2$ we have

$$\begin{aligned}\dot{\rho}_{\alpha\beta}(0) &= \dot{b}_{\alpha\beta}(0) = \big(\boldsymbol{n}'\cdot\boldsymbol{r}'_{,\alpha\beta}\big)^{\cdot}(0) = \dot{\boldsymbol{n}}'(0)\cdot\boldsymbol{r}'_{,\alpha\beta}(0) + \boldsymbol{n}'(0)\cdot\dot{\boldsymbol{r}}'_{,\alpha\beta}(0)\\ &= \boldsymbol{n}\cdot\dot{\boldsymbol{u}}_{,\alpha\beta}(0) + \dot{\boldsymbol{n}}'(0)\cdot\boldsymbol{a}_{\alpha,\beta}.\end{aligned} \tag{5.171}$$

The last term can be transformed in the following way

$$\dot{\boldsymbol{n}}'(0)\cdot \boldsymbol{a}_{\alpha,\beta} = \big[\boldsymbol{n}'\cdot \boldsymbol{I}\boldsymbol{a}_{\alpha,\beta}\big]^{\cdot}(0) = \big[\boldsymbol{n}'\cdot(\boldsymbol{n}\otimes\boldsymbol{n}+\boldsymbol{a}_\gamma\otimes\boldsymbol{a}^\gamma)\boldsymbol{a}_{\alpha,\beta}\big]^{\cdot}(0)$$
$$= \big[(\boldsymbol{n}'\cdot\boldsymbol{n})\kappa_{\alpha\beta} + (\boldsymbol{n}'\cdot\boldsymbol{a}_\gamma)(\boldsymbol{a}^\gamma\cdot\boldsymbol{a}_{\alpha,\beta})\big]^{\cdot}(0).$$

Taking into account that $\boldsymbol{a}^\gamma\cdot\boldsymbol{a}_{\alpha,\beta} = \Gamma^\gamma_{\alpha\beta}$ and

$$\dot{\boldsymbol{n}}'(0)\cdot\boldsymbol{n} = \frac{1}{2}\big[\boldsymbol{n}'\cdot\boldsymbol{n}'\big]^{\cdot}(0) = 0 \qquad \text{and} \qquad \boldsymbol{n}'\cdot\boldsymbol{a}_\gamma = \boldsymbol{n}'\cdot(\boldsymbol{a}'_\gamma - \boldsymbol{u}_{,\gamma}) = -\boldsymbol{n}'\cdot\boldsymbol{u}_{,\gamma},$$

the above relation reduces to

$$\dot{\boldsymbol{n}}'(0)\cdot\boldsymbol{a}_{\alpha,\beta} = \big[-(\boldsymbol{n}'\cdot\boldsymbol{u}_{,\gamma})\Gamma^\gamma_{\alpha\beta}\big]^{\cdot}(0) = -\boldsymbol{n}\cdot\big(\boldsymbol{u}_{,\gamma}\,\Gamma^\gamma_{\alpha\beta}\big)^{\cdot}(0). \tag{5.172}$$

Inserting (5.172) into (5.171) we get

$$\dot{\rho}_{\alpha\beta}(0) = \boldsymbol{n}\cdot\big[\boldsymbol{u}_{,\alpha\beta} - \boldsymbol{u}_{,\gamma}\,\Gamma^\gamma_{\alpha\beta}\big]^{\cdot}(0) = \boldsymbol{n}\cdot\dot{\boldsymbol{u}}_{;\alpha\beta}(0) = \big(\boldsymbol{n}\cdot\boldsymbol{u}_{;\alpha\beta}\big)^{\cdot}(0). \tag{5.173}$$

Since the fields $\rho_{\alpha\beta}$ and $\boldsymbol{u}$ vanish for $\varepsilon = 0$ and we neglect the terms of order $O(\varepsilon^2)$ in the linear theory, the last relation infer that the Eq. (5.170) holds true.

Using the vector field $\boldsymbol{\alpha} = -(\nabla w + \boldsymbol{\kappa}\,\boldsymbol{u})$ defined in (5.98), we can express the tensor $\rho_{\alpha\beta}$ as follows

$$\rho_{\alpha\beta} = -\alpha_{\alpha;\beta} + \kappa^\gamma_\beta\,\big(u_{\gamma;\alpha} - w\,\kappa_{\gamma\alpha}\big), \tag{5.174}$$

where

$$\alpha_\alpha = \boldsymbol{\alpha}\cdot\boldsymbol{a}_\alpha = -\big(w_{,\alpha} + \kappa^\gamma_\alpha\,u_\gamma\big). \tag{5.175}$$

Indeed, to prove (5.174), we rewrite the Eq. (5.170) as follows

$$\rho_{\alpha\beta} = \boldsymbol{n}\cdot\boldsymbol{u}_{;\alpha\beta} = \big(\boldsymbol{n}\cdot\boldsymbol{u}_{,\alpha}\big)_{;\beta} - \boldsymbol{n}_{,\beta}\cdot\boldsymbol{u}_{,\alpha}\,. \tag{5.176}$$

In view of the formula (2.82) and (5.175), we find

$$\begin{aligned} \boldsymbol{n}\cdot\boldsymbol{u}_{,\alpha} &= w_{,\alpha} + \kappa^\beta_\alpha\,u_\beta = -\alpha_\alpha \\ \boldsymbol{n}_{,\beta}\cdot\boldsymbol{u}_{,\alpha} &= \big(-\kappa^\lambda_\beta\,\boldsymbol{a}_\lambda\big)\cdot\big[(u_{\gamma;\alpha} - w\,\kappa_{\gamma\alpha})\boldsymbol{a}^\gamma\big] = -\kappa^\gamma_\beta\,\big(u_{\gamma;\alpha} - w\,\kappa_{\gamma\alpha}\big). \end{aligned} \tag{5.177}$$

If we substitute (5.177) into (5.176), then we obtain that the relation (5.174) holds.

Further, we can express the bending tensor $\rho_{\alpha\beta}$ directly in terms of the displacement $\boldsymbol{u}$. Using (5.175) into (5.174), we derive

$$\rho_{\alpha\beta} = \big(w_{,\alpha} + \kappa^\gamma_\alpha\,u_\gamma\big)_{;\beta} + \kappa^\gamma_\beta\,\big(u_{\gamma;\alpha} - w\,\kappa_{\gamma\alpha}\big),$$

so

$$\rho_{\alpha\beta} = w_{;\alpha\beta} + \kappa^\gamma_\alpha\,u_{\gamma;\beta} + \kappa^\gamma_\beta\,u_{\gamma;\alpha} - w\,\kappa^\gamma_\beta\,\kappa_{\gamma\alpha} + \kappa^\gamma_{\alpha;\beta}\,u_\gamma, \tag{5.178}$$

where

$$
\begin{aligned}
w_{;\alpha\beta} = w_{,\alpha\beta} - w_{,\gamma}\,\Gamma^{\gamma}_{\alpha\beta}, \qquad & u_{\gamma;\alpha} = u_{\gamma,\alpha} - u_{\lambda}\,\Gamma^{\lambda}_{\alpha\gamma} \\
\text{and} \qquad \kappa^{\gamma}_{\alpha;\beta} = \kappa^{\gamma}_{\alpha,\beta} + & \kappa^{\lambda}_{\alpha}\,\Gamma^{\gamma}_{\lambda\beta} - \kappa^{\gamma}_{\lambda}\,\Gamma^{\lambda}_{\alpha\beta}\,.
\end{aligned}
\tag{5.179}
$$

From (5.178) we see that the relation

$$
\rho_{\alpha\beta} = \rho_{\beta\alpha} \tag{5.180}
$$

holds, in view of the Mainardi-Codazzi equations $\kappa^{\gamma}_{\alpha;\beta} = \kappa^{\gamma}_{\beta;\alpha}$ (cf. (2.121)) and the symmetry of the tensor $\boldsymbol{\kappa}^2 = \kappa^{\gamma}_{\alpha}\,\kappa_{\gamma\beta}\,\boldsymbol{a}^{\alpha}\otimes\boldsymbol{a}^{\beta}$.

For the tensor $S^{\lambda}_{\alpha\beta}$ defined in $(5.161)_3$ we can prove the relation

$$
S^{\lambda}_{\alpha\beta} \;=\; \boldsymbol{a}^{\lambda}\cdot\boldsymbol{u}_{;\alpha\beta} - \alpha^{\lambda}\,\kappa_{\alpha\beta} \;=\; \boldsymbol{a}^{\lambda}\cdot\big(\boldsymbol{u}_{;\alpha\beta} - \boldsymbol{\alpha}\,\kappa_{\alpha\beta}\big). \tag{5.181}
$$

Indeed, in view of $(5.166)_3$ we have

$$
\dot{S}^{\lambda}_{\alpha\beta}(0) = \dot{\Gamma}'^{\lambda}_{\alpha\beta}(0) = \big(\boldsymbol{a}'^{\lambda}\cdot\boldsymbol{a}'_{\alpha,\beta}\big)^{\dot{}}(0) = \big[(a'^{\lambda\gamma}\boldsymbol{a}'_{\gamma})\cdot\boldsymbol{a}'_{\alpha,\beta}\big]^{\dot{}}(0),
$$

and using the product rule for differentiation we get

$$
\dot{S}^{\lambda}_{\alpha\beta}(0) = \big(\dot{a}'^{\lambda\gamma}(0)\,\boldsymbol{a}_{\gamma} + a^{\lambda\gamma}\,\dot{\boldsymbol{u}}_{,\gamma}(0)\big)\cdot\boldsymbol{a}_{\alpha,\beta} + \boldsymbol{a}^{\lambda}\cdot\dot{\boldsymbol{u}}_{,\alpha\beta}(0). \tag{5.182}
$$

In order to express the factor $\dot{a}'^{\lambda\gamma}(0)$, we use the relations

$$
\dot{a}'^{\lambda\gamma} = -a'^{\lambda\alpha}\,a'^{\gamma\beta}\,\dot{a}'_{\alpha\beta} \tag{5.183}
$$

and

$$
\dot{a}'_{\alpha\beta}(0) = \big(\boldsymbol{a}'_{\alpha}\cdot\boldsymbol{a}'_{\beta}\big)^{\dot{}}(0) = \dot{\boldsymbol{u}}_{,\alpha}(0)\cdot\boldsymbol{a}_{\beta} + \boldsymbol{a}_{\alpha}\cdot\dot{\boldsymbol{u}}_{,\beta}(0). \tag{5.184}
$$

Using (5.184) into (5.183) we find

$$
\dot{a}'^{\lambda\gamma}(0) = -a^{\lambda\alpha}\,a^{\gamma\beta}\big(\dot{\boldsymbol{u}}_{,\alpha}(0)\cdot\boldsymbol{a}_{\beta} + \boldsymbol{a}_{\alpha}\cdot\dot{\boldsymbol{u}}_{,\beta}(0)\big) = -a^{\lambda\delta}\,\boldsymbol{a}^{\gamma}\cdot\dot{\boldsymbol{u}}_{,\delta}(0) - a^{\gamma\delta}\,\boldsymbol{a}^{\lambda}\cdot\dot{\boldsymbol{u}}_{,\delta}(0)
$$

and substituting this into (5.182) we derive

$$
\begin{aligned}
\dot{S}^{\lambda}_{\alpha\beta}(0) &= \big[-\big(\boldsymbol{a}^{\lambda}\cdot\dot{\boldsymbol{u}}_{,\delta}(0)\big)\boldsymbol{a}^{\delta} - a^{\lambda\delta}\big(\boldsymbol{a}^{\gamma}\cdot\dot{\boldsymbol{u}}_{,\delta}(0)\big)\boldsymbol{a}_{\gamma} + a^{\lambda\gamma}\,\dot{\boldsymbol{u}}_{,\gamma}(0)\big]\cdot\boldsymbol{a}_{\alpha,\beta} \\
&\quad + \boldsymbol{a}^{\lambda}\cdot\dot{\boldsymbol{u}}_{,\alpha\beta}(0) \\
&= \boldsymbol{a}^{\lambda}\cdot\big[\dot{\boldsymbol{u}}_{,\alpha\beta}(0) - \dot{\boldsymbol{u}}_{,\delta}(0)\cdot\big(\boldsymbol{a}^{\delta}\cdot\boldsymbol{a}_{\alpha,\beta}\big)\big] \\
&\quad + a^{\lambda\delta}\big[\dot{\boldsymbol{u}}_{,\delta}(0) - \big(\boldsymbol{a}_{\gamma}\otimes\boldsymbol{a}^{\gamma}\big)\dot{\boldsymbol{u}}_{,\delta}(0)\big]\cdot\boldsymbol{a}_{\alpha,\beta} \\
&= \boldsymbol{a}^{\lambda}\cdot\dot{\boldsymbol{u}}_{;\alpha\beta}(0) + a^{\lambda\delta}\big[\big(\boldsymbol{n}\otimes\boldsymbol{n}\big)\dot{\boldsymbol{u}}_{,\delta}(0)\big]\cdot\boldsymbol{a}_{\alpha,\beta} \\
&= \boldsymbol{a}^{\lambda}\cdot\dot{\boldsymbol{u}}_{;\alpha\beta}(0) + a^{\lambda\delta}\big(\boldsymbol{n}\cdot\dot{\boldsymbol{u}}_{,\delta}(0)\big)\big(\boldsymbol{n}\cdot\boldsymbol{a}_{\alpha,\beta}\big)
\end{aligned}
$$

Using $(5.177)_1$ and $\boldsymbol{n}\cdot\boldsymbol{a}_{\alpha,\beta}=\kappa_{\alpha\beta}$ we deduce

$$\dot{S}^{\lambda}_{\alpha\beta}(0) \;=\; \boldsymbol{a}^{\lambda}\cdot\dot{\boldsymbol{u}}_{;\alpha\beta}(0)-\dot{\alpha}^{\lambda}(0)\kappa_{\alpha\beta} \;=\; \boldsymbol{a}^{\lambda}\cdot\big(\boldsymbol{u}_{;\alpha\beta}-\boldsymbol{\alpha}\,\kappa_{\alpha\beta}\big)^{\dot{}}(0). \tag{5.185}$$

Thus, since the fields $S^{\lambda}_{\alpha\beta}$, $\boldsymbol{u}$ and $\boldsymbol{\alpha}$ vanish for $\varepsilon=0$, we obtain in the linear theory (neglecting terms of order $O(\varepsilon^2)$) that the relation (5.181) holds true.

On the basis of equations (5.170) and (5.181) we can express $\boldsymbol{u}_{;\alpha\beta}$ in the form

$$\boldsymbol{u}_{;\alpha\beta} \;=\; S^{\lambda}_{\alpha\beta}\,\boldsymbol{a}_{\lambda}+\kappa_{\alpha\beta}\,\boldsymbol{\alpha}+\rho_{\alpha\beta}\,\boldsymbol{n}\,. \tag{5.186}$$

Remark The tensor $S^{\lambda}_{\alpha\beta}$ can be expressed in terms of $\varepsilon_{\alpha\beta}$ by the following relation

$$S^{\lambda}_{\alpha\beta}=a^{\lambda\gamma}\,S_{\gamma\alpha\beta} \qquad \text{with} \qquad S_{\gamma\alpha\beta} \;=\; \varepsilon_{\gamma\alpha;\beta}+\varepsilon_{\gamma\beta;\alpha}-\varepsilon_{\alpha\beta;\gamma}\,, \tag{5.187}$$

where we have

$$\varepsilon_{\alpha\beta;\gamma} \;=\; \varepsilon_{\alpha\beta,\gamma}-\varepsilon_{\beta\lambda}\,\Gamma^{\lambda}_{\alpha\gamma}\;-\varepsilon_{\alpha\lambda}\,\Gamma^{\lambda}_{\beta\gamma}\,. \tag{5.188}$$

Let us prove the relation (5.187): from $S^{\lambda}_{\alpha\beta}=a^{\lambda\gamma}\,S_{\gamma\alpha\beta}$ and (5.181) we get

$$S_{\gamma\alpha\beta} \;=\; a_{\lambda\gamma}\,S^{\lambda}_{\alpha\beta} \;=\; \boldsymbol{a}_{\gamma}\cdot\big(\boldsymbol{u}_{;\alpha\beta}-\boldsymbol{\alpha}\,\kappa_{\alpha\beta}\big) \;=\; \boldsymbol{a}_{\gamma}\cdot\boldsymbol{u}_{;\alpha\beta}+(\boldsymbol{u}_{,\gamma}\cdot\boldsymbol{n})\kappa_{\alpha\beta}\,, \tag{5.189}$$

in view of $(5.177)_1$. On the other hand, using (5.188) we can write by a straightforward calculation

$$\begin{aligned}
\varepsilon_{\gamma\alpha;\beta}+\varepsilon_{\gamma\beta;\alpha}-\varepsilon_{\alpha\beta;\gamma} \;&=\; \big(\varepsilon_{\gamma\alpha,\beta}-\varepsilon_{\gamma\lambda}\,\Gamma^{\lambda}_{\alpha\beta}-\varepsilon_{\alpha\lambda}\,\Gamma^{\lambda}_{\gamma\beta}\big)\\
&\quad+\big(\varepsilon_{\gamma\beta,\alpha}-\varepsilon_{\gamma\lambda}\,\Gamma^{\lambda}_{\beta\alpha}-\varepsilon_{\beta\lambda}\,\Gamma^{\lambda}_{\gamma\alpha}\big)-\big(\varepsilon_{\alpha\beta,\gamma}-\varepsilon_{\alpha\lambda}\,\Gamma^{\lambda}_{\beta\gamma}-\varepsilon_{\beta\lambda}\,\Gamma^{\lambda}_{\alpha\gamma}\big)\\
&=\tfrac{1}{2}\big(\boldsymbol{u}_{,\gamma}\cdot\boldsymbol{a}_{\alpha}+\boldsymbol{u}_{,\alpha}\cdot\boldsymbol{a}_{\gamma}\big)_{,\beta}+\tfrac{1}{2}\big(\boldsymbol{u}_{,\gamma}\cdot\boldsymbol{a}_{\beta}+\boldsymbol{u}_{,\beta}\cdot\boldsymbol{a}_{\gamma}\big)_{,\alpha}-\tfrac{1}{2}\big(\boldsymbol{u}_{,\alpha}\cdot\boldsymbol{a}_{\beta}+\boldsymbol{u}_{,\beta}\cdot\boldsymbol{a}_{\alpha}\big)_{,\gamma}\\
&\quad-\big(\boldsymbol{u}_{,\lambda}\cdot\boldsymbol{a}_{\gamma}+\boldsymbol{u}_{,\gamma}\cdot\boldsymbol{a}_{\lambda}\big)\Gamma^{\lambda}_{\alpha\beta}\\
&=\boldsymbol{a}_{\gamma}\cdot\boldsymbol{u}_{,\alpha\beta}-\boldsymbol{a}_{\gamma}\cdot\boldsymbol{u}_{,\lambda}\,\Gamma^{\lambda}_{\alpha\beta}+\boldsymbol{u}_{,\gamma}\cdot\boldsymbol{a}_{\alpha,\beta}-\boldsymbol{u}_{,\gamma}\cdot\boldsymbol{a}_{\lambda}\,\Gamma^{\lambda}_{\alpha\beta}\\
&=\boldsymbol{a}_{\gamma}\cdot\boldsymbol{u}_{;\alpha\beta}+\boldsymbol{u}_{,\gamma}\cdot(\boldsymbol{n}\,\kappa_{\alpha\beta}),
\end{aligned}$$

where we have taken into account (2.76). If we compare this expression with the relation (5.189), we deduce that the Eq. (5.187) holds. □

Finally, we derive with the help of (5.155) and $(5.45)_1$ that

$$\begin{aligned}
\nabla\bar{\boldsymbol{a}}=\bar{\boldsymbol{a}}_{,\beta}\otimes\boldsymbol{a}^{\beta} \;&=\; -\boldsymbol{A}_n^{-1}\big(\underline{\boldsymbol{C}}\big[(\nabla\boldsymbol{u})_{,\beta}+\bar{\boldsymbol{a}}\otimes\boldsymbol{n}_{,\beta}\big]\big)\boldsymbol{n}\otimes\boldsymbol{a}^{\beta}\\
&\quad-\boldsymbol{A}_n^{-1}\big(\underline{\boldsymbol{C}}_{,\beta}[\bar{\boldsymbol{H}}_0]\big)\boldsymbol{n}\otimes\boldsymbol{a}^{\beta}\;+\;\boldsymbol{A}_n^{-1}\big(\underline{\boldsymbol{C}}[\bar{\boldsymbol{H}}_0]\big)\boldsymbol{\kappa}\,.
\end{aligned} \tag{5.190}$$

For the first term in the right-hand side we use the Eq. (5.160) to get

$$\begin{aligned} \boldsymbol{A}_n^{-1}\big(\underline{\boldsymbol{C}}\big[(\nabla \boldsymbol{u})_{,\beta}\big]\big)\boldsymbol{n}\otimes \boldsymbol{a}^\beta &= \boldsymbol{A}_n^{-1}\big(\underline{\boldsymbol{C}}\big[\boldsymbol{u}_{;\alpha\beta}\otimes \boldsymbol{a}^\alpha\big]\big)\boldsymbol{n}\otimes \boldsymbol{a}^\beta + \big[(\nabla \boldsymbol{u})\boldsymbol{\kappa}\big]\boldsymbol{a}_\beta\otimes \boldsymbol{a}^\beta \\ &= \boldsymbol{A}_n^{-1}\big(\underline{\boldsymbol{C}}\big[\boldsymbol{u}_{;\alpha\beta}\otimes \boldsymbol{a}^\alpha\big]\big)\boldsymbol{n}\otimes \boldsymbol{a}^\beta + (\nabla \boldsymbol{u})\boldsymbol{\kappa}\,. \end{aligned} \tag{5.191}$$

Thus, inserting (5.191) into (5.190) we deduce

$$\begin{aligned} \nabla\bar{\boldsymbol{a}} + (\nabla \boldsymbol{u})\boldsymbol{\kappa} = &-\boldsymbol{A}_n^{-1}\big(\underline{\boldsymbol{C}}\big[\boldsymbol{u}_{;\alpha\beta}\otimes \boldsymbol{a}^\alpha + \bar{\boldsymbol{a}}\otimes \boldsymbol{n}_{,\beta}\big]\big)\boldsymbol{n}\otimes \boldsymbol{a}^\beta \\ &-\boldsymbol{A}_n^{-1}\big(\underline{\boldsymbol{C}}_{,\beta}[\bar{\boldsymbol{H}}_0]\big)\boldsymbol{n}\otimes \boldsymbol{a}^\beta \; + \; \boldsymbol{A}_n^{-1}\big(\underline{\boldsymbol{C}}[\bar{\boldsymbol{H}}_0]\big)\boldsymbol{\kappa}\,. \end{aligned} \tag{5.192}$$

For the first term in the right-hand side we employ the relations $\boldsymbol{n}_{,\beta} = -\kappa_{\alpha\beta}\,\boldsymbol{a}^\alpha$ and (5.186) to write

$$\begin{aligned} \boldsymbol{A}_n^{-1}\big(\underline{\boldsymbol{C}}\big[\boldsymbol{u}_{;\alpha\beta}\otimes \boldsymbol{a}^\alpha + \bar{\boldsymbol{a}}\otimes \boldsymbol{n}_{,\beta}\big]\big)\boldsymbol{n}\otimes \boldsymbol{a}^\beta &= S_{\lambda\alpha\beta}\,\boldsymbol{A}_n^{-1}\big(\underline{\boldsymbol{C}}\big[\boldsymbol{a}^\lambda\otimes \boldsymbol{a}^\alpha\big]\big)\boldsymbol{n}\otimes \boldsymbol{a}^\beta \\ +\rho_{\alpha\beta}\,\boldsymbol{A}_n^{-1}\big(\underline{\boldsymbol{C}}[\boldsymbol{n}\otimes \boldsymbol{a}^\alpha]\big)\boldsymbol{n}\otimes \boldsymbol{a}^\beta \; &+ \; \kappa_{\alpha\beta}\,\boldsymbol{A}_n^{-1}\big(\underline{\boldsymbol{C}}[(\boldsymbol{\alpha}-\bar{\boldsymbol{a}})\otimes \boldsymbol{a}^\alpha]\big)\boldsymbol{n}\otimes \boldsymbol{a}^\beta\,. \end{aligned} \tag{5.193}$$

Note that

$$\rho_{\alpha\beta}\,\boldsymbol{A}_n^{-1}\big(\underline{\boldsymbol{C}}[\boldsymbol{n}\otimes \boldsymbol{a}^\alpha]\big)\boldsymbol{n}\otimes \boldsymbol{a}^\beta \;=\; \rho_{\alpha\beta}\,\boldsymbol{A}_n^{-1}\big(\underline{\boldsymbol{C}}[\boldsymbol{a}^\alpha\otimes \boldsymbol{n}]\big)\boldsymbol{n}\otimes \boldsymbol{a}^\beta \;=\; \rho_{\alpha\beta}\,\boldsymbol{a}^\alpha\otimes \boldsymbol{a}^\beta = \boldsymbol{\rho}, \tag{5.194}$$

in view of the relation (5.76) and the minor symmetries of $\underline{\boldsymbol{C}}$. Let us denote by $\boldsymbol{a}^*$ the vector

$$\boldsymbol{a}^* \;:=\; \bar{\boldsymbol{a}} - \boldsymbol{\alpha} \;=\; -\boldsymbol{A}_n^{-1}\big(\underline{\boldsymbol{C}}[\boldsymbol{\varepsilon}]\big)\boldsymbol{n}, \tag{5.195}$$

where we have used (5.104). Taking into account the relations (5.193)–(5.195), the Eq. (5.192) becomes

$$\begin{aligned} \nabla\bar{\boldsymbol{a}} + (\nabla \boldsymbol{u})\boldsymbol{\kappa} = -\boldsymbol{\rho} - S_{\lambda\alpha\beta}\,\boldsymbol{A}_n^{-1}\big(\underline{\boldsymbol{C}}\big[\boldsymbol{a}^\lambda\otimes \boldsymbol{a}^\alpha\big]\big)\boldsymbol{n}\otimes \boldsymbol{a}^\beta \;-\; \boldsymbol{A}_n^{-1}\big(\underline{\boldsymbol{C}}_{,\beta}[\bar{\boldsymbol{H}}_0]\big)\boldsymbol{n}\otimes \boldsymbol{a}^\beta \\ +\; \boldsymbol{A}_n^{-1}\big(\underline{\boldsymbol{C}}[\bar{\boldsymbol{H}}_0]\big)\boldsymbol{\kappa} \;+\; \kappa_{\alpha\beta}\,\boldsymbol{A}_n^{-1}\big(\underline{\boldsymbol{C}}[\boldsymbol{a}^*\otimes \boldsymbol{a}^\alpha]\big)\boldsymbol{n}\otimes \boldsymbol{a}^\beta\,. \end{aligned} \tag{5.196}$$

Hence, we obtain the vector $\bar{\boldsymbol{b}}$ by using the relation $(5.75)_2$. Then, we can determine $\bar{\boldsymbol{H}}_0'$ from the Eq. (5.150).

In the relation (5.196) we can use

$$\big(\underline{\boldsymbol{C}}\big[\boldsymbol{a}^\lambda\otimes \boldsymbol{a}^\alpha\big]\big)\boldsymbol{n} \;=\; C^{\gamma 3\lambda\alpha}\boldsymbol{a}_\gamma + C^{33\lambda\alpha}\boldsymbol{n}\,. \tag{5.197}$$

Indeed, if we denote the components by

$$\underline{\boldsymbol{C}} \;=\; C^{ijkl}\bar{\boldsymbol{g}}_i\otimes\bar{\boldsymbol{g}}_j\otimes\bar{\boldsymbol{g}}_k\otimes\bar{\boldsymbol{g}}_l \qquad \text{with} \qquad \bar{\boldsymbol{g}}_\alpha = \boldsymbol{a}_\alpha \;\text{ and }\; \bar{\boldsymbol{g}}_3 = \boldsymbol{n} \tag{5.198}$$

on the midsurface, then

$$\big(\underline{\boldsymbol{C}}\big[\boldsymbol{a}^\lambda\otimes \boldsymbol{a}^\alpha\big]\big)\boldsymbol{n} \;=\; \big(C^{ij\lambda\alpha}\bar{\boldsymbol{g}}_i\otimes\bar{\boldsymbol{g}}_j\big)\boldsymbol{n} \;=\; C^{\gamma 3\lambda\alpha}\boldsymbol{a}_\gamma + C^{33\lambda\alpha}\boldsymbol{n},$$

so we have shown the relation (5.197). Also, we can write

$$\underline{C}_{,\beta} \;=\; C^{ijkl}{}_{;\beta}\, \bar{g}_i \otimes \bar{g}_j \otimes \bar{g}_k \otimes \bar{g}_l \,+\, O(\kappa), \tag{5.199}$$

where the part $O(\kappa)$ includes terms containing factors of the type $a_{\alpha;\beta} = \kappa_{\alpha\beta}\boldsymbol{n}$ or $\boldsymbol{n}_{,\beta} = -\kappa_{\alpha\beta}\boldsymbol{a}^\alpha$.

Recall that the coupling energy term W_3 has been estimated in Sect. 5.4.3 and note that

$$|\,W_3|\,/\,W_2 \;=\; O(\kappa\varepsilon), \qquad \text{where} \;\; \varepsilon = |\boldsymbol{\varepsilon}|\,.$$

Since we neglected W_3 it is thus consistent to neglect all terms of order $O(\kappa\varepsilon)$.

Suppose that $C^{ijkl}{}_{;\beta} = 0$ (this holds for instance in the case of uniform isotropy, which will be discussed in the next section). Then, from (5.196) it follows that

$$\nabla\bar{\boldsymbol{a}} + (\nabla\boldsymbol{u})\boldsymbol{\kappa} = -\boldsymbol{\rho} - S_{\lambda\alpha\beta}\, \boldsymbol{A}_n^{-1}\big(\underline{C}\big[\boldsymbol{a}^\lambda \otimes \boldsymbol{a}^\alpha\big]\big)\boldsymbol{n} \otimes \boldsymbol{a}^\beta \,+\, O(\kappa\varepsilon). \tag{5.200}$$

Inserting this in (5.150), we obtain

$$\bar{\boldsymbol{H}}_0' \;=\; \bar{\boldsymbol{K}}_0 \,+\, O(\kappa\varepsilon), \tag{5.201}$$

with

$$\bar{\boldsymbol{K}}_0 \,:=\, -\boldsymbol{\rho} - S_{\lambda\alpha\beta}\, \boldsymbol{A}_n^{-1}\big(\underline{C}\big[\boldsymbol{a}^\lambda \otimes \boldsymbol{a}^\alpha\big]\big)\boldsymbol{n} \otimes \boldsymbol{a}^\beta \,+\, \bar{\boldsymbol{b}} \otimes \boldsymbol{n}, \tag{5.202}$$

where, to consistent order, we have

$$\begin{aligned} \bar{\boldsymbol{b}} = S_{\lambda\alpha\beta}\, \boldsymbol{A}_n^{-1}\big\{C^{\gamma3\lambda\alpha}\big(\underline{C}\big[\boldsymbol{A}_n^{-1}\boldsymbol{a}_\gamma \otimes \boldsymbol{a}^\beta\big]\big)\boldsymbol{n} + C^{33\lambda\alpha}\big(\underline{C}\big[\boldsymbol{A}_n^{-1}\boldsymbol{n} \otimes \boldsymbol{a}^\beta\big]\big)\boldsymbol{n}\big\} \\ + \boldsymbol{A}_n^{-1}\big(\underline{C}\big[\boldsymbol{\rho}\big]\big)\boldsymbol{n}\,. \end{aligned} \tag{5.203}$$

Let us prove the relation (5.203): in view of the Eqs. (5.200) and $(5.75)_2$, we have

$$\begin{aligned} \boldsymbol{A}_n\bar{\boldsymbol{b}} \;=\; -\big(\underline{C}[\nabla\bar{\boldsymbol{a}} + (\nabla\boldsymbol{u})\boldsymbol{\kappa}]\big)\boldsymbol{n} \;=\; \big(\underline{C}[\boldsymbol{\rho}]\big)\boldsymbol{n} \\ + S_{\lambda\alpha\beta}\Big(\underline{C}\Big[\boldsymbol{A}_n^{-1}\big(\underline{C}[\boldsymbol{a}^\lambda \otimes \boldsymbol{a}^\alpha]\big)\boldsymbol{n} \otimes \boldsymbol{a}^\beta\Big]\Big)\boldsymbol{n} + O(\kappa\varepsilon). \end{aligned}$$

Substituting here (5.197) we get

$$\begin{aligned} \boldsymbol{A}_n\bar{\boldsymbol{b}} \;=\; \big(\underline{C}[\boldsymbol{\rho}]\big)\boldsymbol{n} \\ + S_{\lambda\alpha\beta}\,\big\{C^{\gamma3\lambda\alpha}\big(\underline{C}\big[\boldsymbol{A}_n^{-1}\boldsymbol{a}_\gamma \otimes \boldsymbol{a}^\beta\big]\big)\boldsymbol{n} + C^{33\lambda\alpha}\big(\underline{C}\big[\boldsymbol{A}_n^{-1}\boldsymbol{n} \otimes \boldsymbol{a}^\beta\big]\big)\boldsymbol{n}\big\} + O(\kappa\varepsilon), \end{aligned}$$

i.e., the Eq. (5.203) holds.

Thus, by using (5.201) in (5.146) and neglecting all the terms of order $O(\kappa\varepsilon)$, we obtain the strain energy density in the form

$$W = \frac{1}{2} h\, \bar{\boldsymbol{H}}_0 \cdot \underline{\boldsymbol{C}}[\bar{\boldsymbol{H}}_0] + \frac{1}{24} h^3\, \bar{\boldsymbol{K}}_0 \cdot \underline{\boldsymbol{C}}[\bar{\boldsymbol{K}}_0], \tag{5.204}$$

where $\bar{\boldsymbol{K}}_0$ is given by (5.202).

5.6 Isotropy

In case of isotropy, the tensor of elastic moduli $\underline{\boldsymbol{C}}$ satisfies

$$\underline{\boldsymbol{C}}[\boldsymbol{T}] = \lambda(\operatorname{tr} \boldsymbol{T})\, \boldsymbol{I} + 2\mu \operatorname{sym} \boldsymbol{T}, \tag{5.205}$$

for any second-order tensor $\boldsymbol{T}$. Then, on the basis of (5.76), we can perform similar calculations as in (4.106) and derive

$$\boldsymbol{A}_n \boldsymbol{w} = \big(\underline{\boldsymbol{C}}[\boldsymbol{w} \otimes \boldsymbol{n}]\big)\boldsymbol{n} = \big[(\lambda + 2\mu)\boldsymbol{n} \otimes \boldsymbol{n} + \mu\boldsymbol{1}\big]\boldsymbol{w},$$

for any vector $\boldsymbol{w}$. Thus, we obtain the acoustic tensor in the form

$$\boldsymbol{A}_n = C\, \boldsymbol{n} \otimes \boldsymbol{n} + \mu\boldsymbol{1}, \qquad \text{where} \quad C = \lambda + 2\mu \tag{5.206}$$

and λ, μ are the Lamé constants of the elastic material. Hence, we have

$$\boldsymbol{A}_n^{-1} = C^{-1}\boldsymbol{n} \otimes \boldsymbol{n} + \mu^{-1}\boldsymbol{1}. \tag{5.207}$$

Also,

$$\underline{\boldsymbol{C}}[\boldsymbol{\varepsilon}] = \lambda(\operatorname{tr} \boldsymbol{\varepsilon})\, \boldsymbol{I} + 2\mu \operatorname{sym} \boldsymbol{\varepsilon} = \lambda(\operatorname{tr} \boldsymbol{\varepsilon})\, \boldsymbol{I} + 2\mu \boldsymbol{\varepsilon} \tag{5.208}$$

and for the components of $\underline{\boldsymbol{C}}$ given by (5.198) at $\zeta = 0$ it holds

$$\begin{aligned} &C^{ijkl} = \lambda\, \bar{g}^{ij} \bar{g}^{kl} + \mu\big(\bar{g}^{ik} \bar{g}^{jl} + \bar{g}^{il} \bar{g}^{kj}\big), \qquad \text{where} \\ &\bar{g}^{\alpha\beta} = a^{\alpha\beta}, \qquad \bar{g}^{33} = 1, \qquad \bar{g}^{\alpha 3} = \bar{g}^{3\alpha} = 0. \end{aligned} \tag{5.209}$$

Notice that in this case we have indeed $C^{ijkl}{}_{;\beta} = 0$, according to (2.88). From (5.209) we deduce that

$$C^{\alpha\beta\gamma\delta} = \lambda\, a^{\alpha\beta} a^{\gamma\delta} + \mu\big(a^{\alpha\gamma} a^{\beta\delta} + a^{\alpha\delta} a^{\beta\gamma}\big) \tag{5.210}$$

and

$$\begin{aligned} C^{33\gamma\delta} &= \lambda\, \bar{g}^{33} a^{\gamma\delta} + \mu\big(\bar{g}^{3\gamma}\,\bar{g}^{3\delta} + \bar{g}^{3\delta}\bar{g}^{\gamma 3}\big) = \lambda\, a^{\gamma\delta},\\ C^{3\beta 3\delta} &= \lambda\, \bar{g}^{3\beta}\bar{g}^{3\delta} + \mu\big(\bar{g}^{33} a^{\beta\delta} + \bar{g}^{3\delta}\bar{g}^{3\beta}\big) = \mu\, a^{\beta\delta},\\ C^{\alpha 3\gamma\delta} &= 0, \qquad C^{333\delta} = 0. \end{aligned} \tag{5.211}$$

Taking into account (5.207), (5.208) and (5.211), we have

$$\underline{\boldsymbol{C}}[\boldsymbol{\varepsilon}]\boldsymbol{n}\otimes\boldsymbol{n} \;=\; C^{33\alpha\beta}\varepsilon_{\alpha\beta}\boldsymbol{n}\otimes\boldsymbol{n} \qquad \text{and} \qquad \boldsymbol{A}_n^{-1}\boldsymbol{n} \;=\; C^{-1}\boldsymbol{n}. \tag{5.212}$$

Then, from (5.105) we get

$$\begin{aligned} \operatorname{sym}\bar{\boldsymbol{H}}_0 \;&=\; \boldsymbol{\varepsilon} \;-\; C^{-1}\, C^{33\alpha\beta}\varepsilon_{\alpha\beta}\boldsymbol{n}\otimes\boldsymbol{n}\\ &=\; \boldsymbol{\varepsilon} \;-\; \frac{\lambda}{\lambda+2\mu}\,(\operatorname{tr}\boldsymbol{\varepsilon})\,\boldsymbol{n}\otimes\boldsymbol{n}\,. \end{aligned} \tag{5.213}$$

Further, we can write

$$\underline{\boldsymbol{C}}[\bar{\boldsymbol{H}}_0] \;=\; \big(\underline{\boldsymbol{C}}[\bar{\boldsymbol{H}}_0]\big)\mathbf{1} \;=\; \underline{\mathscr{M}}[\boldsymbol{\varepsilon}], \tag{5.214}$$

where $\underline{\mathscr{M}}$ is the fourt-order tensor of the plane–stress moduli given by

$$\underline{\mathscr{M}} \;=\; \mathscr{M}^{\alpha\beta\gamma\delta}\,\boldsymbol{a}_\alpha\otimes\boldsymbol{a}_\beta\otimes\boldsymbol{a}_\gamma\otimes\boldsymbol{a}_\delta \tag{5.215}$$

with

$$\mathscr{M}^{\alpha\beta\gamma\delta} \;=\; C^{\alpha\beta\gamma\delta} \;-\; C^{-1}\, C^{\alpha\beta 33}\, C^{\gamma\delta 33}\,. \tag{5.216}$$

Indeed, to verify these relations, we use the Eqs. (5.59) and (5.77) to write

$$\underline{\boldsymbol{C}}[\bar{\boldsymbol{H}}_0] \;=\; \bar{\boldsymbol{P}}_0 \;=\; \bar{\boldsymbol{P}}_0\mathbf{1} \;=\; \big(\underline{\boldsymbol{C}}[\bar{\boldsymbol{H}}_0]\big)\mathbf{1}$$

and from (5.213) and (5.211) we get

$$\begin{aligned} \big(\underline{\boldsymbol{C}}[\bar{\boldsymbol{H}}_0]\big)\mathbf{1} \;&=\; \big(\underline{\boldsymbol{C}}[\operatorname{sym}\bar{\boldsymbol{H}}_0]\big)\mathbf{1} \;=\; \big(\underline{\boldsymbol{C}}[\boldsymbol{\varepsilon} \;-\; C^{-1}\, C^{33\gamma\delta}\varepsilon_{\gamma\delta}\boldsymbol{n}\otimes\boldsymbol{n}]\big)\mathbf{1}\\ &=\; \big(\underline{\boldsymbol{C}}[\boldsymbol{\varepsilon}]\big)\mathbf{1} \;-\; C^{-1}\, C^{33\gamma\delta}\varepsilon_{\gamma\delta}\big(\underline{\boldsymbol{C}}[\boldsymbol{n}\otimes\boldsymbol{n}]\big)\mathbf{1}\\ &=\; C^{\alpha\beta\gamma\delta}\,\varepsilon_{\gamma\delta}\,\boldsymbol{a}_\alpha\otimes\boldsymbol{a}_\beta \;-\; C^{-1}\, C^{33\gamma\delta}\varepsilon_{\gamma\delta}C^{\alpha\beta 33}\,\boldsymbol{a}_\alpha\otimes\boldsymbol{a}_\beta\\ &=\; \big[\big(C^{\alpha\beta\gamma\delta} \;-\; C^{-1}\, C^{\alpha\beta 33}\, C^{\gamma\delta 33}\big)\boldsymbol{a}_\alpha\otimes\boldsymbol{a}_\beta\otimes\boldsymbol{a}_\gamma\otimes\boldsymbol{a}_\delta\big][\boldsymbol{\varepsilon}] \;=\; \underline{\mathscr{M}}[\boldsymbol{\varepsilon}], \end{aligned}$$

so the relations (5.214)–(5.216) hold true. Inserting (5.210) and (5.211) into (5.216), we deduce the following expression for the plane–stress moduli

$$\mathscr{M}^{\alpha\beta\gamma\delta} \;=\; \frac{2\lambda\mu}{\lambda+2\mu}\; a^{\alpha\beta}a^{\gamma\delta} + \mu\big(a^{\alpha\gamma}a^{\beta\delta} + a^{\alpha\delta}a^{\beta\gamma}\big). \tag{5.217}$$

By virtue of (5.213) and (5.214), we can write for the membrane part of the energy

$$\bar{\boldsymbol{H}}_0 \cdot \underline{\boldsymbol{C}}[\bar{\boldsymbol{H}}_0] \;=\; \boldsymbol{\varepsilon} \cdot \underline{\mathscr{M}}[\boldsymbol{\varepsilon}]\,. \tag{5.218}$$

For the bending part of the energy we proceed similarly. From (5.207) and (5.211), we derive that

$$\boldsymbol{A}_n^{-1}\big(\underline{\boldsymbol{C}}\big[\boldsymbol{a}^\lambda \otimes \boldsymbol{a}^\alpha\big]\big)\boldsymbol{n} \;=\; \boldsymbol{A}_n^{-1}\big(C^{33\lambda\alpha}\boldsymbol{n}\big) \;=\; C^{-1}\, C^{33\lambda\alpha}\, \boldsymbol{n} \tag{5.219}$$

and

$$\begin{aligned}\big(\underline{\boldsymbol{C}}\big[\boldsymbol{A}_n^{-1}\boldsymbol{n} \otimes \boldsymbol{a}^\beta\big]\big)\boldsymbol{n} \;&=\; C^{-1}\big(\underline{\boldsymbol{C}}\big[\boldsymbol{n} \otimes \boldsymbol{a}^\beta\big]\big)\boldsymbol{n} \;=\; C^{-1}\big(C^{3\gamma 3\beta}\, \boldsymbol{a}_\gamma\big)\\ &=\; C^{-1}\, \mu\, a^{\gamma\beta}\, \boldsymbol{a}_\gamma \;=\; C^{-1}\, \mu\, \boldsymbol{a}^\beta\,.\end{aligned} \tag{5.220}$$

Using the last relation in (5.203) we get

$$\bar{\boldsymbol{b}} \;=\; S_{\lambda\alpha\beta}\, C^{-1} C^{33\lambda\alpha}\, \boldsymbol{a}^\beta \;+\; \boldsymbol{A}_n^{-1}\big(\underline{\boldsymbol{C}}\big[\boldsymbol{\rho}\big]\big)\boldsymbol{n}\,. \tag{5.221}$$

If we substitute (5.219) and (5.221) into (5.202), then we deduce

$$\bar{\boldsymbol{K}}_0 \;=\; -\boldsymbol{\rho} \;-\; C^{-1} C^{33\lambda\alpha}\, S_{\lambda\alpha\beta}\, \big(\boldsymbol{n} \otimes \boldsymbol{a}^\beta - \boldsymbol{a}^\beta \otimes \boldsymbol{n}\big) \;+\; \boldsymbol{A}_n^{-1}\big(\underline{\boldsymbol{C}}\big[\boldsymbol{\rho}\big]\big)\boldsymbol{n} \otimes \boldsymbol{n}\,. \tag{5.222}$$

Taking into account that

$$\boldsymbol{A}_n^{-1}\big(\underline{\boldsymbol{C}}\big[\boldsymbol{\rho}\big]\big)\boldsymbol{n} \otimes \boldsymbol{n} \;=\; \boldsymbol{A}_n^{-1}\big(C^{33\alpha\beta}\, \rho_{\alpha\beta}\, \boldsymbol{n} \otimes \boldsymbol{n}\big) \;=\; C^{-1} C^{33\alpha\beta}\, \rho_{\alpha\beta}\, \boldsymbol{n} \otimes \boldsymbol{n},$$

from the relation (5.222) and the symmetry of $\boldsymbol{\rho}$ we infer

$$\begin{aligned}\operatorname{sym} \bar{\boldsymbol{K}}_0 \;&=\; -\boldsymbol{\rho} \;+\; C^{-1} C^{33\alpha\beta}\, \rho_{\alpha\beta}\, \boldsymbol{n} \otimes \boldsymbol{n}\\ &=\; -\boldsymbol{\rho} \;+\; \frac{\lambda}{\lambda + 2\mu}\, (\operatorname{tr} \boldsymbol{\rho})\, \boldsymbol{n} \otimes \boldsymbol{n}\,.\end{aligned} \tag{5.223}$$

Notice that this relation depends only on the bending measure $\boldsymbol{\rho}$ and has the same structure as Eq. (5.213).

On the basis of relations (5.201), (5.59) and (5.77) we can write

$$\begin{aligned}\underline{\boldsymbol{C}}[\bar{\boldsymbol{K}}_0] \;&=\; \underline{\boldsymbol{C}}[\bar{\boldsymbol{H}}_0'] + O(\kappa\varepsilon) \;=\; \bar{\boldsymbol{P}}_0' + O(\kappa\varepsilon) \;=\; \bar{P}_0'\mathbf{1} + O(\kappa\varepsilon)\\ &=\; \big(\underline{\boldsymbol{C}}[\bar{\boldsymbol{H}}_0']\big)\mathbf{1} + O(\kappa\varepsilon) \;=\; \big(\underline{\boldsymbol{C}}[\bar{\boldsymbol{K}}_0]\big)\mathbf{1} + O(\kappa\varepsilon).\end{aligned} \tag{5.224}$$

Then, to consistent order, we derive that

$$\bar{\boldsymbol{K}}_0 \cdot \underline{\boldsymbol{C}}[\bar{\boldsymbol{K}}_0] \;=\; \boldsymbol{\rho} \cdot \underline{\mathscr{M}}[\boldsymbol{\rho}], \tag{5.225}$$

where $\underline{\mathcal{M}}$ is given by (5.215). This relation can be proved on the basis of (5.223) and (5.224), in the same way as the Eq. (5.218).

Thus, by virtue of (5.218) and (5.225), the strain energy density (5.204) can be written in the form

$$W = \frac{1}{2} h\, \boldsymbol{\varepsilon} \cdot \underline{\mathcal{M}}[\boldsymbol{\varepsilon}] + \frac{1}{24} h^3\, \boldsymbol{\rho} \cdot \underline{\mathcal{M}}[\boldsymbol{\rho}], \tag{5.226}$$

where the infinitesimal strain tensor $\boldsymbol{\varepsilon}$ is given by (5.169) and the bending tensor $\boldsymbol{\rho}$ by (5.174) or (5.178).

5.7 Equilibrium Equations and Boundary Conditions

Taking into account (5.226) with (5.168) and (5.170), we can put the strain energy density W in the form

$$W = F\big(\boldsymbol{u}_{,\alpha},\ \boldsymbol{u}_{;\alpha\beta}\big), \tag{5.227}$$

for some function F. To derive the equilibrium equations using the same method as in Sect. 4.3, we need to compute the variation

$$\dot{W} = \dot{F} = \boldsymbol{N}^{\alpha} \cdot \dot{\boldsymbol{u}}_{,\alpha} + \boldsymbol{M}^{\alpha\beta} \cdot \dot{\boldsymbol{u}}_{;\alpha\beta}, \tag{5.228}$$

where

$$\begin{aligned} \boldsymbol{N}^{\alpha} &:= \frac{\partial F}{\partial \boldsymbol{u}_{,\alpha}} = \frac{\partial F}{\partial u_{i,\alpha}}\, \boldsymbol{e}_i, \\ \boldsymbol{M}^{\alpha\beta} &:= \frac{\partial F}{\partial \boldsymbol{u}_{;\alpha\beta}} = \frac{\partial F}{\partial u_{i;\alpha\beta}}\, \boldsymbol{e}_i = \boldsymbol{M}^{\beta\alpha}, \end{aligned} \tag{5.229}$$

with $\boldsymbol{u} = u_i \boldsymbol{e}_i$. The fields $\boldsymbol{N}^{\alpha}$ and $\boldsymbol{M}^{\alpha\beta}$ can be obtained from the explicit expression (5.226), as we shall see later in (5.249).

Note that the variation can be written in the alternative form

$$\dot{W} = \dot{F} = \varphi^{\alpha}{}_{;\alpha} - \dot{\boldsymbol{u}} \cdot \boldsymbol{T}^{\alpha}{}_{;\alpha}\,, \tag{5.230}$$

where we have denoted by

$$\boldsymbol{T}^{\alpha} := \boldsymbol{N}^{\alpha} - \boldsymbol{M}^{\beta\alpha}{}_{;\beta} \quad \text{and} \quad \varphi^{\alpha} := \boldsymbol{T}^{\alpha} \cdot \dot{\boldsymbol{u}} + \boldsymbol{M}^{\alpha\beta} \cdot \dot{\boldsymbol{u}}_{,\beta}\,. \tag{5.231}$$

Here, we employ the usual notations

$$\begin{aligned} \boldsymbol{M}^{\beta\alpha}{}_{;\beta} &= \boldsymbol{M}^{\beta\alpha}{}_{,\beta} + \boldsymbol{M}^{\beta\alpha}\,\Gamma^{\lambda}_{\lambda\beta} + \boldsymbol{M}^{\beta\lambda}\,\Gamma^{\alpha}_{\lambda\beta}\,, \\ \boldsymbol{T}^{\alpha}{}_{;\alpha} &= \boldsymbol{T}^{\alpha}{}_{,\alpha} + \boldsymbol{T}^{\beta}\,\Gamma^{\alpha}_{\alpha\beta} = \frac{1}{\sqrt{a}}\big(\sqrt{a}\,\boldsymbol{T}^{\alpha}\big)_{,\alpha}\,, \end{aligned} \tag{5.232}$$

according to (2.87) and (2.90). Indeed, let us prove the relation (5.230): in view of (5.228) and (5.231) we have

$$\begin{aligned} \dot{F} &= \big(\boldsymbol{T}^{\alpha} + \boldsymbol{M}^{\beta\alpha}{}_{;\beta}\big)\cdot\dot{\boldsymbol{u}}_{,\alpha} + \boldsymbol{M}^{\alpha\beta}\cdot\dot{\boldsymbol{u}}_{;\alpha\beta} = \big(\boldsymbol{T}^{\alpha}\cdot\dot{\boldsymbol{u}}\big)_{;\alpha} - \boldsymbol{T}^{\alpha}{}_{;\alpha}\cdot\dot{\boldsymbol{u}} + \big(\boldsymbol{M}^{\alpha\beta}\cdot\dot{\boldsymbol{u}}_{,\beta}\big)_{;\alpha} \\ &= \big(\boldsymbol{T}^{\alpha}\cdot\dot{\boldsymbol{u}} + \boldsymbol{M}^{\alpha\beta}\cdot\dot{\boldsymbol{u}}_{,\beta}\big)_{;\alpha} - \boldsymbol{T}^{\alpha}{}_{;\alpha}\cdot\dot{\boldsymbol{u}} = \varphi^{\alpha}{}_{;\alpha} - \dot{\boldsymbol{u}}\cdot\boldsymbol{T}^{\alpha}{}_{;\alpha}\,, \end{aligned}$$

so the Eq. (5.230) holds true.

For the potential energy E we have the form (cf. (5.147))

$$E = \int_{\Omega}\big(W - \boldsymbol{g}\cdot\boldsymbol{u}\big)\,\mathrm{d}a - \int_{\partial\Omega_t}\big(\boldsymbol{p}_u\cdot\boldsymbol{u} + \boldsymbol{p}_a\cdot\boldsymbol{a}\big)\,\mathrm{d}s\,. \tag{5.233}$$

Using (5.233) and (5.230), we obtain for an equilibrium state

$$\begin{aligned} 0 = \dot{E} &= \int_{\Omega}\big[\varphi^{\alpha}{}_{;\alpha} - \dot{\boldsymbol{u}}\cdot\big(\boldsymbol{T}^{\alpha}{}_{;\alpha} + \boldsymbol{g}\big)\big]\,\mathrm{d}a - \int_{\partial\Omega_t}\big(\boldsymbol{p}_u\cdot\dot{\boldsymbol{u}} + \boldsymbol{p}_a\cdot\dot{\boldsymbol{a}}\big)\,\mathrm{d}s \\ &= \int_{\partial\Omega}\varphi^{\alpha}\,\nu_{\alpha}\,\mathrm{d}s - \int_{\partial\Omega_t}\big(\boldsymbol{p}_u\cdot\dot{\boldsymbol{u}} + \boldsymbol{p}_a\cdot\dot{\boldsymbol{a}}\big)\,\mathrm{d}s - \int_{\Omega}\dot{\boldsymbol{u}}\cdot\big(\boldsymbol{T}^{\alpha}{}_{;\alpha} + \boldsymbol{g}\big)\,\mathrm{d}a\,. \end{aligned} \tag{5.234}$$

This relation gives the equilibrium equation in vector form

$$\boldsymbol{T}^{\alpha}{}_{;\alpha} + \boldsymbol{g} = \boldsymbol{0} \quad \text{in } \Omega. \tag{5.235}$$

To deconstruct this, let us denote the components of $\boldsymbol{N}^{\alpha}$ and $\boldsymbol{M}^{\alpha\beta}$ as follows

$$\boldsymbol{N}^{\alpha} = N^{\beta\alpha}\boldsymbol{a}_{\beta} + N^{\alpha}\boldsymbol{n}, \qquad \boldsymbol{M}^{\alpha\beta} = M^{\lambda\alpha\beta}\boldsymbol{a}_{\lambda} + M^{\alpha\beta}\boldsymbol{n}\,. \tag{5.236}$$

We insert this in $(5.231)_1$ and invoke relations of the type $\boldsymbol{n}_{,\alpha} = -\kappa^{\beta}_{\alpha}\,\boldsymbol{a}_{\beta}$ and $\boldsymbol{a}_{\beta;\alpha} = \kappa_{\beta\alpha}\,\boldsymbol{n}$ (see Gauss equations (2.76)) to write

$$\begin{aligned} \boldsymbol{T}^{\alpha} &= \big(N^{\lambda\alpha}\boldsymbol{a}_{\lambda} + N^{\alpha}\boldsymbol{n}\big) - \big(M^{\lambda\beta\alpha}\boldsymbol{a}_{\lambda} + M^{\beta\alpha}\boldsymbol{n}\big)_{;\beta} \\ &= \big(N^{\lambda\alpha} + M^{\beta\alpha}\,\kappa^{\lambda}_{\beta} - M^{\lambda\beta\alpha}{}_{;\beta}\big)\boldsymbol{a}_{\lambda} + \big(N^{\alpha} - M^{\beta\alpha}{}_{;\beta} - M^{\lambda\beta\alpha}\,\kappa_{\lambda\beta}\big)\boldsymbol{n}\,. \end{aligned} \tag{5.237}$$

Substituting $\boldsymbol{T}^{\alpha}$ into (5.235) and projecting the vectorial equation onto the tangent plane T_{Ω} and the normal direction $\boldsymbol{n}$, we obtain the equilibrium equations in component form

$$\big(N^{\gamma\alpha} + M^{\beta\alpha}\,\kappa^{\gamma}_{\beta} - M^{\gamma\beta\alpha}{}_{;\beta}\big)_{;\alpha} + \big(M^{\beta\alpha}{}_{;\beta} + M^{\lambda\beta\alpha}\,\kappa_{\lambda\beta} - N^{\alpha}\big)\kappa^{\gamma}_{\alpha} + g^{\gamma} = 0,$$
$$\big(N^{\alpha} - M^{\beta\alpha}{}_{;\beta} - M^{\lambda\beta\alpha}\,\kappa_{\lambda\beta}\big)_{;\alpha} + \big(N^{\beta\alpha} + M^{\lambda\alpha}\,\kappa^{\beta}_{\lambda} - M^{\beta\gamma\alpha}{}_{;\gamma}\big)\kappa_{\beta\alpha} + g = 0, \tag{5.238}$$

where $\boldsymbol{g} = g^{\gamma}\boldsymbol{a}_{\gamma} + g\,\boldsymbol{n}$.

To derive the boundary conditions, we employ (5.235) in (5.234) and get

$$0 = \dot{E} = \int_{\partial\Omega} \varphi^{\alpha}\,\nu_{\alpha}\,\mathrm{d}s - \int_{\partial\Omega_t} \big(\boldsymbol{p}_u\cdot\dot{\boldsymbol{u}} + \boldsymbol{p}_a\cdot\dot{\boldsymbol{a}}\big)\,\mathrm{d}s\,. \tag{5.239}$$

Here, we have

$$\dot{\boldsymbol{a}} = -\boldsymbol{A}_n^{-1}\big(\underline{\boldsymbol{C}}[\nabla\dot{\boldsymbol{u}}]\big)\boldsymbol{n} \qquad \text{with} \qquad \nabla\dot{\boldsymbol{u}} = \dot{\boldsymbol{u}}_{,s}\otimes\boldsymbol{\tau} + \dot{\boldsymbol{u}}_{,\nu}\otimes\boldsymbol{\nu},$$

according to (5.94). Then, if we integrate by parts in (5.239) and take into account that $\dot{\boldsymbol{u}} = \boldsymbol{0}$, $\dot{\boldsymbol{u}}_{,\nu} = \boldsymbol{0}$ on $\partial\Omega_u$, then we obtain in a similar way as in the case of plates (see (4.79)–(4.81)) that

$$\int_{\partial\Omega_t} (\boldsymbol{p}_u\cdot\dot{\boldsymbol{u}} + \boldsymbol{p}_a\cdot\dot{\boldsymbol{a}})\,\mathrm{d}s = \int_{\partial\Omega_t} \big(\boldsymbol{f}\cdot\dot{\boldsymbol{u}} + \boldsymbol{c}\cdot\dot{\boldsymbol{u}}_{,\nu}\big)\,\mathrm{d}s, \tag{5.240}$$

for some given vector fields $\boldsymbol{f}$, $\boldsymbol{c}$, which are independent of $\boldsymbol{u}$, $\boldsymbol{u}_{,\nu}$, and are regarded as assigned.

Remark To derive the relation (5.240) we have assumed that the boundary $\partial\Omega$ is *smooth*. Otherwise, we have to consider "corner forces". □

In view of definition $(5.231)_2$, the first integral in (5.239) can be written as

$$\int_{\partial\Omega} \varphi^{\alpha}\,\nu_{\alpha}\,\mathrm{d}s = \int_{\partial\Omega} \big(\dot{\boldsymbol{u}}\cdot\boldsymbol{T}^{\alpha}\,\nu_{\alpha} + \dot{\boldsymbol{u}}_{,\beta}\cdot\boldsymbol{M}^{\alpha\beta}\,\nu_{\alpha}\big)\,\mathrm{d}s, \tag{5.241}$$

where

$$\dot{\boldsymbol{u}}_{,\beta} = (\nabla\dot{\boldsymbol{u}})\boldsymbol{a}_{\beta} = \big(\dot{\boldsymbol{u}}_{,s}\otimes\boldsymbol{\tau} + \dot{\boldsymbol{u}}_{,\nu}\otimes\boldsymbol{\nu}\big)\boldsymbol{a}_{\beta} = \dot{\boldsymbol{u}}_{,s}\,\tau_{\beta} + \dot{\boldsymbol{u}}_{,\nu}\,\nu_{\beta}\,.$$

Inserting the latter into (5.241) and integrating by parts, we derive as before (see (4.101) for plates)

$$\int_{\partial\Omega} \varphi^{\alpha}\,\nu_{\alpha}\,\mathrm{d}s = \int_{\partial\Omega} \big\{\big[\boldsymbol{T}^{\alpha}\nu_{\alpha} - \big(\boldsymbol{M}^{\alpha\beta}\nu_{\alpha}\tau_{\beta}\big)_{,s}\big]\cdot\dot{\boldsymbol{u}} + \big(\boldsymbol{M}^{\alpha\beta}\nu_{\alpha}\nu_{\beta}\big)\cdot\dot{\boldsymbol{u}}_{,\beta}\big\}\mathrm{d}s, \tag{5.242}$$

for a smooth boundary $\partial\Omega$. By virtue of (5.240) and (5.242), the relation (5.239) yields

$$\int_{\partial\Omega_t} \{[\boldsymbol{T}^\alpha \nu_\alpha - (\boldsymbol{M}^{\alpha\beta}\nu_\alpha\tau_\beta)_{,s}] \cdot \dot{\boldsymbol{u}} + (\boldsymbol{M}^{\alpha\beta}\nu_\alpha\nu_\beta)\cdot \dot{\boldsymbol{u}}_{,\beta}\}\mathrm{d}s \\ = \int_{\partial\Omega_t} (\boldsymbol{f}\cdot\dot{\boldsymbol{u}} + \boldsymbol{c}\cdot\dot{\boldsymbol{u}}_{,\nu})\,\mathrm{d}s, \tag{5.243}$$

since $\dot{\boldsymbol{u}} = \boldsymbol{0}$, $\dot{\boldsymbol{u}}_{,\nu} = \boldsymbol{0}$ on $\partial\Omega_u$. This equation gives the boundary conditions in the following form

$$\boldsymbol{T}^\alpha\nu_\alpha - (\boldsymbol{M}^{\alpha\beta}\nu_\alpha\tau_\beta)_{,s} = \boldsymbol{f} \quad \text{and} \quad \boldsymbol{M}^{\alpha\beta}\nu_\alpha\nu_\beta = \boldsymbol{c}. \tag{5.244}$$

Next, we want to express the fields $\boldsymbol{N}^\alpha$ and $\boldsymbol{M}^{\alpha\beta}$ in terms of kinematics. In view of (5.226) and (5.228) we have

$$\boldsymbol{N}^\alpha\cdot\dot{\boldsymbol{u}}_{,\alpha} + \boldsymbol{M}^{\alpha\beta}\cdot\dot{\boldsymbol{u}}_{;\alpha\beta} = \dot{W} = h\,\underline{\mathscr{M}}[\boldsymbol{\varepsilon}]\cdot\dot{\boldsymbol{\varepsilon}} + \frac{1}{12}h^3\,\underline{\mathscr{M}}[\boldsymbol{\rho}]\cdot\dot{\boldsymbol{\rho}}, \tag{5.245}$$

where $\dot{\boldsymbol{\varepsilon}} = \dot{\varepsilon}_{\alpha\beta}\boldsymbol{a}^\alpha\otimes\boldsymbol{a}^\beta$ and $\dot{\boldsymbol{\rho}} = \dot{\rho}_{\alpha\beta}\boldsymbol{a}^\alpha\otimes\boldsymbol{a}^\beta$ with

$$\dot{\varepsilon}_{\alpha\beta} = \frac{1}{2}(\boldsymbol{a}_\alpha\cdot\dot{\boldsymbol{u}}_{,\beta} + \boldsymbol{a}_\beta\cdot\dot{\boldsymbol{u}}_{,\alpha}), \qquad \dot{\rho}_{\alpha\beta} = \boldsymbol{n}\cdot\dot{\boldsymbol{u}}_{;\alpha\beta}. \tag{5.246}$$

Thus,

$$\underline{\mathscr{M}}[\boldsymbol{\rho}]\cdot\dot{\boldsymbol{\rho}} = \underline{\mathscr{M}}[\boldsymbol{\rho}]\cdot(\dot{\rho}_{\alpha\beta}\boldsymbol{a}^\alpha\otimes\boldsymbol{a}^\beta) = \Big(\boldsymbol{a}^\alpha\cdot(\underline{\mathscr{M}}[\boldsymbol{\rho}])\boldsymbol{a}^\beta\Big)\dot{\rho}_{\alpha\beta} \\ = \Big(\boldsymbol{a}^\alpha\cdot(\underline{\mathscr{M}}[\boldsymbol{\rho}])\boldsymbol{a}^\beta\Big)\boldsymbol{n}\cdot\dot{\boldsymbol{u}}_{;\alpha\beta}. \tag{5.247}$$

Also, by the minor symmetries of $\underline{\mathscr{M}}$, we can write

$$\begin{aligned} \underline{\mathscr{M}}[\boldsymbol{\varepsilon}]\cdot\dot{\boldsymbol{\varepsilon}} &= \frac{1}{2}\Big(\underline{\mathscr{M}}[\boldsymbol{\varepsilon}]\cdot(\boldsymbol{a}^\alpha\otimes\boldsymbol{a}^\beta)\Big)(\boldsymbol{a}_\alpha\cdot\dot{\boldsymbol{u}}_{,\beta} + \boldsymbol{a}_\beta\cdot\dot{\boldsymbol{u}}_{,\alpha}) \\ &= \frac{1}{2}\Big(\underline{\mathscr{M}}[\boldsymbol{\varepsilon}]\cdot\mathrm{sym}(\boldsymbol{a}^\alpha\otimes\boldsymbol{a}^\beta)\Big)(\boldsymbol{a}_\alpha\cdot\dot{\boldsymbol{u}}_{,\beta} + \boldsymbol{a}_\beta\cdot\dot{\boldsymbol{u}}_{,\alpha}) \\ &= \Big(\underline{\mathscr{M}}[\boldsymbol{\varepsilon}]\cdot\mathrm{sym}(\boldsymbol{a}^\alpha\otimes\boldsymbol{a}^\beta)\Big)(\boldsymbol{a}_\beta\cdot\dot{\boldsymbol{u}}_{,\alpha}) \\ &= \Big(\boldsymbol{a}^\alpha\cdot(\underline{\mathscr{M}}[\boldsymbol{\varepsilon}])\boldsymbol{a}^\beta\Big)\boldsymbol{a}_\beta\cdot\dot{\boldsymbol{u}}_{,\alpha}. \end{aligned} \tag{5.248}$$

Hence, using (5.247) and (5.248) in (5.245), we deduce

$$\begin{aligned} \boldsymbol{N}^\alpha &= h\Big(\boldsymbol{a}^\alpha\cdot(\underline{\mathscr{M}}[\boldsymbol{\varepsilon}])\boldsymbol{a}^\beta\Big)\boldsymbol{a}_\beta, \\ \boldsymbol{M}^{\alpha\beta} &= \frac{1}{12}h^3\Big(\boldsymbol{a}^\alpha\cdot(\underline{\mathscr{M}}[\boldsymbol{\rho}])\boldsymbol{a}^\beta\Big)\boldsymbol{n}. \end{aligned} \tag{5.249}$$

Taking into account the component representations $\boldsymbol{N}^{\alpha} = N^{\beta\alpha}\boldsymbol{a}_{\beta} + N^{\alpha}\boldsymbol{n}$ and $\boldsymbol{M}^{\alpha\beta} = M^{\lambda\alpha\beta}\boldsymbol{a}_{\lambda} + M^{\alpha\beta}\boldsymbol{n}$ introduced in (5.236), we obtain from (5.249) that

$$\begin{aligned} \boldsymbol{N}^{\alpha} &= N^{\beta\alpha}\boldsymbol{a}_{\beta} \text{ (and } N^{\alpha} = 0), \\ \boldsymbol{M}^{\alpha\beta} &= M^{\alpha\beta}\boldsymbol{n} \text{ (and } M^{\lambda\alpha\beta} = 0), \end{aligned} \tag{5.250}$$

where

$$\begin{aligned} N^{\alpha\beta} &= N^{\beta\alpha} = h\,\boldsymbol{a}^{\alpha}\cdot\big(\mathcal{M}[\boldsymbol{\varepsilon}]\big)\boldsymbol{a}^{\beta} \quad \text{and} \\ M^{\alpha\beta} &= M^{\beta\alpha} = \frac{1}{12}h^{3}\,\boldsymbol{a}^{\alpha}\cdot\big(\mathcal{M}[\boldsymbol{\rho}]\big)\boldsymbol{a}^{\beta}. \end{aligned} \tag{5.251}$$

Notice that $N^{\alpha\beta}$ is a function of $\boldsymbol{\varepsilon}$, while $M^{\alpha\beta}$ is a function of $\boldsymbol{\rho}$.

Remark The components N^{α} and $M^{\lambda\alpha\beta}$ vanish in the isotropic case, cf. (5.250). In general, note that N^{α} and $M^{\lambda\alpha\beta}$ are non-zero in the absence of reflection symmetry, see the paper [5]. □

Due to the relations (5.250), the equilibrium equations (5.238) simplify in the case of isotropic shells and reduce to

$$\begin{aligned} \big(N^{\gamma\alpha} + M^{\beta\alpha}\kappa_{\beta}^{\gamma}\big)_{;\alpha} + M^{\beta\alpha}{}_{;\beta}\,\kappa_{\alpha}^{\gamma} + g^{\gamma} &= 0, \\ \big(N^{\beta\alpha} + M^{\lambda\alpha}\kappa_{\lambda}^{\beta}\big)\kappa_{\beta\alpha} - M^{\beta\alpha}{}_{;\beta\alpha} + g &= 0. \end{aligned} \tag{5.252}$$

These equilibrium equations are also called the *Koiter equations* for linear shells, see e.g., [3, p. 341].

Note that the boundary conditions (5.244) can also be simplified in this case, as we shall see later in Sect. 5.9.

5.8 Classical Membrane Theory

The linear membrane theory retains only the terms of order $O(h)$, while the moments $M^{\alpha\beta}$ are neglected.

In this case, the equilibrium equations (5.252) reduce to

$$N^{\gamma\alpha}{}_{;\alpha} + g^{\gamma} = 0 \quad \text{and} \quad N^{\beta\alpha}\kappa_{\beta\alpha} + g = 0, \tag{5.253}$$

which can be written in the vectorial form

$$\boldsymbol{N}^{\alpha}{}_{;\alpha} + \boldsymbol{g} = \boldsymbol{0}. \tag{5.254}$$

Indeed, in view of (5.250) and (2.76),

$$\boldsymbol{N}^{\alpha}{}_{;\alpha} = \big(N^{\beta\alpha}\boldsymbol{a}_{\beta}\big)_{;\alpha} = N^{\beta\alpha}{}_{;\alpha}\,\boldsymbol{a}_{\beta} + N^{\beta\alpha}\boldsymbol{a}_{\beta;\alpha} = N^{\beta\alpha}{}_{;\alpha}\,\boldsymbol{a}_{\beta} + N^{\beta\alpha}\kappa_{\beta\alpha}\,\boldsymbol{n}.$$

Also, the symmetry condition $N^{\alpha\beta} = N^{\beta\alpha}$ can be written in the vectorial form

$$\boldsymbol{N}^{\alpha} \times \boldsymbol{a}_{\alpha} = \boldsymbol{0}, \tag{5.255}$$

since

$$\boldsymbol{N}^{\alpha} \times \boldsymbol{a}_{\alpha} = N^{\beta\alpha}\boldsymbol{a}_{\beta} \times \boldsymbol{a}_{\alpha} = N^{\beta\alpha}\,\tilde{\varepsilon}_{\beta\alpha}\,\boldsymbol{n} = \sqrt{a}\left(N^{12} - N^{21}\right)\boldsymbol{n}.$$

In the absence of distributed body loads, the above equilibrium equations can be reduced to the search of a single stress function, see Exercise 5.2.

The boundary conditions $(5.244)_1$ for linear membrane shells reduce to

$$\boldsymbol{N}^{\alpha}\nu_{\alpha} = \boldsymbol{f}, \qquad \text{i.e.} \qquad N^{\beta\alpha}\nu_{\alpha} = f^{\beta}. \tag{5.256}$$

The basic kinematic variable is

$$\varepsilon_{\alpha\beta} = \frac{1}{2}\left(\boldsymbol{a}_{\alpha}\cdot\boldsymbol{u}_{,\beta} + \boldsymbol{a}_{\beta}\cdot\boldsymbol{u}_{,\alpha}\right), \tag{5.257}$$

where $\boldsymbol{u}(\theta^{\alpha}) = u_{\alpha}\boldsymbol{a}^{\alpha} + w\,\boldsymbol{n}$ is the displacement field. Then, as before (cf. (5.101))

$$\varepsilon_{\alpha\beta} = \frac{1}{2}\left(u_{\alpha;\beta} + u_{\beta;\alpha}\right) - w\,\kappa_{\alpha\beta}\,. \tag{5.258}$$

In the isotropic case, we can use the foregoing relations to obtain

$$N^{\alpha\beta} = \frac{Eh}{1-\nu^2}\left[\nu\varepsilon_{\lambda}^{\lambda}\,a^{\alpha\beta} + (1-\nu)\varepsilon^{\alpha\beta}\right], \tag{5.259}$$

where E is the Young modulus, ν the Poisson ratio of the material, and $\varepsilon_{\beta}^{\alpha} = a^{\alpha\gamma}\varepsilon_{\gamma\beta}$, $\varepsilon^{\alpha\beta} = a^{\alpha\gamma}a^{\beta\delta}\varepsilon_{\gamma\delta}$ (raising the index, see (1.18)). Indeed, to verify (5.259) we write $(5.249)_1$ in the form

$$N^{\alpha\beta} = h\,\boldsymbol{a}^{\beta}\cdot\left(\underline{\mathscr{M}}[\boldsymbol{\varepsilon}]\right)\boldsymbol{a}^{\alpha}$$

and using (5.217) we find

$$N^{\alpha\beta} = h\left(\frac{2\lambda\mu}{\lambda+2\mu}\,\varepsilon_{\lambda}^{\lambda}\,a^{\alpha\beta} + 2\mu\varepsilon^{\alpha\beta}\right), \tag{5.260}$$

so the relation (5.259) holds, since $\frac{2\lambda\mu}{\lambda+2\mu} = \frac{E\nu}{1-\nu^2}$ and $\mu = \frac{E}{2(1+\nu)}$, see (4.135). Note that $\varepsilon_{\lambda}^{\lambda} = \text{tr}\boldsymbol{\varepsilon}$.

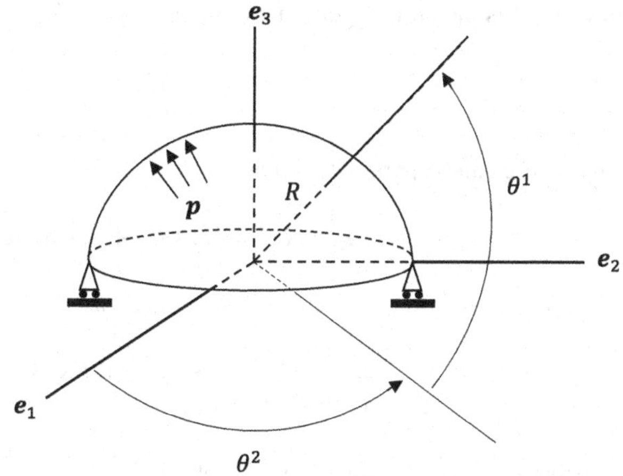

Fig. 5.4 Hemispherical membrane shell under constant lateral pressure

5.8.1 Equilibrium Under Normal Pressure

Let us consider that the applied distributed loads have the form $\boldsymbol{g} = p\,\boldsymbol{n}$ (normal pressure). Then, the relevant equilibrium equations are

$$N^{\alpha\beta}{}_{;\alpha} = 0, \qquad N^{\alpha\beta}\kappa_{\alpha\beta} + p = 0, \qquad \left(N^{\alpha\beta} = N^{\beta\alpha}\right). \tag{5.261}$$

Observe that, for $\kappa_{\alpha\beta} \neq 0$, these equations furnish a *statically determinate* system for $N^{\alpha\beta}\left(= N^{\beta\alpha}\right)$. This unusual feature is an artifact of linearization about a stress-free state.

When working in *orthogonal coordinates*, the mixed components N^{α}_{β} are also the *physical components* of the tensor (see (1.86)), so it is convenient to work with the system

$$N^{\alpha}_{\beta\,;\alpha} = 0, \qquad N^{\alpha}_{\beta}\,\kappa^{\beta}_{\alpha} + p = 0, \tag{5.262}$$

where we have clearly $N^{\alpha}_{\beta} = a_{\beta\gamma}\,N^{\alpha\gamma}$ and $\kappa^{\alpha}_{\beta} = a^{\alpha\gamma}\,\kappa_{\gamma\beta}$.

Example: Hemispherical shell under lateral pressure

A hemispherical shell with radius R that is roller-pinned to equator is under lateral constant pressure p (Fig. 5.4). Let us determine the displacement components u, w and the mixed stress components N^{α}_{β}.

We assume axisymmetry of the displacement field $\boldsymbol{u}$, i.e.,

$$\boldsymbol{u} = u\left(\theta^1\right)\boldsymbol{t} + w(\theta^1)\boldsymbol{n}, \tag{5.263}$$

in which $\boldsymbol{t}$ is the unit tangent to meridian. Thus

$$\boldsymbol{t} = \frac{\boldsymbol{a}_1}{|\boldsymbol{a}_1|} \,. \tag{5.264}$$

This sphere can be parametrized as

$$\boldsymbol{r} = R\cos\theta^1 \boldsymbol{e}_r\left(\theta^2\right) + R\sin\theta^1 \boldsymbol{e}_3 \tag{5.265}$$

and from this relation, we can derive $\boldsymbol{a}_1$ and $\boldsymbol{a}_2$ as

$$\begin{aligned} \boldsymbol{a}_1 &= -R\sin\theta^1 \boldsymbol{e}_r\left(\theta^2\right) + R\cos\theta^1 \boldsymbol{e}_3, \\ \boldsymbol{a}_2 &= R\cos\theta^1 \boldsymbol{e}_\theta\left(\theta^2\right). \end{aligned} \tag{5.266}$$

By using $\boldsymbol{a}_1$ and $\boldsymbol{a}_2$ we can find $a_{\alpha\beta},\ a^{\alpha\beta}$ and a as

$$a_{\alpha\beta} = R^2 \begin{pmatrix} 1 & 0 \\ 0 & \cos^2\theta^1 \end{pmatrix}, \qquad a^{\alpha\beta} = R^{-2} \begin{pmatrix} 1 & 0 \\ 0 & \frac{1}{\cos^2\theta^1} \end{pmatrix}, \qquad \sqrt{a} = R^2\cos\theta^1$$

and

$$\sqrt{a}\,\boldsymbol{n} = \boldsymbol{a}_1 \times \boldsymbol{a}_2, \quad \text{so} \quad \boldsymbol{n} = -\cos\theta^1 \boldsymbol{e}_r - \sin\theta^1 \boldsymbol{e}_3 \,. \tag{5.267}$$

Moreover, we have

$$\begin{aligned} \boldsymbol{a}_{1,1} &= -R(\cos\theta^1 \boldsymbol{e}_r + \sin\theta^1 \boldsymbol{e}_3), \\ \boldsymbol{a}_{1,2} &= -R\sin\theta^1 \boldsymbol{e}_\theta, \\ \boldsymbol{a}_{2,2} &= -R\cos\theta^1 \boldsymbol{e}_r, \end{aligned} \tag{5.268}$$

which can be used to get

$$\kappa_{\alpha\beta} = R \begin{pmatrix} 1 & 0 \\ 0 & \cos^2\theta^1 \end{pmatrix}. \tag{5.269}$$

Also,

$$\begin{aligned} \boldsymbol{a}^1 &= -R^{-1}\sin\theta^1 \boldsymbol{e}_r + R^{-1}\cos\theta^1 \boldsymbol{e}_3, \\ \boldsymbol{a}^2 &= \tfrac{1}{R\cos\theta^1} \boldsymbol{e}_\theta \,. \end{aligned} \tag{5.270}$$

By using $\Gamma^\alpha_{\beta\gamma} = \boldsymbol{a}^\alpha \cdot \boldsymbol{a}_{\beta\gamma}$ we find

$$\begin{aligned} \Gamma^1_{22} &= \sin\theta^1\cos\theta^1, \\ \Gamma^2_{12} &= \Gamma^2_{21} = -\tan\theta^1 \end{aligned} \tag{5.271}$$

and the rest are all zero. With

$$\boldsymbol{t} = \frac{\boldsymbol{a}}{|\boldsymbol{a}_1|} = R^{-1}\boldsymbol{a}_1 = R\,\boldsymbol{a}^1 \quad \text{and} \quad \boldsymbol{u} = u\,\boldsymbol{t} + w\,\boldsymbol{n} = u_\alpha \boldsymbol{a}^\alpha + w\,\boldsymbol{n}, \tag{5.272}$$

we get

$$u_1 = R\,u\left(\theta^1\right), \qquad \text{and} \qquad u_2 = 0. \tag{5.273}$$

By using

$$u_{\alpha;\beta} = u_{\alpha,\beta} - u_\gamma \Gamma^\gamma_{\alpha\beta} \tag{5.274}$$

we find

$$\begin{aligned} u_{1;1} &= R\,u'(\theta^1), \\ u_{1;2} &= u_{2;1} = 0, \\ u_{2;2} &= -R \sin\theta^1 \cos\theta^1 u(\theta^1), \end{aligned} \tag{5.275}$$

which along with Eq. (5.258) yield

$$\begin{aligned} \varepsilon_{11} &= R(u' - w), \\ \varepsilon_{12} &= \varepsilon_{21} = 0, \\ \varepsilon_{22} &= -R\cos\theta^1 \left(u \sin\theta^1 + w\cos\theta^1\right). \end{aligned} \tag{5.276}$$

In view of $\varepsilon^\alpha_\beta = a^{\alpha\gamma}\varepsilon_{\beta\gamma}$ we obtain

$$\begin{aligned} \varepsilon^1_1 &= R^{-1}(u' - w), \\ \varepsilon^2_2 &= -R^{-1}(u\tan\theta^1 + w) \end{aligned} \tag{5.277}$$

and the other components are all zero. From

$$N^\alpha_\beta = k\left[\nu\varepsilon^\lambda_\lambda\delta^\alpha_\beta + (1-\nu)\varepsilon^\alpha_\beta\right], \qquad \text{where} \qquad k = \frac{Eh}{1-\nu^2}, \tag{5.278}$$

we can conclude that only N^1_1 and N^2_2 are non-zero. We have

$$\begin{aligned} N^\alpha_{\beta;\alpha} &= 0, \\ N^\alpha_\beta \kappa^\beta_\alpha &= -p = q, \end{aligned} \tag{5.279}$$

with

$$\begin{aligned} N^1_{1;1} &= N^1_{1,1}, \\ N^1_{2;1} &= 0, \\ N^2_{1;2} &= -\tan\theta^1\left(N^1_1 - N^2_2\right), \\ N^2_{2;2} &= 0. \end{aligned} \tag{5.280}$$

Then from (5.279) we find

$$N^1_{1,1} - \tan\theta^1(N^1_1 - N^2_2) = 0, \tag{5.281}$$

with

$$\kappa^{\alpha}_{\beta} = a^{\alpha\gamma}\kappa_{\gamma\beta} = R^{-1}\begin{pmatrix} 1 & 0 \\ 0 & 1 \end{pmatrix} \tag{5.282}$$

and from Eq. (5.279) we get

$$N^1_1 + N^2_2 = qR\,. \tag{5.283}$$

The last relation can be used along with Eq. (5.281) to get

$$N^1_{1,1} - 2\tan\theta^1 N^1_1 = -qR\tan\theta^1, \tag{5.284}$$

or

$$\frac{1}{g}\left(gN^1_1\right)_{,1} = -qR\tan\theta^1, \tag{5.285}$$

in which g is defined as

$$g = \exp\left[\int_0^{\theta^1}(-2\tan\phi)\mathrm{d}\phi\right] = \exp\left[\ln(\cos^2\theta^1)\right] = \cos^2\theta^1. \tag{5.286}$$

Therefore, we can write

$$\frac{\mathrm{d}}{\mathrm{d}\theta^1}(\cos^2\theta^1 N^1_1) = -qR\sin\theta^1\cos\theta^1. \tag{5.287}$$

By integrating from θ^1 to $\frac{\pi}{2}$ we find

$$-(\cos^2\theta^1)N^1_1(\theta^1) = \frac{-qR}{2}\sin^2\phi\Big|_{\theta^1}^{\frac{\pi}{2}} = -\frac{qR}{2}\cos^2\theta^1 \tag{5.288}$$

and as a result, we find N^1_1 as

$$N^1_1 = \frac{qR}{2}\,. \tag{5.289}$$

Then, from Eq. (5.283) we can find N^2_2 as

$$N^2_2 = \frac{qR}{2}\,. \tag{5.290}$$

In order to find the displacements, we start from

$$\begin{aligned} \frac{qR}{2} &= N^1_1 = k(\varepsilon^1_1 + \nu\varepsilon^2_2) = \frac{k}{R}[(u' - w) - \nu(u\tan\theta^1 + w)], \\ \frac{qR}{2} &= N^2_2 = k(\varepsilon^2_2 + \nu\varepsilon^1_1) = \frac{k}{R}[\nu(u' - w) - (u\tan\theta^1 + w)]. \end{aligned} \tag{5.291}$$

By adding these two equations we find

$$\frac{qR^2}{k(1+\nu)} = u' - u\tan\theta^1 - 2w \tag{5.292}$$

and subtracting them will result in

$$u' + u\tan\theta^1 = 0. \tag{5.293}$$

By multiplying equation (5.293) with $\cos\theta^1$ we can rewrite this equation as

$$\frac{d}{d\theta^1}\left(u\cos\theta^1\right) = 0, \quad \text{so} \quad u(\theta^1)\cos\theta^1 = C. \tag{5.294}$$

From the boundary condition that the sphere is pinned at equator, i.e., $u\,(0) = 0$, we find $C = 0$ and consequently

$$u(\theta^1) \equiv 0. \tag{5.295}$$

Then, Eq. (5.292) will give

$$w = \frac{-qR^2}{2k(1+\nu)} = -\frac{qR^2(1-\nu)}{2Eh}\,. \tag{5.296}$$

We have $w < 0$ because $\boldsymbol{n}$ is inward.

5.9 Bending Theory

The boundary conditions are (cf. (5.244))

$$\boldsymbol{f} = \boldsymbol{T}^\alpha\nu_\alpha - \left(\boldsymbol{M}^{\alpha\beta}\nu_\alpha\tau_\beta\right)_{,s} \quad \text{and} \quad \boldsymbol{c} = \boldsymbol{M}^{\alpha\beta}\nu_\alpha\nu_\beta\,. \tag{5.297}$$

where $\boldsymbol{T}^\alpha$ is given by (5.231), i.e.

$$\boldsymbol{T}^\alpha = \boldsymbol{N}^\alpha - \boldsymbol{M}^{\beta\alpha}{}_{;\beta}\,. \tag{5.298}$$

In view of (5.250), we can replace here $\boldsymbol{N}^\alpha = N^{\lambda\alpha}\boldsymbol{a}_\lambda$ and $\boldsymbol{M}^{\beta\alpha} = M^{\beta\alpha}\boldsymbol{n}$ and deduce

$$\boldsymbol{T}^\alpha = N^{\lambda\alpha}\boldsymbol{a}_\lambda - \left(M^{\beta\alpha}\boldsymbol{n}\right)_{;\beta} = \left(N^{\lambda\alpha} + M^{\beta\alpha}\kappa_\beta^\lambda\right)\boldsymbol{a}_\lambda - M^{\beta\alpha}{}_{;\beta}\,\boldsymbol{n}\,. \tag{5.299}$$

The last relation can be written in the form

$$\boldsymbol{T}^\alpha = P^{\lambda\alpha}\boldsymbol{a}_\lambda + S^\alpha\boldsymbol{n}, \tag{5.300}$$

where we denote by

$$\begin{aligned} P^{\lambda\alpha} &:= N^{\lambda\alpha} + M^{\beta\alpha}\kappa_\beta^\lambda \ \text{(tangential stresses)},\\ S^\alpha &:= -M^{\beta\alpha}{}_{;\beta} \qquad\quad \text{(transversal shear tractions)}. \end{aligned} \tag{5.301}$$

Thus, the boundary conditions (5.297) read

$$\begin{aligned} &\boldsymbol{f} = S^\alpha \nu_\alpha \boldsymbol{n} + P^{\lambda\alpha}\nu_\alpha \boldsymbol{a}_\lambda - \left(M^{\alpha\beta}\nu_\alpha\tau_\beta \boldsymbol{n}\right)_{,s}\\ \text{and}\quad &\boldsymbol{c} = M^{\alpha\beta}\nu_\alpha\nu_\beta \boldsymbol{n}. \end{aligned} \tag{5.302}$$

Recall the equations of equilibrium (cf. (5.252))

$$\begin{aligned} \left(N^{\mu\alpha} + M^{\beta\alpha}\kappa_\beta^\mu\right)_{;\alpha} + M^{\beta\alpha}{}_{;\beta}\,\kappa_\alpha^\mu + g^\mu &= 0,\\ \left(N^{\beta\alpha} + M^{\lambda\alpha}\kappa_\lambda^\beta\right)\kappa_{\beta\alpha} - M^{\beta\alpha}{}_{;\beta\alpha} + g &= 0. \end{aligned} \tag{5.303}$$

With the help of (5.301), these equations can be written in the equivalent form

$$\begin{aligned} P^{\beta\alpha}{}_{;\alpha} - \kappa_\alpha^\beta S^\alpha + g^\beta &= 0,\\ S^\alpha{}_{;\alpha} + \kappa_{\beta\alpha} P^{\beta\alpha} + g &= 0. \end{aligned} \tag{5.304}$$

For the case of isotropic, uniform materials, we can express the tensors $N^{\alpha\beta}$ and $M^{\alpha\beta}$ in terms of the strain and bending measures, respectively, as follows

$$\begin{aligned} N^{\alpha\beta} &= k\left[\nu\varepsilon_\lambda^\lambda a^{\alpha\beta} + (1-\nu)\varepsilon^{\alpha\beta}\right], \qquad C = \frac{Eh}{1-\nu^2},\\ M^{\alpha\beta} &= D\left[\nu\,\rho_\lambda^\lambda a^{\alpha\beta} + (1-\nu)\rho^{\alpha\beta}\right], \qquad D = \frac{Eh^3}{12(1-\nu^2)}, \end{aligned} \tag{5.305}$$

where $\rho^{\alpha\beta} = a^{\alpha\gamma}a^{\beta\delta}\rho_{\gamma\delta}$ and $\rho_{\alpha\beta}$ is given by (5.170) (or equivalently (5.178)). Indeed, in view of (5.217) and (5.249) we derive

$$\begin{aligned} N^{\alpha\beta} &= h\left(\frac{2\lambda\mu}{\lambda+2\mu}\,\varepsilon_\lambda^\lambda a^{\alpha\beta} + 2\mu\varepsilon^{\alpha\beta}\right),\\ M^{\alpha\beta} &= \frac{1}{12}h^3\left(\frac{2\lambda\mu}{\lambda+2\mu}\,\rho_\lambda^\lambda a^{\alpha\beta} + 2\mu\,\rho^{\alpha\beta}\right), \end{aligned} \tag{5.306}$$

and using (4.134) we obtain that the relations (5.305) hold true.

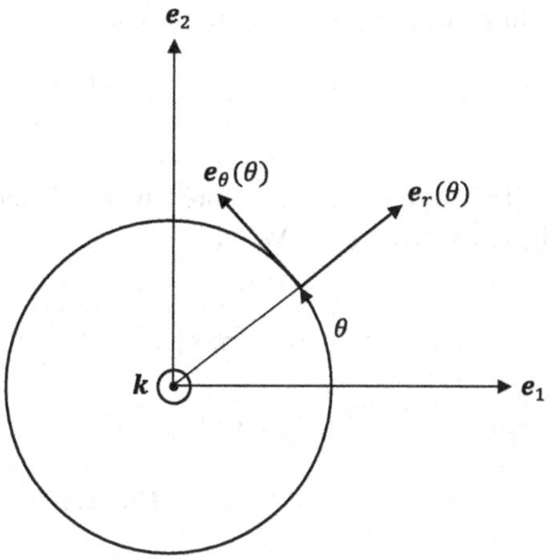

Fig. 5.5 Cross-section curve of the cylindrical midsurface

5.9.1 *Circular Cylindrical Shells*

Let us consider the equations for circular cylindrical shells in detail. The midsurface of the shell is described by the position vector

$$\boldsymbol{r} = R\,\boldsymbol{e}_r(\theta) + z\,\boldsymbol{k} \qquad \text{with} \qquad \boldsymbol{e}_r(\theta) = \cos\theta\,\boldsymbol{e}_1 + \sin\theta\,\boldsymbol{e}_2, \tag{5.307}$$

where θ is the angle at the cylinder axis, z is the coordinate in the direction of $\boldsymbol{k} = \boldsymbol{e}_3$ and the constant R is the radius. We also use

$$\boldsymbol{e}_\theta(\theta) = -\sin\theta\,\boldsymbol{e}_1 + \cos\theta\,\boldsymbol{e}_2, \tag{5.308}$$

which is a tangent unit vector to the cross-section curve, see Fig. 5.5.

We take the curvilinear coordinates $\theta^1 = R\,\theta$ (arclength parameter along the cross-section curve) and $\theta^2 = z$ (axial coordinate). The natural basis $\{\boldsymbol{a}_\alpha\}$ is given by

$$\boldsymbol{a}_1 = R\,\boldsymbol{e}_r'(\theta)\,\theta_{,1} = \boldsymbol{e}_r'(\theta) = \boldsymbol{e}_\theta(\theta) \qquad \text{and} \qquad \boldsymbol{a}_2 = \boldsymbol{k}\,. \tag{5.309}$$

Thus, we have

$$(a_{\alpha\beta}) = \begin{pmatrix} 1 & 0 \\ 0 & 1 \end{pmatrix} = \left(a^{\alpha\beta}\right) \qquad \text{and} \qquad \sqrt{a} = 1\,. \tag{5.310}$$

The dual basis $\{\boldsymbol{a}^\alpha\}$ in the tangent plane is the same, i.e.

$$\boldsymbol{a}^1 = \boldsymbol{e}_\theta \qquad \text{and} \qquad \boldsymbol{a}^2 = \boldsymbol{k}, \tag{5.311}$$

while the unit normal to the surface is

$$n = \frac{a_1 \times a_2}{\sqrt{a}} = e_r(\theta). \tag{5.312}$$

To determine the curvature tensor $\kappa_{\alpha\beta}$ and the Christoffel symbols, let us compute the derivatives $a_{\alpha,\beta}$. We get

$$a_{1,1} = \frac{\partial}{\partial\theta^1}\, e_\theta(\theta) = \frac{1}{R}\, e'_\theta(\theta) = -\frac{1}{R}\, e_r \quad \text{and} \quad a_{1,2} = a_{2,\alpha} = \mathbf{0}. \tag{5.313}$$

Then, in view of (2.75) we find

$$\Gamma^{\gamma}_{\alpha\beta} = a^{\gamma} \cdot a_{\alpha,\beta} = 0\,. \tag{5.314}$$

From the relation $\kappa_{\alpha\beta} = n \cdot a_{\alpha,\beta}$ (cf. (2.70)) we obtain

$$\kappa_{11} = n \cdot a_{1,1} = -\frac{1}{R} \quad \text{and} \quad \kappa_{12} = \kappa_{2\alpha} = 0 \tag{5.315}$$

and similarly

$$\kappa^1_1 = -\frac{1}{R}\,, \quad \kappa^1_2 = \kappa^2_\alpha = 0 \quad \text{and} \quad \kappa^{11} = -\frac{1}{R}\,, \quad \kappa^{12} = \kappa^{2\alpha} = 0. \tag{5.316}$$

Axisymmetric deformations

In case of axisymmetry, the displacement field u admits the following decomposition in the orthonormal local basis $\{e_\theta,\, k,\, e_r\}$

$$u = u(z)\, k + w(z)\, e_r(\theta) + v(z)\, e_\theta(\theta), \tag{5.317}$$

where $u(z)$ is the axial displacement, $w(z)$ the radial displacement and $v(z)$ stands for the twist of the cylindrical shell. Then, u. w and v are physical components (see Sect. 1.4).

Let us consider the case of axisymmetric deformations with *no twist*. Then, we have $v(z) = 0$ and the representation (5.317) becomes

$$u = u(z)\, k + w(z)\, e_r(\theta). \tag{5.318}$$

Using the latter relation, we shall express the strain tensor $\varepsilon_{\alpha\beta}$ and the bending tensor $\rho_{\alpha\beta}$, as well as the stress tensors. To this aim, we compute the derivatives

$$u_{,1} = \frac{1}{R}\frac{\partial u}{\partial\theta} = \frac{w}{R}\, e_\theta\,, \qquad u_{,2} = \frac{\partial u}{\partial z} = u'(z)\, k + w'(z)\, e_r$$

and

$$\boldsymbol{u}_{,11} = -\frac{w}{R^2}\,\boldsymbol{e}_r\,, \qquad \boldsymbol{u}_{,12} = \boldsymbol{u}_{,21} = \frac{w'}{R}\,\boldsymbol{e}_\theta\,, \qquad \boldsymbol{u}_{,22} = u''(z)\,\boldsymbol{k} + w''(z)\,\boldsymbol{e}_r\,.$$

We employ the relations (cf. (5.168))

$$\varepsilon_{\alpha\beta} = \frac{1}{2}\left(\boldsymbol{u}_{,\alpha}\cdot\boldsymbol{a}_\beta + \boldsymbol{u}_{,\beta}\cdot\boldsymbol{a}_\alpha\right),$$

where

$$\boldsymbol{a}_1\cdot\boldsymbol{u}_{,1} = \frac{w}{R}\,, \qquad \boldsymbol{a}_1\cdot\boldsymbol{u}_{,2} = 0 = \boldsymbol{a}_2\cdot\boldsymbol{u}_{,1}\,, \qquad \boldsymbol{a}_2\cdot\boldsymbol{u}_{,2} = u'(z).$$

Then, we obtain the components of the infinitesimal strain tensor $\boldsymbol{\varepsilon}$

$$\begin{aligned}
\varepsilon_{11} &= \frac{w}{R} = \varepsilon^{11} \quad \text{(circumferential)}\\
\varepsilon_{12} &= \varepsilon_{21} = 0 \quad \text{(shear)}\\
\varepsilon_{22} &= u'(z) = \varepsilon^{22} \ \text{(axial)}.
\end{aligned}$$

Thus, we have

$$\boldsymbol{\varepsilon} = \frac{w}{R}\,\boldsymbol{e}_\theta\otimes\boldsymbol{e}_\theta + u'(z)\,\boldsymbol{k}\otimes\boldsymbol{k}\,, \qquad \operatorname{tr}\boldsymbol{\varepsilon} = u' + \frac{w}{R}\,. \tag{5.319}$$

The latter relations and (5.305) yield

$$\begin{aligned}
N^{11} &= C\left[\nu\left(u' + \frac{w}{R}\right) + (1-\nu)\,\frac{w}{R}\right] = C\left(\frac{w}{R} + \nu\,u'\right),\\
N^{22} &= C\left[\nu\left(u' + \frac{w}{R}\right) + (1-\nu)\,u'\right] = C\left(u' + \nu\,\frac{w}{R}\right),\\
N^{12} &= 0 = N^{21}\,.
\end{aligned} \tag{5.320}$$

To find $M^{\alpha\beta}$ we require (cf. (5.170))

$$\rho_{\alpha\beta} = \boldsymbol{n}\cdot\boldsymbol{u}_{;\alpha\beta} = \boldsymbol{e}_r\cdot\boldsymbol{u}_{,\alpha\beta}\,.$$

Thus, we get

$$\rho_{11} = -\frac{w}{R^2} = \rho^{11}, \qquad \rho_{22} = w'' = \rho^{22}, \qquad \rho_{12} = 0 = \rho_{21}\,.$$

which yield

$$\boldsymbol{\rho} = -\frac{w}{R^2}\,\boldsymbol{e}_\theta\otimes\boldsymbol{e}_\theta + w''\boldsymbol{k}\otimes\boldsymbol{k}\,, \qquad \operatorname{tr}\boldsymbol{\rho} = w'' - \frac{w}{R^2}\,.$$

According to (5.305),

$$M^{\alpha\beta} = D\big[\nu\,(\mathrm{tr}\,\boldsymbol{\rho})\,a^{\alpha\beta} + (1-\nu)\,\rho^{\alpha\beta}\big],$$

so we have

$$\begin{aligned} M^{11} &= D\Big[\nu\Big(w'' - \frac{w}{R^2}\Big) + (1-\nu)\Big(-\frac{w}{R^2}\Big)\Big] = D\Big(\nu\,w'' - \frac{w}{R^2}\Big),\\ M^{22} &= D\Big[\nu\Big(w'' - \frac{w}{R^2}\Big) + (1-\nu)\,w''\Big] = D\Big(w'' - \nu\,\frac{w}{R^2}\Big),\\ M^{12} &= 0 = M^{21}\,. \end{aligned} \tag{5.321}$$

Next, we obtain the tangential tractions using the relations

$$P^{\alpha\beta} = N^{\alpha\beta} + M^{\alpha\gamma}\,\kappa_\gamma^\beta\,,$$

yielding

$$\begin{aligned} P^{11} &= N^{11} + \kappa_\gamma^1\,M^{1\gamma} = N^{11} + \kappa_1^1\,M^{11} = N^{11} - \frac{1}{r}\,M^{11},\\ P^{22} &= N^{22} + \kappa_\gamma^2\,M^{2\gamma} = N^{22}\,,\\ P^{12} &= N^{12} + \kappa_\gamma^1\,M^{2\gamma} = \underbrace{N^{12}}_{=0} + \kappa_1^1\,\underbrace{M^{21}}_{=0} + \underbrace{\kappa_2^1}_{=0}\,M^{22} = 0,\\ P^{21} &= N^{21} + \kappa_\gamma^2\,M^{1\gamma} = N^{21} = 0\,. \end{aligned}$$

Similarly, we compute the transverse shear tractions from the relations

$$S^\alpha = -M^{\beta\alpha}{}_{;\beta} = -M^{\beta\alpha}{}_{,\beta}\,,$$

so

$$\begin{aligned} S^1 &= -M^{\beta 1}{}_{,\beta} = -M^{11}{}_{,1} - \underbrace{M^{21}{}_{,2}}_{=0} = -\frac{1}{R}\,\frac{\partial}{\partial\theta}\,M^{11} = 0,\\ S^2 &= -M^{\beta 2}{}_{,\beta} = -\underbrace{M^{12}{}_{,1}}_{=0} - M^{22}{}_{,2} = -\big(M^{22}\big)' = -D\Big(w''' - \nu\,\frac{w'}{R^2}\Big). \end{aligned}$$

The equilibrium equations (5.304) in this case are

$$P^{\beta\alpha}{}_{;\alpha} - \kappa_\alpha^\beta\,S^\alpha = 0 \qquad (g^\beta = 0), \tag{5.322}$$

and

$$S^\alpha{}_{;\alpha} + \kappa_{\beta\alpha}\,P^{\beta\alpha} + g = 0. \tag{5.323}$$

In (5.322) we have

$$P^{1\alpha}{}_{;\alpha} = P^{1\alpha}{}_{,\alpha} = P^{11}{}_{,1} + \underbrace{P^{12}{}_{,2}}_{=0} = \frac{1}{R}\frac{\partial}{\partial\theta} P^{11} = 0,$$

$$P^{2\alpha}{}_{;\alpha} = P^{2\alpha}{}_{,\alpha} = \underbrace{P^{21}{}_{,1}}_{=0} + P^{22}{}_{,2} = \left(P^{22}\right)' = \left(N^{22}\right)',$$

$$\kappa^1_\alpha S^\alpha = \kappa^1_1 \underbrace{S^1}_{=0} + \underbrace{\kappa^1_2}_{=0} S^2 = 0, \qquad \kappa^2_\alpha S^\alpha = 0\,.$$

Thus, Eq. (5.322) reduce to

$$C\left(u' + \nu\frac{w}{R}\right) = N^{22} = \text{constant}. \tag{5.324}$$

Next, we compute the terms appearing in (5.323)

$$S^\alpha{}_{;\alpha} = S^\alpha{}_{,\alpha} = \left(S^2\right)' = -\left(M^{22}\right)'' = -D\left(w^{IV} - \nu\frac{w''}{R^2}\right)$$

and

$$\kappa_{\beta\alpha} P^{\beta\alpha} = \kappa_{11} P^{11} = -\frac{1}{R}\left(N^{11} - \frac{1}{R} M^{11}\right).$$

Then, the equilibrium equations (5.323) can be written

$$-R\left(M^{22}\right)'' - \left(N^{11} - \frac{1}{R} M^{11}\right) + R g = 0. \tag{5.325}$$

Here, we have

$$N^{11} = C\left(\frac{w}{R} + \nu u'\right).$$

If we eliminate u' from (5.324), then the latter relation becomes

$$N^{11} = \nu N^{22} + C(1-\nu^2)\frac{w}{R}\,. \tag{5.326}$$

Then, in view of (5.326) and $(5.321)_1$, we get

$$N^{11} - \frac{1}{R} M^{11} = \nu N^{22} + C\left[(1-\nu^2) + \frac{1}{12}\left(\frac{h}{R}\right)^2\right]\frac{w}{R} - \nu\frac{D}{R} w'', \tag{5.327}$$

since $D = \frac{1}{12} h^2 C$. In case of thin shells, we can neglect the term $\frac{1}{12}\left(\frac{h}{R}\right)^2$ in the square brackets and write

$$N^{11} - \frac{1}{R}\, M^{11} \;\simeq\; \nu\, N^{22} + C(1-\nu^2)\,\frac{w}{R} - \nu\,\frac{D}{R}\, w'' \qquad (\text{for } \nu^2 \neq 1). \tag{5.328}$$

Hence, the equilibrium equation (5.325) reduces to the form

$$w^{IV} + \frac{12(1-\nu^2)}{h^2 R^2}\, w - \frac{2\nu}{R^2}\, w'' \;=\; -\frac{\nu\, N^{22}}{R\, D} + \frac{g}{D}\,. \tag{5.329}$$

Homogeneous solutions

Let us denote by

$$\zeta := \frac{z}{R} \qquad \text{and} \qquad \omega := \frac{w(z)}{R} = \frac{w(R\zeta)}{R} = \omega(\zeta).$$

Then, we have $w = \omega\, R$ and the derivatives are

$$w'(z) = \dot{\omega}(\zeta), \quad w''(z) = \frac{1}{R}\,\ddot{\omega}(\zeta), \quad w'''(z) = \frac{1}{R^2}\,\dddot{\omega}\,(\zeta), \quad w^{IV}(z) = \frac{1}{R^3}\,\omega^{(4)}(\zeta),$$

where a superposed dot designates the derivative with respect to ζ.

Hence, the homogeneous equations corresponding to (5.329), multiplied by R^3, is

$$\omega^{(4)} + 12(1-\nu^2)\left(\frac{R}{h}\right)^2 \omega - 2\nu\,\ddot{\omega} \;=\; 0. \tag{5.330}$$

We seek for solutions of the form $\omega(\zeta) = e^{\lambda\zeta}$. From Eq. (5.330) we get

$$\lambda^4 + 4\alpha^4 - 2\nu\,\lambda^2 \;=\; 0, \tag{5.331}$$

where we have denoted by

$$\alpha = \left[3(1-\nu^2)\left(\frac{R}{h}\right)^2\right]^{1/4}, \qquad \text{so} \qquad \alpha^4 \sim \left(\frac{R}{h}\right)^2 \gg 1.$$

So we expect that the Eq. (5.331) has complex roots and

$$\lambda^4 + 4\alpha^4 \;\simeq\; 0. \tag{5.332}$$

We can write the latter relation as

$$\left(\lambda^2 + 2\alpha^2\right)^2 - 4\lambda^2\alpha^2 \;=\; 0,$$

so

$$\lambda^2 + 2\alpha^2 \;=\; \pm 2\lambda\,\alpha,$$

which represent two quadratic equations with the roots

$$\lambda = \pm\alpha \pm i\,\alpha\,.$$

In each case we have

$$\left|\lambda^2\right| \sim \alpha^2 \sim \frac{R}{h} \quad \text{and} \quad \frac{\left|\lambda^2\right|}{\left|\lambda^4\right|} \sim \frac{h}{R} \ll 1,$$

so this is consistent with the simplification in (5.332).

This is tantamount to suppressing w'' in the original differential equation (5.329) and writing

$$w^{IV} + \frac{12(1-\nu^2)}{h^2 R^2}\, w = -\frac{\nu\, N^{22}}{R\, D} + \frac{g}{D}, \qquad \text{for} \quad \frac{h}{R} \ll 1\,. \tag{5.333}$$

Then, we obtain the general solution

$$w(z) = e^{-\beta z}(\bar{A}\cos\beta z + \bar{B}\sin\beta z) + e^{\beta z}(\bar{C}\cos\beta z + \bar{D}\sin\beta z) \,+\, w_p(z), \tag{5.334}$$

where $\bar{A}, \bar{B}, \bar{C}, \bar{D}$ are arbitrary constants, w_p is a particular solution of the inhomogeneous equation, and we denote

$$\beta = \left(\frac{3(1-\nu^2)}{R^2\, h^2}\right)^{1/4} = \frac{\alpha}{R} \qquad (\text{with} \quad \beta\, z = \alpha\, \zeta\,).$$

Now, we can integrate the Eq. (5.324) to find the solution $u(z)$. In this way, we determine the displacement field (5.318).

Boundary Data at $z = 0$

Let us consider the end edge of the cylindrical shell at $z = 0$ and write the boundary conditions.

At this edge, the boundary curve is given by the position vector $\boldsymbol{r}(\theta) = R\, \boldsymbol{e}_r(\theta)$, so the outward unit normal to the curve (lying in the tangent plane) is

$$\boldsymbol{\nu} = -\boldsymbol{k} = -\boldsymbol{a}_2\,,$$

see Fig. 5.6. Also, the unit tangent vector to the boundary curve is given by

$$\boldsymbol{\tau} = \frac{\partial \boldsymbol{r}}{\partial \theta^1} = \boldsymbol{a}_1 = \boldsymbol{e}_\theta\,,$$

since $\theta^1 = R\,\theta$ is the arclength parameter.

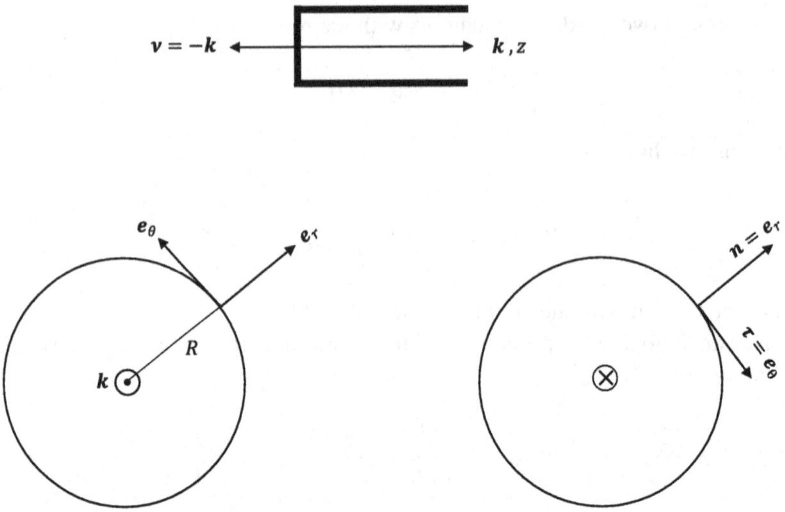

Fig. 5.6 Boundary data at the end edge curve $z = 0$

The unit normal vector to the midsurface is $\boldsymbol{n} = \boldsymbol{e}_r(\theta)$. Indeed, we have

$$\boldsymbol{n} = \frac{\boldsymbol{a}_1 \times \boldsymbol{a}_2}{|\boldsymbol{a}_1 \times \boldsymbol{a}_2|} = \boldsymbol{\tau} \times (-\boldsymbol{\nu}) = \boldsymbol{e}_\theta \times \boldsymbol{k} = \boldsymbol{e}_r \tag{5.335}$$

and the vector basis $\{\boldsymbol{\nu}, \boldsymbol{\tau}, \boldsymbol{n}\}$ is ortonormal and right-handed.

In view of $\boldsymbol{\tau} = \boldsymbol{a}_1$ and $\boldsymbol{\nu} = -\boldsymbol{a}_2$, we have the components

$$\tau_1 = \tau^1 = 1, \quad \tau_2 = \tau^2 = 0 \quad \text{and} \quad \nu_1 = \nu^1 = 0, \quad \nu_2 = \nu^2 = -1. \tag{5.336}$$

The first boundary condition in (5.302) can be written as

$$\boldsymbol{f} = \boldsymbol{T}^\alpha \nu_\alpha - \frac{\partial}{\partial \theta^1}\left(M^{\alpha\beta} \nu_\alpha \tau_\beta \, \boldsymbol{n}\right), \tag{5.337}$$

where $\boldsymbol{T}^\alpha = P^{\beta\alpha} \boldsymbol{a}_\beta + S^\alpha \boldsymbol{n}$. Substituting (5.335) and (5.336) into (5.337) we get

$$\boldsymbol{f} = -\boldsymbol{T}^2 + \frac{1}{R}\frac{\mathrm{d}}{\mathrm{d}\theta}\big(\underbrace{M^{21}}_{=0} \boldsymbol{e}_r\big) = -\boldsymbol{T}^2 = -P^{\beta 2} \boldsymbol{a}_\beta - S^2 \boldsymbol{n}\,.$$

Using here $P^{12} = 0$ and $P^{22} = N^{22}$ we find

$$\boldsymbol{f} = -N^{22} \boldsymbol{k} - S^2 \boldsymbol{e}_r\,. \tag{5.338}$$

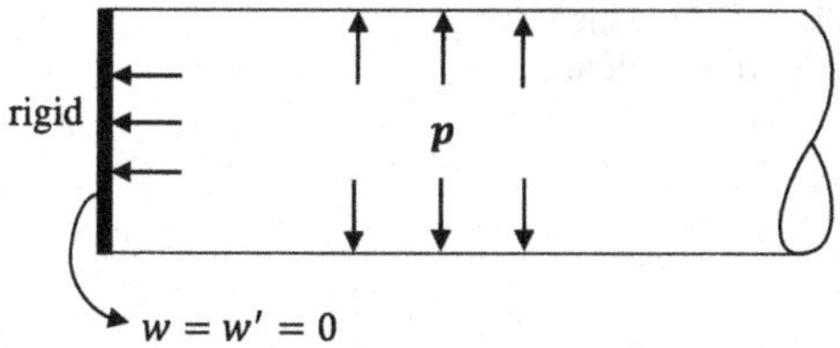

Fig. 5.7 Clamped cylindrical shell under internal pressure

Also, the second boundary condition in (5.302) has the form

$$\boldsymbol{c} = M^{\alpha\beta} \nu_\alpha \nu_\beta \, \boldsymbol{n} = M^{22} \boldsymbol{e}_r \, . \tag{5.339}$$

Then, from (5.318), (5.338) and (5.339) we obtain

$$\boldsymbol{f} \cdot \boldsymbol{u} = -N^{22} u - S^2 w \tag{5.340}$$

and

$$\boldsymbol{c} \cdot \boldsymbol{u}_{,\nu} = (M^{22} \boldsymbol{e}_r) \cdot \left(-\frac{\partial \boldsymbol{u}}{\partial z} \right) = (M^{22} \boldsymbol{e}_r) \cdot (-u' \boldsymbol{k} - w' \boldsymbol{e}_r) = -M^{22} w', \tag{5.341}$$

since $\boldsymbol{u}_{,\nu} = \boldsymbol{u}_{,\alpha} \nu^\alpha = -\frac{\partial \boldsymbol{u}}{\partial z}$.

The terms $\boldsymbol{f} \cdot \boldsymbol{u}$ and $\boldsymbol{c} \cdot \boldsymbol{u}_{,\nu}$ appear in the expression of the potential energy, see (5.233) with (5.240).

Example 1

Consider a circular cylindrical shell clamped at its left end under an internal pressure p, see Fig. 5.7. We will determine the solution in the vicinity of the clamped end.

In this problem, S^2 and M^{22} are not assigned. At $z = 0$ we have

$$\boldsymbol{k} \cdot \boldsymbol{f} = -\frac{p\pi R^2}{2\pi R} = -\frac{1}{2} pR, \tag{5.342}$$

which can be used to realize that

$$N^{22} = \frac{1}{2} pR \ (= \text{const.}) \, . \tag{5.343}$$

On the other hand, Eq. (5.329) can be written as

$$w^{IV} + 4\beta^4 w - 2\nu \frac{w''}{R^2} = \frac{p}{2D} (2 - \nu) \tag{5.344}$$

and as a result, we find the particular solution

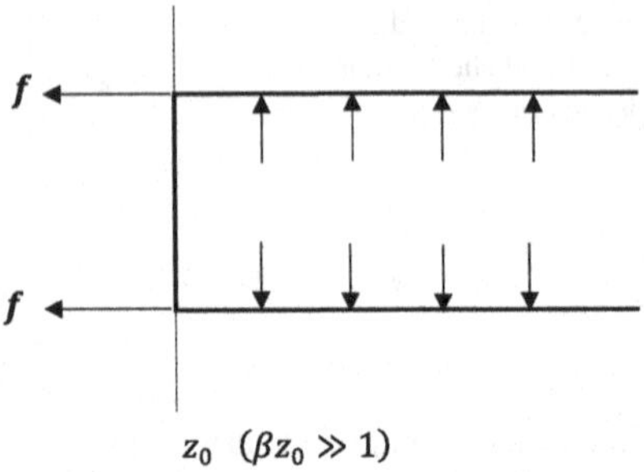

Fig. 5.8 Pure membrane cylinfrical shell under constant pressure

$$w_p(z) = \frac{p}{8D\beta^4}(2-\nu)\,. \tag{5.345}$$

Suppose that the cylinder is long, i.e., $L/R \gg 1$. We seek a solution that remains bounded as $\zeta = z/R \to \infty$. Then, we find $C = D = 0$ and the general solution

$$w\,(z) = e^{-\beta z}(A\cos\beta z + B\sin\beta z) + w_p(z). \tag{5.346}$$

This solution must satisfy the boundary conditions

$$w(0) = w'(0) = 0 \qquad \text{at} \qquad z = 0. \tag{5.347}$$

By enforcing these boundary conditions, we find

$$w(z) = \frac{p}{8D\beta^4}(2-\nu)[1 - e^{-\beta z}(\cos\beta z + \sin\beta z)]. \tag{5.348}$$

The traction at the section $z = z_0$ is

$$\boldsymbol{f}_{|z_0} = -N^{22}\boldsymbol{k} - S^2\boldsymbol{e}_r, \tag{5.349}$$

where

$$S^2 = -(M^{22})' = D(\nu\frac{w'}{R^2} - w'') \;\sim\; e^{-\beta z_0}. \tag{5.350}$$

For $\beta z_0 \gg 1$, we obtain

$$\boldsymbol{f}_{|z_0} \;\sim\; -\frac{1}{2}\,p\,R\,\boldsymbol{k}, \tag{5.351}$$

which is the classical pressure vessel theory, i.e., pure membrane result as can be seen in the Fig. 5.8.

The traction on a generator is shown in the Fig. 5.9, from which we can find

$$\nu^1 = \nu_1 = 1, \qquad \nu^2 = \nu_2 = 0, \qquad \tau^1 = \tau_1 = 0, \qquad \tau^2 = \tau_2 = 1. \tag{5.352}$$

Fig. 5.9 Traction on a generator of the cylindrical shell

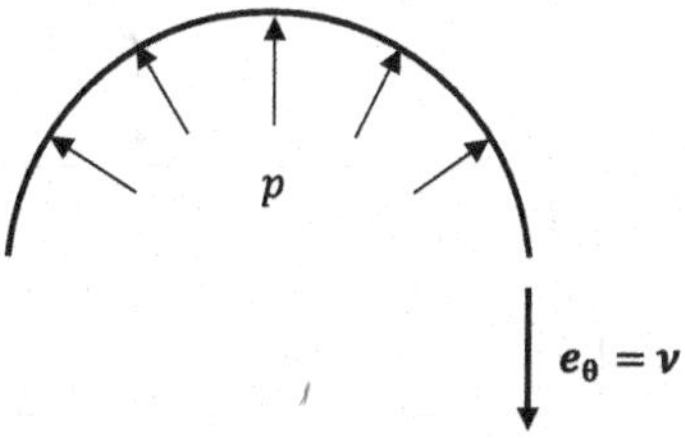

Then, it follows

$$M^{\alpha\beta}\nu_\alpha\tau_\beta = M^{12} = 0 \tag{5.353}$$

and as a result, we find

$$f = T^\alpha\nu_\alpha = T^1 = P^{\beta 1}a_\beta + \underbrace{S^1}_{=0}n = P^{11}e_\theta = (N^{11} - \frac{1}{R}M^{11})e_\theta\,. \tag{5.354}$$

But, we have

$$N^{11} - \frac{1}{R}M^{11} = Rp - r(M^{22})'' \tag{5.355}$$

and from Eq. (5.325), where

$$(M^{22})'' = D(w^{IV} - \nu\frac{w''}{R^2}) \sim e^{-\beta z_0}, \tag{5.356}$$

we conclude that

$$f \sim R\,p\,e_\theta \quad \text{for} \quad z_0 \gg \frac{1}{\beta}\,. \tag{5.357}$$

Therefore, hoop and axial tractions will converge to the value of these quantities in the theory of pure membrane as $\beta z_0 \to \infty$.

Remark Bending effects are important and localized near the boundaries, as needed to satisfy boundary conditions (here, on w') that cannot be accommodated by membrane theory alone, while away from such boundaries, membrane theory (the leading order model) dominates. □

Example 2

Let us consider a pinched cylindrical shell under a ring load ($p = 0 = N^{22}$) with open ends, see Fig. 5.10. We will use the symmetry of the problem an replace it with

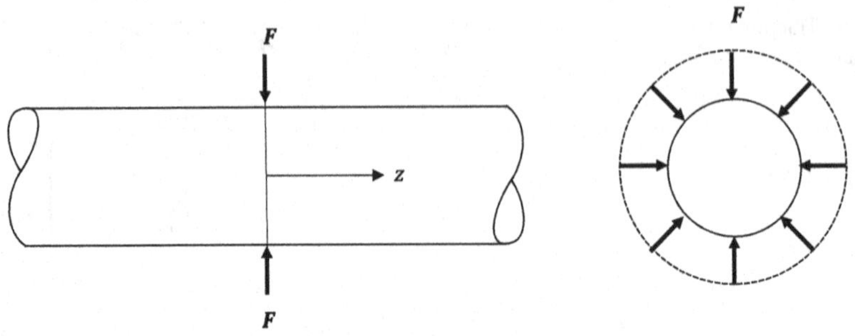

Fig. 5.10 Pinched cylindrical shell under a ring load

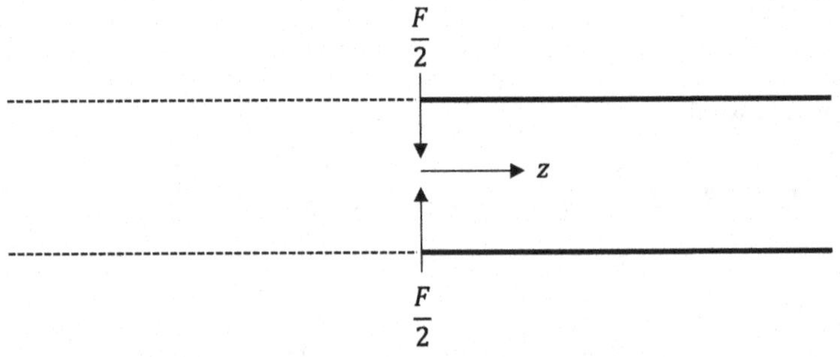

Fig. 5.11 The equivalent problem to a pinched cylindrical shell

an equivalent problem. We are investigating the solution in the immediate vicinity of the load.

The equivalent problem is shown below in Fig. 5.11. In this problem we consider only the half shell in the range $z \geq 0$, together with the boundary conditions $w'(0) = 0$, but w and M^{22} are not prescribed whereas S^2 is prescribed.

The traction force at $z = 0$ is

$$-\frac{F}{2}\,\boldsymbol{e}_r = \boldsymbol{f} = -S^2\boldsymbol{e}_r, \qquad \text{so} \qquad S^2 = \frac{F}{2}\,, \tag{5.358}$$

where

$$S^2 = D\big(\nu\frac{w'}{r^2} - w''\big). \tag{5.359}$$

Then, by using the boundary condition $w'(0) = 0$ we find

$$w''(0) = -\frac{S^2}{D} = -\frac{F}{2D}\,. \tag{5.360}$$

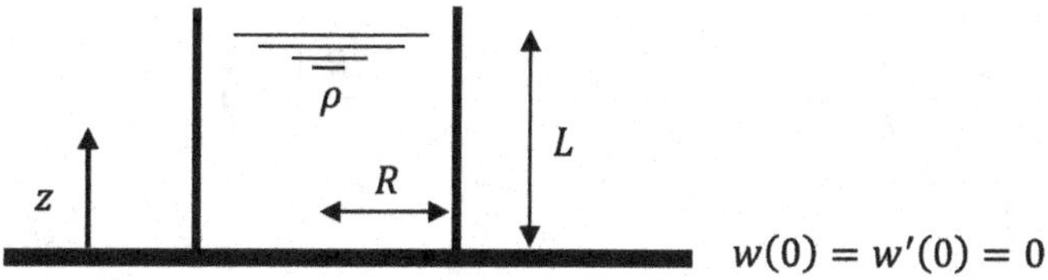

Fig. 5.12 Cylindrical shell under a hydrostatic pressure (water tank)

In this example, we again suppose that $L/R \gg 1$ and we seek for the bounded solution for $\zeta \rightarrow \infty$ as:

$$w\,(z) = e^{-\beta z}(A\cos\beta z + B\sin\beta z). \tag{5.361}$$

By using boundary conditions, we find

$$A = B = -\frac{F}{8\beta^3 D} \tag{5.362}$$

and as a result, we find the displacement

$$w\,(z) = -\frac{F}{8\beta^3 D}e^{-\beta z}(\cos\beta z + \sin\beta z). \tag{5.363}$$

Example 3

In the last example we study the deformation of a cylindrical shell under a hydrostatic pressure (water tank), see Fig. 5.12. For simplicity we suppose that the length L of the shell is much larger than the radius R and we study the solution in the vicinity of the ground (the plane $z = 0$). The bounded solution is determined.

We have the hydrostatic pressure

$$p = \rho g(L - z), \tag{5.364}$$

in which g is the gravitational constant. Then, the right hand side of Eq. (5.329) is $\frac{\rho g}{D}(L - z)$, and a particular solution to this differential equation is

$$w_p(z) = \frac{\rho g}{4\beta^4 D}(L - z). \tag{5.365}$$

For $L/R \gg 1$, the bounded solution is

$$\frac{4\beta^2 D}{\rho g}w(z) = L - z - e^{-\beta z}\big[L\cos\beta z + (L - \frac{1}{\beta})\sin\beta z\big], \tag{5.366}$$

where

$$L - \frac{1}{\beta} = L\Big(1 - \frac{1}{L\beta}\Big) = L\Big(1 - \underbrace{\frac{R}{L}\frac{1}{\alpha}}_{\sim\frac{R}{L}\sqrt{\frac{h}{R}}\ll 1}\Big) \sim L\,. \tag{5.367}$$

Remark For the case of finite height L of the cylinder, one can determine the solution of this problem by taking into account appropriate boundary conditions at the edge $z = l$, see Exercise 5.4. □

5.10 Exercises

5.1 Show that the term $\frac{1}{24}h^3\,\boldsymbol{\rho}\cdot\mathscr{M}[\boldsymbol{\rho}]$ from the approximate theory for combined bending and stretching coincides with the rigorously derived leading-order strain energy in the case of *pure bending*, see (5.126).

5.2 (a) Recall the *statically determinate* system associated with linear membrane theory for infinitesimal equilibrium deformations from a stress-free reference configuration. In the absence of distributed loads this is:

$$N^{\alpha\beta}{}_{;\alpha} = 0, \qquad N^{\alpha\beta}\kappa_{\alpha\beta} = 0, \qquad N^{\alpha\beta} = N^{\beta\alpha}\,.$$

For the special case of a developable surface (i.e., zero Gaussian curvature) we can reduce this problem to the search of a *stress function* $F(\theta^\alpha)$ as follows:

Show that to satisfy the first equation it is sufficient that

$$N^{\alpha\beta} = \tilde{\varepsilon}^{\alpha\lambda}\tilde{\varepsilon}^{\beta\gamma}\,F_{;\gamma\lambda}\,,$$

where $\tilde{\varepsilon}^{\alpha\beta}$ is the alternator tensor. This expression for the stress is also necessary if the domain is simply connected.

Hint: Use $F_{;\gamma\delta} = F_{,\gamma\delta} - \Gamma^{\lambda}_{\gamma\delta}\,F_{,\lambda}$ to prove that $F_{;\gamma\delta\nu} - F_{;\gamma\nu\delta} = R^{\mu}{}_{\cdot\gamma\delta\nu}\,F_{,\mu}$, the right-hand side of which involves the Gaussian curvature.

Show that the second equilibrium equation then yields a second order partial differential equation to be solved for F provided that the surface is not plane, namely

$$\kappa_{\alpha\beta}\,\tilde{\varepsilon}^{\alpha\gamma}\tilde{\varepsilon}^{\beta\delta}\,F_{;\gamma\delta} = 0.$$

(b) Generalize the foregoing procedure to surfaces with non-zero Gaussian curvature, as follows: Use the vector form of the equilibrium equations, $\boldsymbol{N}^{\alpha}{}_{;\alpha} = \boldsymbol{0}$ and $\boldsymbol{N}^{\alpha}\times\boldsymbol{a}_{\alpha} = \boldsymbol{0}$. To satisfy the first equation it is sufficient (and in simply-connected regions, necessary) that $\boldsymbol{N}^{\alpha} = \tilde{\varepsilon}^{\alpha\lambda}\,\boldsymbol{W}_{,\lambda}$ for some vector-valued 'stress function' $\boldsymbol{W}(\theta^\alpha)$. Use the second equation to obtain a coupled system of partial differential equations for the components of $\boldsymbol{W} = W^{\alpha}\boldsymbol{a}_{\alpha} + W\,\boldsymbol{n}$. Show that if the Gaussian curvature is non-zero then the tangential components W^{α} can be

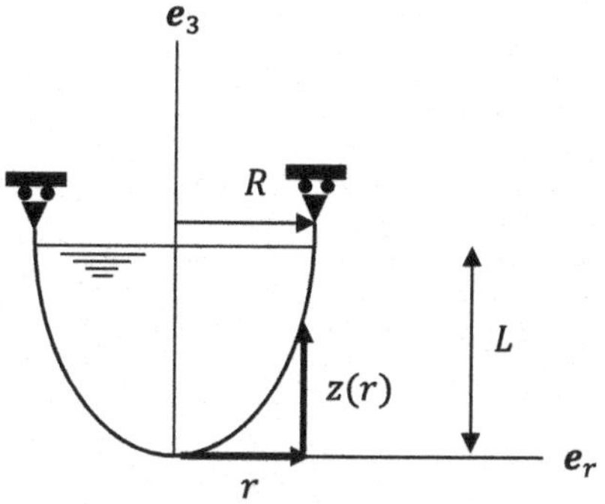

Fig. 5.13 Paraboloid of revolution filled with water (see Exercise 5.3)

eliminated from this system, yielding a single partial differential equation for $\boldsymbol{W}(\theta^\alpha)$. Obtain this equation explicitly, and show that the stress $N^{\alpha\beta}$ is determined by solutions of the partial differential equation.

Finally, show that the formulation of part (a), for the case of a surface with zero Gaussian curvature, may be recovered by taking $\boldsymbol{W} = W^\alpha \boldsymbol{a}_\alpha$ and retracing the present analysis.

5.3 For this problem use isotropic membrane theory. Assume an axisymmetric displacement field of the form $\boldsymbol{u} = u\,\boldsymbol{t} + w\,\boldsymbol{n}$, where $\boldsymbol{t}$ is the unit tangent to a meridian and u, w are independent of the azimuth θ. A paraboloid of revolution (see Fig. 5.13), described by

$$\boldsymbol{r} = r\,\boldsymbol{e}_r(\theta) + z(r)\,\boldsymbol{e}_3 \qquad \text{with} \qquad z(r) = L\left(r/R\right)^2,$$

is filled with water of density ρ and gravity is acting. The lateral pressure is $p = \rho\, g\,(L - z)$. Find the mixed components N^α_β of the membrane stress at $r = 0, R$.

5.4 Solve the problem of a cylindrical liquid-filled tank under gravity that we considered in Example 3 of Sect. 5.9. Assume clamping conditions at the base and impose appropriate boundary conditions at the top of the shell.

References

1. Chadwick, P.: Continuum Mechanics: Concise Theory and Problems. Dover, New York (1976)
2. Ciarlet, P.G.: An Introduction to Differential Geometry with Applications to Elasticity. Springer, Dordrecht (2005)
3. Ciarlet, P.G.: Mathematical Elasticity, vol. III. Theory of Shells, North-Holland, Amsterdam (2000)
4. Naghdi, P.M.: The theory of shells and plates. In: Flügge, W. (ed.) Handbuch der Physik, vol. VIa/2, pp. 425–640. Springer, Berlin (1972)
5. Steigmann, D.: Extension of Koiter's linear shell theory to materials exhibiting arbitrary symmetry. Int. J. Eng. Sci. **51**, 216–232 (2012)

Chapter 6
Nonlinear Equations for Plates and Shells

Abstract Here we develop the nonlinear theories of plates and shells, and show how Koiter's shell theory emerges in the framework of our dimension reduction procedure for nonlinearly elastic materials.

Preliminary Remarks

Contemporary research on the theoretical foundation of theories of thin plates and shells emphasizes their relationship to three-dimensional finite elasticity theory. These efforts are typically based on the method of gamma convergence [9], concerned with the limiting variational problem for small thickness, or on asymptotic analysis of the weak forms of the equilibrium equations [2, 8]. However, neither method has generated a model that accommodates bending and stretching in a single framework. The current state of the art in the rigorous derivation of plate theory by gamma convergence is illustrated by [9], which concludes with the observation: "A wide open problem is the question of whether we can rigorously justify theories which are two-dimensional but still involve the small thickness parameter.[...] A typical case involves boundary conditions that cause part of the shell to stretch, but another part to bend with no stretching". Indeed, such problems are of primary interest in applications. Evidently, then, at present there exists no rigorously derived model for combined bending and stretching.

In contrast, the work of Hilgers and Pipkin [12, 13, 15], inspired by the need for a regularization of membrane theory for problems in which membrane theory has no solution, furnishes the first careful consideration of the relationship between plate theory and modern three-dimensional nonlinear elasticity in the presence of combined bending and stretching. This work subsumes the models obtained by asymptotic analysis and gamma convergence, and furnishes an extension of Koiter's small-strain model [16, 17] to large midsurface strains.

A parallel approach based on asymptotic expansion of the local differential equations has recently been pursued by Dai and co-workers [21]. An interesting open question, originally posed by Koiter [18], concerns the relationship between the

D. J. Steigmann et al., *Lecture Notes on the Theory of Plates and Shells*,
Solid Mechanics and Its Applications 274,
https://doi.org/10.1007/978-3-031-25674-5_6

equations generated by this procedure and the Euler-Lagrange equations associated with the energies generated by the present approach.

In this chapter we review the Hilgers-Pipkin model from the point of view developed in [24, 25] for thin elastic bodies. Attention is confined in Sect. 6.1 to plates. This allows us to illustrate the main ideas as simply as possible while avoiding the less important details associated with the differential geometry of shells. Extensions of these ideas to shells are discussed in Sect. 6.2.

6.1 Asymptotic Derivation of Nonlinear Plate Models

Standard notation is adopted. If $\mathscr{M}$ is a fourth-order tensor and $\boldsymbol{T}$ a second-order tensor, then $\mathscr{M}[\boldsymbol{T}]$ is the second-order tensor with Cartesian components $\mathscr{M}_{iAjB}T_{jB}$. We use Div to denote the three-dimensional divergence operator, and div its two-dimensional counterpart. For example, $\text{Div}\,\boldsymbol{A} = A_{iA,A}\boldsymbol{e}_i$ and $\text{div}\,\boldsymbol{A} = A_{i\alpha,\alpha}\boldsymbol{e}_i$, where $\{\boldsymbol{e}_i\}$ is an orthonormal basis and subscripts preceded by commas are used to denote partial derivatives with respect to Cartesian coordinates. We also use ∇ to denote the two-dimensional gradient. The unit vector $\boldsymbol{k} = \boldsymbol{e}_3$ identifies the orientation of the plate midplane prior to deformation.

6.1.1 *Background from Three-Dimensional Nonlinear Elasticity Theory*

We recall briefly some basic equations from three-dimensional nonlinear elasticity theory, which were presented in Chaps. 2 and 3 in more details. In the purely mechanical theory of nonlinear elasticity upon which the considerations of this chapter are based, the Piola stress $\tilde{\boldsymbol{P}}$ of the three-dimensional theory is given by the values of the function

$$\tilde{\boldsymbol{P}}(\tilde{\boldsymbol{F}}) = \mathscr{W}_{\tilde{\boldsymbol{F}}}\,, \tag{6.1}$$

the derivative with respect to the deformation gradient $\tilde{\boldsymbol{F}}$ of the strain energy $\mathscr{W}(\tilde{\boldsymbol{F}})$ per unit reference volume. The material is assumed to be uniform for the sake of simplicity, so that the strain-energy function does *not* depend explicitly on position $\boldsymbol{x}$ in a reference configuration κ. Superposed tildes are used to denote three-dimensional quantities. The same symbols, without tildes, are used to denote their midplane values.

The force per unit area transmitted across a surface with unit normal $\boldsymbol{N}$ in κ is

$$\tilde{\boldsymbol{p}} = \tilde{\boldsymbol{P}}(\tilde{\boldsymbol{F}})\boldsymbol{N}\,. \tag{6.2}$$

It is well known that this, together with the equilibrium equation

$$\mathrm{Div}\, \tilde{\boldsymbol{P}} = \boldsymbol{0}, \tag{6.3}$$

are the natural boundary condition and Euler equation for energy-minimizing deformations under conditions of conservative loading without body force, holding on a subset of $\partial\kappa$ and in κ respectively.

A plate is a material body identified with κ, which is generated by the parallel translation of a plane region Ω, with piecewise smooth boundary curve $\partial\Omega$, in the direction orthogonal to Ω. The body itself occupies the volume $\bar{\Omega} \times [-h/2, h/2]$, where $\bar{\Omega} = \Omega \cup \partial\Omega$ and h is the (uniform) thickness. Let l be another length scale such as the diameter of Ω or that of an interior hole. We assume that $h/l \ll 1$. Further, we regard l as a fixed scale and adopt it as the measure of length. This allows us to set $l = 1$ and thus to simplify the notation.

Our goal is an optimal expression for the term E in the expansion

$$\mathscr{E} = E + o(h^3) \tag{6.4}$$

of the potential energy $\mathscr{E}$ of the thin three-dimensional body, in which $h \ll 1$. This is shown below to have the form

$$E = h\, E_1 + h^3 E_3\,, \tag{6.5}$$

in which E_1 and E_3 are not explicitly dependent on h. We will show that E_1 is the conventional membrane energy, whereas E_3 is associated with bending effects.

If a particular deformation minimizes the three-dimensional energy; i.e., if it is stable, then the perturbation $\Delta\mathscr{E}$ relative to that deformation satisfies $\Delta\mathscr{E} \geq 0$ for any kinematically possible alternative. Using the expansion $\mathscr{E} = h\, E_1 + o(h)$, this in turn yields $\Delta E_1 + o(h)/h \geq 0$. Passing to the limit, we obtain $\Delta E_1 \geq 0$, and conclude that at leading order in thickness, stable deformations minimize the membrane energy. If attention is restricted to deformations that are strain-free at the midsurface, and if the boundary data are compatible with such deformations, then ΔE_1 vanishes identically and the same argument yields $\Delta E_3 \geq 0$. In this case admissible deformations of the plate correspond to pure bending, and three-dimensional energy minimizers minimize E_3, again at leading order in thickness. These observations underlie the approach to membrane and inextensional bending theory via gamma convergence. However, in the case of combined bending and stretching of a finite-thickness plate in which terms of order h and h^3 are retained simultaneously, the inequality $\Delta\mathscr{E} \geq 0$ satisfied by equilibria in the three-dimensional theory does *not* imply that $\Delta E \geq 0$. This is the reason why the method of gamma convergence, which is concerned exclusively with the derivation of the limiting minimization problem, has not succeeded in generating a single model for combined bending and stretching, except in the fortuitous circumstance—exemplified by special cases of the linear theory—when the two effects decouple at leading order [22]. Accordingly, we do *not* expect E to be minimized at a stable equilibrium state. One may seek to rectify this situation by expanding the energy to higher orders in h. However, in the nonlinear theory it is impractical to do so, as this requires higher-order three-dimensional elastic moduli [8], which are excessively unwieldy for strain-energy functions commonly used in nonlinear elasticity theory.

An interesting exception to the foregoing observation occurs when $E_1 = O(h^2)$. In this case $\mathcal{E}/h^3 = \bar{E} + o(h^3)/h^3$, where $\bar{E}$ does not depend explicitly on h. Passing to the limit, we conclude that minimizers in the exact theory correspond to minimizers in the approximate theory; that is, if a deformation minimizes $\mathcal{E}$, than it also minimizes $\bar{E}$ at leading order, and vice versa. This situation obtains in the case of wrinkling, in which the energies of stretching and bending are of comparable order [1, 11, 28]. We will show that in such circumstances $\bar{E}$ may be identified with Koiter's expression for the energy. This fact lends further support to the widespread view [3] that Koiter's model provides the best 'all-around' theory of plates and shells, despite the fact that it does not emerge as a Gamma-limit or a formal asymptotic limit.

We assume throughout that equilibrium deformations satisfy the strong-ellipticity condition

$$\boldsymbol{a} \otimes \boldsymbol{b} \cdot \mathcal{M}(\tilde{\boldsymbol{F}})[\boldsymbol{a} \otimes \boldsymbol{b}] > 0 \quad \text{for all} \quad \boldsymbol{a} \otimes \boldsymbol{b} \neq \boldsymbol{0}, \tag{6.6}$$

where

$$\mathcal{M}(\tilde{\boldsymbol{F}}) = \mathcal{W}_{\tilde{\boldsymbol{F}}\tilde{\boldsymbol{F}}} \tag{6.7}$$

is the tensor of elastic moduli. It is well known that this condition must hold pointwise in the body if $\tilde{\boldsymbol{F}}(\boldsymbol{x})$ is the gradient of an energy-minimizing deformation.

We shall also make use of the strain-dependent elastic moduli $\mathcal{C}(\tilde{\boldsymbol{E}})$, where

$$\tilde{\boldsymbol{E}} = \frac{1}{2}(\tilde{\boldsymbol{F}}^T\tilde{\boldsymbol{F}} - \boldsymbol{I}) \tag{6.8}$$

is the strain in which $\boldsymbol{I}$ is the identity for 3-space, and

$$\mathcal{C}(\tilde{\boldsymbol{E}}) = \mathcal{U}_{\tilde{\boldsymbol{E}}\tilde{\boldsymbol{E}}}\,, \tag{6.9}$$

in which

$$\mathcal{U}(\tilde{\boldsymbol{E}}) = \mathcal{W}(\tilde{\boldsymbol{F}}) \tag{6.10}$$

is the associated strain-energy function. An application of the chain rule, combined with the minor symmetries of $\mathcal{C}$, furnishes

$$\mathcal{M}(\tilde{\boldsymbol{F}})[\boldsymbol{A}] = \boldsymbol{A}\tilde{\boldsymbol{S}} + \tilde{\boldsymbol{F}}\mathcal{C}(\tilde{\boldsymbol{E}})[\tilde{\boldsymbol{F}}^T\boldsymbol{A}] \tag{6.11}$$

for any tensor $\boldsymbol{A}$, where

$$\tilde{\boldsymbol{S}} = \mathcal{U}_{\tilde{\boldsymbol{E}}} \tag{6.12}$$

is the symmetric second Piola-Kirchhoff stress, given in terms of the Piola stress by

$$\tilde{\boldsymbol{P}} = \tilde{\boldsymbol{F}}\tilde{\boldsymbol{S}}. \tag{6.13}$$

We assume $\mathscr{U}(\cdot)$ to be convex in a neighborhood of the origin in strain space, with the origin furnishing an isolated local minimum. Thus $\tilde{\boldsymbol{S}}$ vanishes at zero strain, and $\mathscr{C}(\mathbf{0})$ is positive definite in the sense that $\boldsymbol{A} \cdot \mathscr{C}(\mathbf{0})[\boldsymbol{A}] > 0$ for all non-zero symmetric $\boldsymbol{A}$. Then,

$$\tilde{\boldsymbol{S}} = \mathscr{C}(\mathbf{0})[\tilde{\boldsymbol{E}}] + o(\left|\tilde{\boldsymbol{E}}\right|). \tag{6.14}$$

It follows from (6.11), (6.14) that

$$\mathscr{M}(\boldsymbol{I})[\boldsymbol{A}] = \mathscr{C}(\mathbf{0})[\boldsymbol{A}] \tag{6.15}$$

and hence that our hypotheses yields strong ellipticity at zero strain, as in classical linear elasticity theory.

6.1.2 Small-Thickness Estimate of the Energy

Position in the reference placement of the plate may be written

$$\boldsymbol{x} = \boldsymbol{u} + \varsigma \boldsymbol{k} \tag{6.16}$$

where $\boldsymbol{u} \in \Omega$ and $\varsigma \in [-h/2, h/2]$. We assume the origin to lie on Ω. The projection

$$\mathbf{1} = \boldsymbol{I} - \boldsymbol{k} \otimes \boldsymbol{k}, \tag{6.17}$$

is the identity on the translation space Ω' of Ω. The three-dimensional deformation gradient satisfies $d\tilde{\boldsymbol{y}} = \tilde{\boldsymbol{F}} d\boldsymbol{x}$, where $\tilde{\boldsymbol{y}} = \tilde{\boldsymbol{\chi}}(\boldsymbol{x})$ is the position after deformation of the material point $\boldsymbol{x}$ and $\tilde{\boldsymbol{\chi}}$ is the deformation function. Using this with $\tilde{\boldsymbol{y}} = \hat{\boldsymbol{y}}(\boldsymbol{u}, \varsigma) = \tilde{\boldsymbol{\chi}}(\boldsymbol{u} + \varsigma \boldsymbol{k})$ and $d\boldsymbol{u} \in \Omega'$ yields the alternative representations

$$(\tilde{\boldsymbol{F}}\mathbf{1})d\boldsymbol{u} + \tilde{\boldsymbol{F}}\boldsymbol{k}\, d\varsigma = d\hat{\boldsymbol{y}} = (\nabla\hat{\boldsymbol{y}})d\boldsymbol{u} + \hat{\boldsymbol{y}}' d\varsigma, \tag{6.18}$$

where ∇ is the (two-dimensional) gradient with respect to $\boldsymbol{u}$ at fixed ς and the notation $(\cdot)'$ is used to denote $\partial(\cdot)/\partial\varsigma$ at fixed $\boldsymbol{u}$. It follows from $\tilde{\boldsymbol{F}} = \tilde{\boldsymbol{F}}\mathbf{1} + \tilde{\boldsymbol{F}}\boldsymbol{k} \otimes \boldsymbol{k}$ that

$$\hat{\boldsymbol{F}} = \nabla\hat{\boldsymbol{y}} + \hat{\boldsymbol{y}}' \otimes \boldsymbol{k}, \tag{6.19}$$

where $\hat{\boldsymbol{F}}(\boldsymbol{u}, \varsigma) = \tilde{\boldsymbol{F}}(\boldsymbol{u} + \varsigma\boldsymbol{k})$.

The total strain energy in a given deformation is

$$\mathscr{S} = \int_{\kappa} \mathscr{W}\left(\tilde{\boldsymbol{F}}(\boldsymbol{x})\right) dv = \int_{\Omega} \int_{-h/2}^{h/2} \mathscr{W}\left(\hat{\boldsymbol{F}}(\boldsymbol{u}, \varsigma)\right) d\varsigma\, da. \tag{6.20}$$

If $\tilde{\chi}(x)$ is sufficiently smooth, then by Leibniz' Rule and Taylor's Theorem, applied to the small parameter h,

$$\int_{-h/2}^{h/2} \mathscr{W}\big(\hat{\boldsymbol{F}}(\boldsymbol{u}, \varsigma)\big)\mathrm{d}\varsigma = h\mathscr{W}(\boldsymbol{F}) + \frac{1}{24}h^3\mathscr{W}'' + \cdots, \tag{6.21}$$

where, by the chain rule,

$$\mathscr{W}' = \boldsymbol{P}\cdot\boldsymbol{F}' \quad \text{and} \quad \mathscr{W}'' = \boldsymbol{P}'\cdot\boldsymbol{F}' + \boldsymbol{P}\cdot\boldsymbol{F}'' \tag{6.22}$$

in which

$$\boldsymbol{F}^{(n)} = \hat{\boldsymbol{F}}^{(n)}_{|\varsigma=0} \tag{6.23}$$

and

$$\boldsymbol{P} = \tilde{\boldsymbol{P}}(\boldsymbol{F}), \qquad \boldsymbol{P}' = \mathscr{M}(\boldsymbol{F})[\boldsymbol{F}']. \tag{6.24}$$

From (6.19) we have

$$\hat{\boldsymbol{F}}' = \nabla\hat{\boldsymbol{y}}' + \hat{\boldsymbol{y}}''\otimes\boldsymbol{k} \quad \text{and} \quad \hat{\boldsymbol{F}}'' = \nabla\hat{\boldsymbol{y}}'' + \hat{\boldsymbol{y}}'''\otimes\boldsymbol{k}. \tag{6.25}$$

It follows that

$$\boldsymbol{F} = \nabla\boldsymbol{r} + \boldsymbol{d}\otimes\boldsymbol{k}, \quad \boldsymbol{F}' = \nabla\boldsymbol{d} + \boldsymbol{g}\otimes\boldsymbol{k}, \quad \text{and} \quad \boldsymbol{F}'' = \nabla\boldsymbol{g} + \boldsymbol{h}\otimes\boldsymbol{k}, \tag{6.26}$$

where

$$\boldsymbol{r} = \boldsymbol{y}, \quad \boldsymbol{d} = \boldsymbol{y}', \quad \boldsymbol{g} = \boldsymbol{y}'' \quad \text{and} \quad \boldsymbol{h} = \boldsymbol{y}''', \tag{6.27}$$

in which

$$\boldsymbol{y}^{(n)} = \hat{\boldsymbol{y}}^{(n)}_{|\varsigma=0} \tag{6.28}$$

are *independent* functions of $\boldsymbol{u} \in \Omega$. These are the coefficient vectors in the order—ς^3 expansion

$$\hat{\boldsymbol{y}}(\boldsymbol{u}, \varsigma) = \boldsymbol{r}(\boldsymbol{u}) + \varsigma\,\boldsymbol{d}(\boldsymbol{u}) + \frac{1}{2}\varsigma^2\boldsymbol{g}(\boldsymbol{u}) + \frac{1}{6}\varsigma^3\boldsymbol{h}(\boldsymbol{u}) + \cdots. \tag{6.29}$$

Here $\boldsymbol{r}(\boldsymbol{u})$ is the position of a material point on the deformed image ω of the midsurface Ω; its gradient $\nabla\boldsymbol{r}$ maps Ω' to the tangent plane T_ω to ω at the material point $\boldsymbol{u}$, and the functions $\boldsymbol{d}(\boldsymbol{u})$, $\boldsymbol{g}(\boldsymbol{u})$ and $\boldsymbol{h}(\boldsymbol{u})$—the *director* fields—provide information about the three-dimensional deformation in the vicinity of the midplane.

The strain energy may be expanded as

$$\mathscr{S} = \int_\kappa \mathscr{W}\big(\tilde{\boldsymbol{F}}(\boldsymbol{x})\big)\mathrm{d}v = S + o(h^3), \tag{6.30}$$

where

$$S = \int_{\Omega} W(\boldsymbol{d}, \boldsymbol{g}, \boldsymbol{h}, \nabla \boldsymbol{r}, \nabla \boldsymbol{d}, \nabla \boldsymbol{g}) da\,, \tag{6.31}$$

in which

$$W = h\mathscr{W}(\nabla \boldsymbol{r} + \boldsymbol{d} \otimes \boldsymbol{k}) + \frac{1}{24}h^3\{\boldsymbol{P} \cdot (\nabla \boldsymbol{g} + \boldsymbol{h} \otimes \boldsymbol{k}) + \boldsymbol{P}' \cdot (\nabla \boldsymbol{d} + \boldsymbol{g} \otimes \boldsymbol{k})\} \tag{6.32}$$

is the order—h^3 strain energy per unit area of Ω. We will see that this formula subsumes the strain energies associated with conventional membrane theory and inextensional bending theory.

We remark that this expression does not furnish the complete strain energy for the order—ς^3 truncation of the three-dimensional deformation. The latter contributes additional terms at higher order in h. However, rather than model a given truncation, our objective here is an accurate order—h^3 expression for the potential energy that is as accurate as possible by the standard of the three-dimensional theory and which yields a meaningful minimization problem in its own right.

To obtain an order—h^3 expansion of the potential energy of the loads, we first consider the simplest case in which $\partial\Omega$ consists of the union of disjoint arcs $\partial\Omega_e$ and $\partial\Omega_n$, where essential and natural boundary conditions, respectively, are specified. For example, suppose three-dimensional position is assigned on $\partial\kappa_{C_e} = \partial\Omega_e \times C$, where $C = [-h/2, h/2]$. We refer to this as a clamped edge. If dead loads are assigned on $\partial\kappa_{C_n} = \partial\Omega_n \times C$, then the potential energy of the three-dimensional body is $\mathscr{E} = \mathscr{S} - \mathscr{L}$, where $\mathscr{S}$ is the total strain energy defined by (6.20), and

$$\mathscr{L} = \int_{\partial\Omega_n} \left(\int_{-h/2}^{h/2} \tilde{\boldsymbol{p}} \cdot \tilde{\boldsymbol{\chi}}\, d\varsigma\right) da \tag{6.33}$$

is the load potential, in which $\tilde{\boldsymbol{p}}(\boldsymbol{x}) = \hat{\boldsymbol{p}}(\boldsymbol{u}, \varsigma)$ is the assigned (three-dimensional) Piola traction. Using a formula like (6.21), we can easily show that

$$\mathscr{L} = L + o(h^3), \tag{6.34}$$

where

$$L = \int_{\partial\Omega_n} \chi(\boldsymbol{r}, \boldsymbol{d}, \boldsymbol{g})\, da \tag{6.35}$$

with

$$\chi(\boldsymbol{r}, \boldsymbol{d}, \boldsymbol{g}) = \boldsymbol{p}_r \cdot \boldsymbol{r} + \boldsymbol{p}_d \cdot \boldsymbol{d} + \boldsymbol{p}_g \cdot \boldsymbol{g}\,, \tag{6.36}$$

and with

$$\boldsymbol{p}_r = h\boldsymbol{p} + \frac{1}{24}h^3\boldsymbol{p}'', \quad \boldsymbol{p}_d = \frac{1}{12}h^3\boldsymbol{p}' \text{ and } \boldsymbol{p}_g = \frac{1}{24}h^3\boldsymbol{p}, \tag{6.37}$$

in which $\boldsymbol{p}_r$, $\boldsymbol{p}_d$ and $\boldsymbol{p}_g$ are assigned and the primes identify derivatives of the three-dimensional traction with respect to ς, evaluated at $\varsigma = 0$. The order—h^3 estimate of the potential energy is thus given by

$$E = \int_\Omega W \,\mathrm{d}a - L\,, \tag{6.38}$$

with

$$L = \int_{\partial\Omega_n} (\boldsymbol{p}_r \cdot \boldsymbol{r} + \boldsymbol{p}_d \cdot \boldsymbol{d} + \boldsymbol{p}_g \cdot \boldsymbol{g})\,\mathrm{d}s, \tag{6.39}$$

Comparison with (6.5) furnishes

$$E_1 = \int_\Omega \mathscr{W}(\nabla \boldsymbol{r} + \boldsymbol{d} \otimes \boldsymbol{k})\,\mathrm{d}a - \int_{\partial\Omega_n} \boldsymbol{p} \cdot \boldsymbol{r}\,\mathrm{d}s \tag{6.40}$$

and

$$24E_3 = \int_\Omega \{\mathscr{M}(\boldsymbol{F})[\nabla \boldsymbol{d} + \boldsymbol{g} \otimes \boldsymbol{k}] \cdot (\nabla \boldsymbol{d} + \boldsymbol{g} \otimes \boldsymbol{k}) + \boldsymbol{P} \cdot (\nabla \boldsymbol{g} + \boldsymbol{h} \otimes \boldsymbol{k})\}\,da$$
$$- \int_{\partial\Omega_n} (\boldsymbol{p}'' \cdot \boldsymbol{r} + 2\boldsymbol{p}' \cdot \boldsymbol{d} + \boldsymbol{p} \cdot \boldsymbol{g})\,\mathrm{d}s. \tag{6.41}$$

Later we will require the Piola-Nanson formula (see also Eq. (2.19))

$$\alpha \boldsymbol{n} = \boldsymbol{F}^* \boldsymbol{k} \tag{6.42}$$

with $\alpha = |\boldsymbol{F}^*\boldsymbol{k}|$ is the areal stretch of the midplane and $\boldsymbol{n}$ is the unit normal to the tangent plane T_ω to ω at the material point $\boldsymbol{u}$. We note that the determinant of the deformation gradient, evaluated at the midplane, is $J = \boldsymbol{F}^*\boldsymbol{k} \cdot \boldsymbol{F}\boldsymbol{k}$. Thus,

$$J = \alpha\, \boldsymbol{n} \cdot \boldsymbol{d}. \tag{6.43}$$

Accordingly, the requirement $J > 0$ typically imposed in continuum mechanics is equivalent to the requirement $\boldsymbol{d} \in S_+$, where S_+ is the half-space

$$S_+ = \{\boldsymbol{v} \,|\, \boldsymbol{v} \cdot \boldsymbol{n} > 0\}. \tag{6.44}$$

We do not impose bulk incompressibility ($J = 1$), although doing so presents no difficulty [31].

6.1.3 Membrane Limit

Membrane theory is associated with the leading order energy in (6.5). Thus,

$$\mathscr{E}/h = E_m + o(h)/h, \tag{6.45}$$

where

$$E_m = \int_\Omega \mathscr{W}(\nabla \boldsymbol{r} + \boldsymbol{d} \otimes \boldsymbol{k})\,\mathrm{d}a - \int_{\partial\Omega_n} \boldsymbol{p} \cdot \boldsymbol{r}\,\mathrm{d}s \tag{6.46}$$

is the membrane energy in the dead-load boundary-value problem.

The energy is stationary with respect to $\boldsymbol{d}$ if and only if the membrane is in a state of plane stress; i.e.,

$$\{\mathscr{W}_F(\nabla \boldsymbol{r} + \boldsymbol{d} \otimes \boldsymbol{k})\}\boldsymbol{k} = \boldsymbol{0}. \tag{6.47}$$

From (6.1) and (6.13) we then have that $\boldsymbol{Sk} = \boldsymbol{0}$, which combines with the symmetry of $\boldsymbol{S}$ to yield

$$\boldsymbol{S} = S_{\alpha\beta}\boldsymbol{e}_\alpha \otimes \boldsymbol{e}_\beta\,. \tag{6.48}$$

To prove that (6.47) may be solved uniquely for $\boldsymbol{d}$, we first show that any solution, $\bar{\boldsymbol{d}}$ say, minimizes $\mathscr{W}$ pointwise. To this end we fix $\nabla \boldsymbol{r}$ and define $M(\boldsymbol{d}) = \mathscr{W}(\nabla \boldsymbol{r} + \boldsymbol{d} \otimes \boldsymbol{k})$. Let $\boldsymbol{d}(u)$ be a twice-differentiable function. The derivatives of $\sigma(u) = M\big(\boldsymbol{d}(u)\big)$ are

$$\dot{\sigma} = \dot{\boldsymbol{d}} \cdot \boldsymbol{P}(\nabla \boldsymbol{r} + \boldsymbol{d} \otimes \boldsymbol{k})\boldsymbol{k} = \dot{\boldsymbol{d}} \cdot M_{\boldsymbol{d}} \tag{6.49}$$

and

$$\begin{aligned} \ddot{\sigma} &= \ddot{\boldsymbol{d}} \cdot \boldsymbol{P}(\nabla \boldsymbol{r} + \boldsymbol{d} \otimes \boldsymbol{k})\boldsymbol{k} + \dot{\boldsymbol{d}} \otimes \boldsymbol{k} \cdot \mathscr{M}(\nabla \boldsymbol{r} + \boldsymbol{d} \otimes \boldsymbol{k})[\dot{\boldsymbol{d}} \otimes \boldsymbol{k}] \\ &= \ddot{\boldsymbol{d}} \cdot M_{\boldsymbol{d}} + \dot{\boldsymbol{d}} \cdot (M_{\boldsymbol{dd}})\dot{\boldsymbol{d}}\,. \end{aligned} \tag{6.50}$$

Thus,

$$M_{\boldsymbol{d}}(\bar{\boldsymbol{d}}) = \boldsymbol{P}(\nabla \boldsymbol{r} + \bar{\boldsymbol{d}} \otimes \boldsymbol{k})\boldsymbol{k} \tag{6.51}$$

vanishes by (6.47), whereas

$$M_{\boldsymbol{dd}}(\bar{\boldsymbol{d}}) = \mathscr{A}(\nabla \boldsymbol{r} + \bar{\boldsymbol{d}} \otimes \boldsymbol{k}), \tag{6.52}$$

where $\mathscr{A}(\boldsymbol{F})$ is the acoustic tensor defined, for any vector $\boldsymbol{v}$, by

$$\mathscr{A}(\boldsymbol{F})\boldsymbol{v} = \{\mathscr{M}(\boldsymbol{F})[\boldsymbol{v} \otimes \boldsymbol{k}]\}\boldsymbol{k}. \tag{6.53}$$

This is positive definite by virtue of the strong ellipticity condition (6.6).

We conclude that $\ddot{\sigma} > 0$ on straight-line paths defined by $\boldsymbol{d}(u) = u\boldsymbol{d}_2 + (1 - u)\boldsymbol{d}_1$ with $\boldsymbol{d}_1, \boldsymbol{d}_2 \in S_+$ fixed and $0 \le u \le 1$. These paths are admissible because the domain S_+ of $M(\cdot)$ is convex. Integrating with respect to u yields $\dot{\sigma}(u) > \dot{\sigma}(0)$ for $u \in (0, 1]$ and hence $\sigma(1) - \sigma(0) > \dot{\sigma}(0)$, proving that $M(\boldsymbol{d})$ is a strictly convex function; i.e.,

$$M(\boldsymbol{d}_2) - M(\boldsymbol{d}_1) > M_{\boldsymbol{d}}(\boldsymbol{d}_1) \cdot (\boldsymbol{d}_2 - \boldsymbol{d}_1), \tag{6.54}$$

for all unequal pairs $\boldsymbol{d}_1, \boldsymbol{d}_2$. It follows that M is minimized absolutely at a stationary point and hence that (6.47) has a unique solution $\bar{\boldsymbol{d}}(\nabla \boldsymbol{r})$.

An interesting corollary is that for a given midplane deformation, the strain energy is minimized absolutely when the midplane is in a state of plane stress.

With this solution incorporated, the membrane energy reduces to the functional of $\boldsymbol{r}$ defined by (6.45) or (6.46) with their integrands replaced by $\mathscr{W}\left(\nabla \boldsymbol{r} + \bar{\boldsymbol{d}}(\nabla \boldsymbol{r}) \otimes \boldsymbol{k}\right)$. However, it transpires that this function fails to satisfy the relevant (two-dimensional) Legendre-Hadamard condition, which must be satisfied pointwise by any stable (i.e., energy-minimizing, see Sect. 3.4) state, even if $\mathscr{W}$ is strongly elliptic in the three-dimensional sense. This is due to the presence of compressive stresses in the plane stress-deformation relation, whereas such stresses are precluded by the Legendre-Hadamard condition.

To verify this claim, we define the membrane strain-energy function

$$W(\nabla \boldsymbol{r}) = \mathscr{W}\left(\nabla \boldsymbol{r} + \bar{\boldsymbol{d}}(\nabla \boldsymbol{r}) \otimes \boldsymbol{k}\right). \tag{6.55}$$

Its derivatives are

$$\partial W / \partial r_{i,\alpha} = \partial \mathscr{W} / \partial F_{i\alpha} + \left(\partial \mathscr{W} / \partial d_j\right) K_{ji\alpha}, \tag{6.56}$$

where $K_{ji\alpha} = \partial \bar{d}_j / \partial F_{i\alpha}$ and F_{iA} are the components of $(6.26)_1$ with $F_{i\alpha} = r_{i,\alpha}$ and $F_{i3} = d_i$. The derivatives $\partial \mathscr{W} / \partial d_j$ vanish identically by (6.47), yielding

$$\partial W / \partial r_{i,\alpha} = P_{i\alpha} . \tag{6.57}$$

The associated moduli are

$$E_{i\alpha j\beta} = \partial^2 W / \partial r_{i,\alpha} \partial r_{j,\beta} . \tag{6.58}$$

The operative Legendre-Hadamard necessary condition for energy minimizers is (see Sect. 3.4 or [14])

$$E_{i\alpha j\beta} x_i x_j y_\alpha y_\beta \geq 0 \quad \text{for all} \quad x_i, y_\alpha. \tag{6.59}$$

For frame-invariant strain energies, this has the interesting consequence that the symmetric (plane) second Piola-Kirchhoff stress $S_{\alpha\beta}$, defined by $P_{i\beta} = F_{i\alpha} S_{\alpha\beta}$, satisfies

$$S_{\alpha\beta} y_\alpha y_\beta \geq 0, \tag{6.60}$$

and is thus positive semi-definite.

To see this we observe that by virtue of frame invariance, W is a function, U say, of the surface metric $a_{\alpha\beta} = F_{i\alpha} F_{i\beta}$, or, equivalently, of the surface strain $\varepsilon_{\alpha\beta} = \frac{1}{2}(F_{i\alpha} F_{i\beta} - \delta_{\alpha\beta})$, where $\delta_{\alpha\beta}$ is the Kronecker delta. The chain rule then yields

$$S_{\alpha\beta} = \partial U / \partial \varepsilon_{\alpha\beta}, \tag{6.61}$$

in which we understand $\varepsilon_{\alpha\beta}$ to be replaced by $\frac{1}{2}(\varepsilon_{\alpha\beta} + \varepsilon_{\beta\alpha})$ in the function U, with $\varepsilon_{\alpha\beta}$ and $\varepsilon_{\beta\alpha}$ being regarded as independent when computing the partial derivative. With this it follows by straightforward application of the chain rule that

$$E_{i\alpha j\beta} = \delta_{ij} S_{\alpha\beta} + F_{i\mu} F_{j\lambda} D_{\alpha\mu\beta\lambda}, \tag{6.62}$$

where

$$D_{\alpha\mu\beta\lambda} = \partial^2 U / \partial \varepsilon_{\alpha\mu} \partial \varepsilon_{\beta\lambda} \tag{6.63}$$

are the plane-stress elastic moduli. These possess the usual major and minor symmetries.

The Legendre-Hadamard condition (6.59) may thus be reduced to

$$x_i\, x_i\, S_{\alpha\beta}\, y_\alpha\, y_\beta + D_{\alpha\mu\beta\lambda}\, y_\alpha\, z_\mu\, y_\beta\, z_\lambda \geq 0, \tag{6.64}$$

where $z_\mu = x_i F_{i\mu}$ is a two-vector on the undeformed midplane. For the choice $x_i = n_i$—the unit normal to the *deformed* midsurface—z_μ vanishes. We then obtain (6.60) and the conclusion that energy minimizers yield a stress field that is pointwise positive semi-definite. This severe restriction means that boundary-value problems based on W will generally fail to have energy minimizing solutions. In such circumstances well-posedness may be restored via *relaxation*, in which the function W is replaced by its *quasiconvexification* [6]; i.e., the largest quasiconvex function not exceeding W anywhere on its domain. The latter automatically satisfies the Legendre-Hadamard inequality at all deformations and provides the foundation for the *tension-field theory* of elastic membranes [23].

6.1.4 Pure Bending

For deformations that generate zero strain at the midplane, our constitutive hypotheses imply that the midplane stress $\boldsymbol{P}$ and edge traction $\boldsymbol{p}$ vanish identically, and hence that

$$\mathscr{E}/h^3 = E_b + o(h^3)/h^3, \tag{6.65}$$

where

$$24E_b = \int_{\Omega} \mathcal{M}(\boldsymbol{F})[\nabla \boldsymbol{d} + \boldsymbol{g} \otimes \boldsymbol{k}] \cdot (\nabla \boldsymbol{d} + \boldsymbol{g} \otimes \boldsymbol{k})\, \mathrm{d}a - \int_{\partial\Omega_n} (\boldsymbol{p}'' \cdot \boldsymbol{r} + 2\boldsymbol{p}' \cdot \boldsymbol{d})\, \mathrm{d}s \tag{6.66}$$

in the case of dead loading. Moreover, the midplane value of the deformation gradient is then a rotation, $\boldsymbol{R}$ say, implying that $\nabla \boldsymbol{r} = \boldsymbol{R}\boldsymbol{1}$ and $\boldsymbol{d} = \boldsymbol{R}\boldsymbol{k} = \boldsymbol{n}$, the unit normal to the deformed midsurface. Thus $\boldsymbol{d}$ is determined by $\nabla \boldsymbol{r}$; we write $\boldsymbol{d} = \bar{\boldsymbol{d}}(\nabla \boldsymbol{r})$ as before. It follows that E_b is a functional of the midplane deformation $\boldsymbol{r}$ and the vector field $\boldsymbol{g}$.

This energy is stationary with respect to $\boldsymbol{g}$ if and only if

$$\{\mathcal{M}(\boldsymbol{F})[\nabla \bar{\boldsymbol{d}} + \boldsymbol{g} \otimes \boldsymbol{k}]\}\boldsymbol{k} = \boldsymbol{0}, \tag{6.67}$$

or

$$\mathcal{A}(\boldsymbol{F})\boldsymbol{g} = -\{\mathcal{M}(\boldsymbol{F})[\nabla \bar{\boldsymbol{d}}]\}\boldsymbol{k}, \tag{6.68}$$

where $\mathcal{A}(\boldsymbol{F})$ is the acoustic tensor defined by (6.53). Thus (6.67) has the unique solution $\boldsymbol{g} = \bar{\boldsymbol{g}}(\nabla \boldsymbol{r}, \nabla\nabla \boldsymbol{r})$, say.

The solution $\bar{\boldsymbol{g}}$ also minimizes the strain energy. To see this we fix $\nabla \boldsymbol{r}$ and define

$$B(\boldsymbol{g}) = \frac{1}{2}\mathcal{M}(\nabla \boldsymbol{r} + \bar{\boldsymbol{d}} \otimes \boldsymbol{k})[\nabla \bar{\boldsymbol{d}} + \boldsymbol{g} \otimes \boldsymbol{k}] \cdot (\nabla \bar{\boldsymbol{d}} + \boldsymbol{g} \otimes \boldsymbol{k}). \tag{6.69}$$

Consider a parametrized path $\boldsymbol{g}(u)$. The derivatives of $\sigma(u) = B(\boldsymbol{g}(u))$ with respect to u are

$$\dot{\sigma} = \dot{\boldsymbol{g}} \cdot \{\mathcal{M}(\nabla \boldsymbol{r} + \bar{\boldsymbol{d}} \otimes \boldsymbol{k})[\nabla \bar{\boldsymbol{d}} + \boldsymbol{g} \otimes \boldsymbol{k}]\}\boldsymbol{k} = \dot{\boldsymbol{g}} \cdot B_g \tag{6.70}$$

and

$$\begin{aligned}\ddot{\sigma} &= \ddot{\boldsymbol{g}} \cdot \{\mathcal{M}(\nabla \boldsymbol{r} + \bar{\boldsymbol{d}} \otimes \boldsymbol{k})[\nabla \bar{\boldsymbol{d}} + \boldsymbol{g} \otimes \boldsymbol{k}]\}\boldsymbol{k} + \dot{\boldsymbol{g}} \otimes \boldsymbol{k} \cdot \mathcal{M}(\nabla \boldsymbol{r} + \bar{\boldsymbol{d}} \otimes \boldsymbol{k})[\dot{\boldsymbol{g}} \otimes \boldsymbol{k}] \\ &= \ddot{\boldsymbol{g}} \cdot B_g + \dot{\boldsymbol{g}} \cdot (B_{gg})\dot{\boldsymbol{g}}\,,\end{aligned} \tag{6.71}$$

where we have used the major symmetry of $\mathcal{M}$. Thus,

$$B_g(\bar{\boldsymbol{g}}) = \{\mathcal{M}(\nabla \boldsymbol{r} + \bar{\boldsymbol{d}} \otimes \boldsymbol{k})[\nabla \bar{\boldsymbol{d}} + \bar{\boldsymbol{g}} \otimes \boldsymbol{k}]\}\boldsymbol{k} \tag{6.72}$$

vanishes by (6.67), whereas

$$B_{gg} = \mathcal{A}(\nabla \boldsymbol{r} + \bar{\boldsymbol{d}} \otimes \boldsymbol{k}). \tag{6.73}$$

Then, $\ddot{\sigma} > 0$ on straight-line paths defined by $\boldsymbol{g}(u) = u\boldsymbol{g}_2 + (1-u)\boldsymbol{g}_1$ with $\boldsymbol{g}_1, \boldsymbol{g}_2$ fixed and $0 \le u \le 1$. These paths belong to the linear space of 3-vectors, a convex set. Integrating with respect to u yields $\dot{\sigma}(u) > \dot{\sigma}(0)$ for $u \in (0, 1]$ and $\sigma(1) - \sigma(0) > \dot{\sigma}(0)$. This establishes the strict convexity of the function $B(\boldsymbol{g})$:

$$B(g_2) - B(g_1) > B_g(g_1) \cdot (g_2 - g_1) \tag{6.74}$$

for all unequal pairs $\boldsymbol{g}_1, \boldsymbol{g}_2$. It follows that B is minimized absolutely at a stationary point and thus that the solution $\bar{\boldsymbol{g}}$ to (6.68) furnishes the optimal order—h^3 energy.

The explicit energy is obtained from (6.66) on noting, from (6.11) with $\boldsymbol{F} = \boldsymbol{R}$, that

$$\mathcal{M}(\boldsymbol{R})[\nabla \boldsymbol{n} + \bar{\boldsymbol{g}} \otimes \boldsymbol{k}] \cdot (\nabla \boldsymbol{n} + \bar{\boldsymbol{g}} \otimes \boldsymbol{k}) = \boldsymbol{B} \cdot \mathcal{C}(\boldsymbol{0})[\boldsymbol{B}], \tag{6.75}$$

where

$$\boldsymbol{B} = \boldsymbol{R}^T[\nabla \boldsymbol{n} + \bar{\boldsymbol{g}} \otimes \boldsymbol{k}] \tag{6.76}$$

is the bending strain [9, 24].

Further, (6.65) implies that E_b furnishes the rigorous leading-order energy for isometric deformations of the midplane in the limit as thickness tends to zero. This result is in precise agreement with that obtained by formal asymptotic expansions [8] and the method of gamma convergence [9]. Nevertheless the result is not entirely satisfactory. For, Gauss' *Theorema Egregium* implies that the deformed midsurface is necessarily developable; i.e., that it is a cylinder or a cone. Accordingly, E_b does not furnish a model of plates that can be used in general applications.

6.1.5 Asymptotic Model for Combined Bending and Stretching

Having derived the order—h^3 expansion of the potential energy for a three-dimensional deformation (cf. (6.5), (6.40), (6.41)), we use it to derive energetically optimal director fields $\boldsymbol{d}$ and $\boldsymbol{g}$ for a given midplane deformation $\boldsymbol{r}$. That is, we minimize the energy with respect to these director fields at a fixed midplane deformation. Accordingly, we impose the stationarity condition

$$h\dot{E}_1 + h^3\dot{E}_3 + o(h^3) = 0, \tag{6.77}$$

in which the superposed dot refers to the variational (or Gateaux) derivative. We regard this as an asymptotic expansion of the three-dimensional equilibrium statement $\mathcal{E}^{\cdot} = 0$. Accordingly, we require

$$\dot{E}_1 = 0 \quad \text{and} \quad \dot{E}_3 = 0. \tag{6.78}$$

The first of these follows simply on dividing (6.77) by h and evaluating the resulting equation in the limit $h \to 0$; the second result then follows on dividing by h^3 and passing to the same limit.

For a fixed midplane deformation ($\dot{\boldsymbol{r}} = \boldsymbol{0}$), Eq. $(6.78)_1$ reduces to

$$\int_{\Omega} \mathscr{W}_{\boldsymbol{F}} \cdot \dot{\boldsymbol{d}} \otimes \boldsymbol{k} \, da = 0\,, \tag{6.79}$$

in which the variation $\dot{\boldsymbol{d}}$ is arbitrary in Ω. Accordingly, by the Fundamental Lemma,

$$\boldsymbol{P}\boldsymbol{k} = \boldsymbol{0} \quad \text{in} \quad \Omega, \tag{6.80}$$

where $\boldsymbol{P} = \tilde{\boldsymbol{P}}(\nabla \boldsymbol{r} + \boldsymbol{d} \otimes \boldsymbol{k})$. This of course is just the plane-stress condition (6.47), yielding $\boldsymbol{d} = \bar{\boldsymbol{d}}(\nabla \boldsymbol{r})$ and thus determining $\boldsymbol{d}$ in terms of the midplane deformation field $\boldsymbol{r}$. Because we are considering the latter to be fixed in the present discussion, Eq. $(6.78)_2$ reduces to

$$\int_{\Omega} \left\{ 2\mathscr{M}(\boldsymbol{F})[\nabla \bar{\boldsymbol{d}} + \boldsymbol{g} \otimes \boldsymbol{k}] \cdot \dot{\boldsymbol{g}} \otimes \boldsymbol{k} + \boldsymbol{P} \cdot (\nabla \dot{\boldsymbol{g}} + \dot{\boldsymbol{h}} \otimes \boldsymbol{k}) \right\} da - \int_{\partial \Omega_n} \boldsymbol{p} \cdot \dot{\boldsymbol{g}} \, ds = 0, \tag{6.81}$$

where the major symmetry of $\mathscr{M}$ has been invoked. Invoking (6.80), integrating the term involving $\nabla \dot{\boldsymbol{g}}$ by parts using Green's theorem, and imposing $\dot{\boldsymbol{g}}_{|\partial \Omega} = \boldsymbol{0}$, this becomes

$$\int_{\Omega} [2\boldsymbol{P}'\boldsymbol{k} - \operatorname{div}(\boldsymbol{P}\boldsymbol{1})] \cdot \dot{\boldsymbol{g}} \, da + \int_{\partial \Omega_n} (\boldsymbol{P}\boldsymbol{1}\boldsymbol{\nu} - \boldsymbol{p}) \cdot \dot{\boldsymbol{g}} \, ds = 0, \tag{6.82}$$

which implies that

$$2\boldsymbol{P}'\boldsymbol{k} = \operatorname{div}(\boldsymbol{P}\boldsymbol{1}) \quad \text{in} \quad \Omega \qquad \text{and} \qquad \boldsymbol{p} = \boldsymbol{P}\boldsymbol{1}\boldsymbol{\nu} \quad \text{on} \quad \partial \Omega_n\,. \tag{6.83}$$

The second of these results is in precise agreement with the three-dimensional theory. However, the first is not. To see this we note that in the three-dimensional theory (6.3) holds at all points of the plate and hence on the midplane in particular, where it reduces to

$$\operatorname{div}(\boldsymbol{P}\boldsymbol{1}) + \boldsymbol{P}'\boldsymbol{k} = \boldsymbol{0}. \tag{6.84}$$

We attempt to reconcile this with $(6.83)_1$ by using the three-dimensional theory to relate $\boldsymbol{P}'\boldsymbol{k}$ to the tractions $\tilde{\boldsymbol{p}}^{\pm}$ at the upper and lower lateral surfaces of the plate. With $\boldsymbol{N} = \pm \boldsymbol{k}$ at the upper and lower surfaces of the plate, respectively, a Taylor expansion of (6.2) gives

$$\tilde{\boldsymbol{p}}^{\pm} = \hat{\boldsymbol{p}}(\boldsymbol{u}, \pm h/2) = \pm \boldsymbol{P}\boldsymbol{k} + (h/2)\boldsymbol{P}'\boldsymbol{k} \pm (h^2/8)\boldsymbol{P}''\boldsymbol{k} \; + \; O(h^3), \tag{6.85}$$

which is equivalent to

$$\tilde{\boldsymbol{p}}^{+} + \tilde{\boldsymbol{p}}^{-} = h\boldsymbol{P}'\boldsymbol{k} \; + \; O(h^3) \qquad \text{and} \qquad \tilde{\boldsymbol{p}}^{+} - \tilde{\boldsymbol{p}}^{-} = 2\boldsymbol{P}\boldsymbol{k} \; + \; O(h^2). \tag{6.86}$$

If $\tilde{\boldsymbol{p}}^{\pm} = O(h^3)$, we conclude that

$$\boldsymbol{Pk} = O(h^2) \quad \text{and} \quad \boldsymbol{P}'\boldsymbol{k} = O(h^2). \tag{6.87}$$

The first of these is consistent with the prediction (6.80), whereas the second implies that $(6.83)_1$ and (6.84) are consistent with each other in the sense that both yield the estimate $\operatorname{div}(\boldsymbol{P1}) = O(h^2)$. With this information we may re-write (6.41) as

$$\begin{aligned} 24E_3 = &\int_{\Omega} \{\mathcal{M}(\boldsymbol{F})[\nabla\boldsymbol{d} + \boldsymbol{g}\otimes\boldsymbol{k}]\cdot(\nabla\boldsymbol{d} + \boldsymbol{g}\otimes\boldsymbol{k})\}\, da + \int_{\partial\Omega_e} \boldsymbol{P1\nu}\cdot\boldsymbol{g}\, ds \\ &- \int_{\partial\Omega_n} (\boldsymbol{p}''\cdot\boldsymbol{r} + 2\boldsymbol{p}'\cdot\boldsymbol{d})\, ds + O(h^2), \end{aligned} \tag{6.88}$$

where we have made use of $(6.83)_2$ and (6.80), the latter implying that $\boldsymbol{d} = \bar{\boldsymbol{d}}(\nabla\boldsymbol{r})$, which we assume to hold on the closure of Ω. Recalling that E_3 is multiplied by h^3 in the expansion (6.5), we are justified in suppressing terms of order $O(h^2)$ in E_3 as this does not affect the accuracy of the expansion. For consistency we must then suppress $\boldsymbol{P}'\boldsymbol{k}$; i.e., we must impose

$$\{\mathcal{M}(\boldsymbol{F})[\nabla\bar{\boldsymbol{d}} + \boldsymbol{g}\otimes\boldsymbol{k}]\}\boldsymbol{k} = \boldsymbol{0}, \tag{6.89}$$

which, as we have seen in the case of pure bending, uniquely determines $\boldsymbol{g} = \bar{\boldsymbol{g}}(\nabla\boldsymbol{r}, \nabla\nabla\boldsymbol{r})$ in Ω. With $\boldsymbol{g}_{|\partial\Omega_e}$ fixed by the data for the three-dimensional parent model, and with $\boldsymbol{P}$ now determined by $\nabla\boldsymbol{r}$, we conclude that if the part $\partial\Omega_e$ of the boundary is clamped; that is, if $\boldsymbol{r}$ and the normal derivative $\boldsymbol{r}_{,\nu}$ are assigned on $\partial\Omega_e$, then the full gradient $\nabla\boldsymbol{r}$, consisting of the normal and tangential derivatives of $\boldsymbol{r}$, is likewise fixed on $\partial\Omega_e$ and hence that the integral $\int_{\partial\Omega_e} \boldsymbol{P1\nu}\cdot\boldsymbol{g}\, ds$ is fixed by the given data in the three-dimensional parent theory. Accordingly it contributes an unimportant constant to the overall energy. This may be suppressed without loss of generality, and so we have

$$24E_3 = \int_{\Omega} \mathcal{M}(\boldsymbol{F})[\nabla\bar{\boldsymbol{d}} + \bar{\boldsymbol{g}}\otimes\boldsymbol{k}]\cdot(\nabla\bar{\boldsymbol{d}} + \bar{\boldsymbol{g}}\otimes\boldsymbol{k})\, da - \int_{\partial\Omega_n} (\boldsymbol{p}''\cdot\boldsymbol{r} + 2\boldsymbol{p}'\cdot\bar{\boldsymbol{d}})\, ds, \tag{6.90}$$

which is a functional of $\boldsymbol{r}$ alone.

With the foregoing results in effect the approximate energy becomes

$$\begin{aligned} E = h\Big\{&\int_{\Omega} \mathcal{W}(\nabla\boldsymbol{r} + \bar{\boldsymbol{d}}\otimes\boldsymbol{k})\, da - \int_{\partial\Omega_n} \boldsymbol{p}\cdot\boldsymbol{r}\, ds\Big\} \\ &+ \frac{1}{24}h^3\Big\{\int_{\Omega} \mathcal{M}(\boldsymbol{F})[\nabla\bar{\boldsymbol{d}} + \bar{\boldsymbol{g}}\otimes\boldsymbol{k}]\cdot(\nabla\bar{\boldsymbol{d}} + \bar{\boldsymbol{g}}\otimes\boldsymbol{k})\, da - \int_{\partial\Omega_n} (\boldsymbol{p}''\cdot\boldsymbol{r} + 2\boldsymbol{p}'\cdot\bar{\boldsymbol{d}})\, ds\Big\}, \end{aligned} \tag{6.91}$$

in the case of dead loading, in which $\boldsymbol{p}$, $\boldsymbol{p}'$ and $\boldsymbol{p}''$ are assigned on $\partial\Omega_n$.

The solution $\bar{\boldsymbol{g}}$ to (6.89) involves the gradient $\nabla\bar{\boldsymbol{d}}$ and may thus be expressed in terms of the first and second gradients, $\nabla\boldsymbol{r}$ and $\nabla\nabla\boldsymbol{r}$ respectively, of the midsurface deformation function $\boldsymbol{r}(\boldsymbol{u})$. To derive the explicit form of this function we observe that the function $\bar{\boldsymbol{d}}(\nabla\boldsymbol{r})$ satisfies (6.47) identically in $\nabla\boldsymbol{r}$. We write the latter in the form

$$\partial\mathscr{W}/\partial F_{i3} \equiv 0, \tag{6.92}$$

where $F_{i3} = d_i$, and differentiate with respect to $\nabla\boldsymbol{r} = r_{i,\alpha}\,\boldsymbol{e}_i \otimes \boldsymbol{e}_\alpha$ (with $\boldsymbol{e}_3 = \boldsymbol{k}$), obtaining

$$\mathscr{M}_{i3j\beta} + \mathscr{A}_{ik}\,K_{kj\beta} = 0, \tag{6.93}$$

where $\mathscr{A}_{ik}$ are the components of the acoustic tensor defined in (6.53) and

$$K_{kj\beta} = \partial\bar{d}_k/\partial r_{j,\beta}\,. \tag{6.94}$$

Accordingly,

$$K_{kj\beta} = -\mathscr{A}_{ki}^{-1}\mathscr{M}_{i3j\beta}\,, \tag{6.95}$$

and the chain rule yields

$$\bar{d}_{i,\alpha} = K_{ij\beta}\,r_{j,\beta\alpha}\,. \tag{6.96}$$

Then, combining (6.68) in the form

$$\mathscr{A}_{ik}g_k = -\mathscr{M}_{i3j\alpha}\,\bar{d}_{j,\alpha} \tag{6.97}$$

with (6.95), we conclude that

$$\bar{g}_i(\nabla\boldsymbol{r}, \nabla\nabla\boldsymbol{r}) = K_{ij\alpha}\,\bar{d}_{j,\alpha} = K_{ij\alpha}\,K_{jk\beta}\,r_{k,\alpha\beta} \tag{6.98}$$

and hence that the second integrand in (6.90) (or (6.91)) is a homogeneous quadratic function of the second derivatives $r_{k,\alpha\beta}$. Moreover, as noted in the discussion of pure-bending theory, $\bar{\boldsymbol{g}}$ minimizes the energy E with respect to $\boldsymbol{g}$.

In the foregoing reduction the deformation $\boldsymbol{r}$ was held fixed ($\dot{\boldsymbol{r}} = \boldsymbol{0}$) in $(6.78)_{1,2}$. The reason is that $(6.78)_1$ would otherwise yield the membrane problem which, as we have seen, fails to possess a solution unless the energy is replaced by its relaxation, whereas the purpose of the order—h^3 expansion is to regularize the membrane problem. In this case it is logical to regard E as the operative approximate energy and to render it stationary with respect to $\boldsymbol{r}$ to derive the relevant equilibrium problem. However, it transpires, rather unexpectedly, that we have failed to cure the ill-posedness of the (unrelaxed) membrane problem! Thus the minimization problem for E is also typically ill-posed; that is, that in general this problem possesses no solution.

6.1.6 Reflection Symmetry and Ill-Posedness

As we have noted, there is no reason to suppose that minimizers of E, if any, are related to those of the three-dimensional energy $\mathscr{E}$. Nevertheless, it is of interest to determine whether or not E admits minimizers. If a deformation $\boldsymbol{r}$ minimizes E, then it satisfies the operative Legendre-Hadamard condition pointwise in Ω (see Sect. 3.4). In the present context this is the requirement that the part of the energy density that is homogeneous quadratic in the second derivatives $r_{i,\alpha\beta}$ be non-negative definite when the $r_{i,\alpha\beta}$ are replaced by $v_i b_\alpha b_\beta$, with v_i an arbitrary 3-vector and b_α an arbitrary 2-vector [14]. This restriction affects only the second integrand in (6.90) (or (6.91)), which has the component form

$$I = \mathscr{M}_{i\alpha j\beta}\, \bar{d}_{i,\alpha}\, \bar{d}_{j,\beta} + 2\mathscr{M}_{i\alpha j3}\, \bar{d}_{i,\alpha}\, \bar{g}_j + \mathscr{M}_{i3j3}\, \bar{g}_i\, \bar{g}_j\, . \tag{6.99}$$

Substituting (6.96) and (6.95), after some algebra we obtain

$$I = G_{i\alpha j\beta}\, K_{ik\lambda}\, K_{jl\mu}\, r_{k,\lambda\alpha}\, r_{l,\beta\mu}\, , \tag{6.100}$$

where

$$G_{i\alpha j\beta} = \mathscr{M}_{i\alpha j\beta} - A_{kl}\, K_{ki\alpha}\, K_{lj\beta}\, . \tag{6.101}$$

The relevant Legendre-Hadamard condition is thus given by

$$G_{i\alpha j\beta}\, a_i\, a_j\, b_\alpha\, b_\beta \geq 0, \quad \text{where} \quad a_i = K_{ij\beta}\, v_j\, b_\beta\, , \tag{6.102}$$

and this must hold for every v_i and b_α.

It transpires that $G_{i\alpha j\beta} = E_{i\alpha j\beta}$, the moduli for pure membrane theory (cf. (6.58)). To see this we compute a further derivative of (6.56), obtaining

$$\begin{aligned} E_{i\alpha j\beta} &= \partial^2 \mathscr{W} / \partial F_{i\alpha} \partial F_{j\beta} + (\partial^2 \mathscr{W} / \partial F_{i\alpha} \partial d_k) K_{kj\beta} \\ &= \mathscr{M}_{i\alpha j\beta} + \mathscr{M}_{i\alpha k3}\, K_{kj\beta} \\ &= \mathscr{M}_{i\alpha j\beta} - A_{kl}\, K_{li\alpha}\, K_{kj\beta}\, , \end{aligned} \tag{6.103}$$

and the claim follows on comparison with (6.101).

In view of the discussion leading to (6.64) and (6.60), we conclude that if a_i can be chosen to be aligned with the normal to the deformed midsurface, then minimizers of E must again deliver a plane second Piola-Kirchhoff stress field that is pointwise positive semi-definite. Hilgers and Pipkin [15] have shown that this situation occurs in the practically important case in which the three-dimensional material possesses reflection symmetry with respect to the midplane Ω; i.e., $\mathscr{U}(\boldsymbol{E}) = \mathscr{U}(\boldsymbol{Q}\boldsymbol{E}\boldsymbol{Q}^T)$, with $\boldsymbol{Q} = \boldsymbol{I} - 2\boldsymbol{k} \otimes \boldsymbol{k}$. Once again this restriction on the state of stress implies that minimizers of E generally fail to exist and therefore that solution procedures relying on the construction of energy-minimizing sequences of deformations, such as the finite-element method, cannot be applied. This is a serious drawback for the practical implementation of the theory.

We conclude that the minimization problem for E is generally ill-posed, despite the fact that the fields $\bar{\boldsymbol{d}}$ and $\bar{\boldsymbol{g}}$ minimize the energy for any given midsurface position field. In [15] this is addressed by introducing ad hoc strain-gradient terms which ensure that the Legendre-Hadamard condition is automatically satisfied without qualification. However, these regularizing terms are unrelated to the three-dimensional parent theory. Accordingly, in [15] the existence issue is addressed at the expense of accuracy.

6.1.7 Koiter's Model

Another way to cure the problematic ill-posedness of E is simply to suppress the contribution of the stress to the order—h^3 term in the energy. To justify this simplification we may suppose that $S_{\alpha\beta} = O(h)$ at the outset. The constitutive hypotheses discussed in Sect. 6.1.1 then imply that the strain $\varepsilon_{\alpha\beta} = O(h)$, so that

$$S_{\alpha\mu} = D_{\alpha\mu\beta\lambda}(\mathbf{0})\varepsilon_{\beta\lambda} + o(h), \tag{6.104}$$

where $D_{\alpha\mu\beta\lambda}(\mathbf{0})$ are the classical plane-stress elastic moduli evaluated at zero strain. In view of (6.11) the stress $\boldsymbol{S}$ may then be suppressed in the second integral in (6.90) (or (6.91)) without affecting the order—h^3 accuracy of E. The operative Legendre-Hadamard condition then becomes

$$D_{\alpha\mu\beta\lambda}(\mathbf{0})\, y_\alpha\, w_\mu\, y_\beta\, w_\lambda \geq 0, \tag{6.105}$$

where $w_\mu = a_i F_{i\mu}$ and a_i is given by $(6.102)_2$. That this inequality is automatically satisfied may be seen on observing that, for any $\boldsymbol{A}$,

$$0 \leq \boldsymbol{A} \cdot \mathscr{C}(\mathbf{0})[\boldsymbol{A}] = D_{\alpha\mu\beta\lambda}(\mathbf{0})\, A_{\alpha\mu}\, A_{\beta\lambda} \tag{6.106}$$

under plane-stress conditions, and choosing $A_{\alpha\mu} = y_\alpha\, w_\mu$.

Thus in the present circumstances the midplane strain energy is approximated by

$$\frac{1}{2} D_{\alpha\mu\beta\lambda}(\mathbf{0})\, \varepsilon_{\alpha\mu}\, \varepsilon_{\beta\lambda} + o(h^2).$$

Because this is multiplied by h in (6.90) (or (6.91)), it then follows that

$$\mathscr{E}/h^3 = \bar{E} + o(h^3)/h^3\,,$$

where $\bar{E}$ involves the sum of the integrals of a homogeneous quadratic function of the surface strain and a homogeneous quadratic function of the bending strain. Thus energy minimizers in the parent theory minimize $\bar{E}$ at leading order in h. Further, the minimization problem for $\bar{E}$ is well posed [3]. The energy $\bar{E}$ is precisely Koiter's energy for combined bending and stretching [17, 27].

The difficulty, of course, is that the magnitude of the stress $S_{\alpha\beta}$ is not known *a priori* and thus that the assumptions underpinning Koiter's model, if true, can only be justified *a posteriori*.

We will present explicit details of this theory in the case of isotropy, including the relevant equilibrium equations, as a special case of those for shells in Sect. 6.2.

6.2 Koiter's Shell Theory

Koiter's nonlinear shell theory, which purports to be valid for isotropic shells undergoing finite deformations with small strains, enjoys a unique status. This stems in large part from the extensive assessment of it in recent years, on the part of Ciarlet and his school, as an approximation to elasticity theory for thin bodies. A large body of evidence is given in [3] in support of the contention that the Koiter model furnishes the 'best' all-around theory, despite the fact that it has not been obtained either as a Gamma-limit or an asymptotic limit of the three-dimensional theory. The important features of Koiter's theory are associated with the areal strain-energy density (see [3, p. 156] and [17, p. 30])

$$W = \frac{1}{2} h\, \boldsymbol{\varepsilon} \cdot \mathscr{D}[\boldsymbol{\varepsilon}] + \frac{1}{24} h^3 \boldsymbol{\rho} \cdot \mathscr{D}[\boldsymbol{\rho}], \tag{6.107}$$

where $\mathscr{D}$ is the positive-definite tensor of linear plane-stress elastic moduli, h is the thickness of the shell prior to deformation, and $\boldsymbol{\varepsilon}$ and $\boldsymbol{\rho}$ respectively are surface tensors that characterize the change in metric and curvature of the shell midsurface induced by deformation. Thus, according to a fundamental theorem of surface theory [3], the energy vanishes for rigid-body motions of the midsurface and is positive definite for any non-rigid motion.

As in prior chapters, commas followed by subscripts are used to denote partial derivatives with respect to convected surface coordinates, while vertical strokes are used for covariant derivatives on the reference surface.

Superposed tildes are again used to denote three-dimensional fields whereas variables appearing without the tilde are the restrictions of these fields to a midsurface Ω embedded in the three-dimensional shell-like body κ. Let h denote the thickness of the shell. Our basic assumption is that the shell is thin in the sense that $h/l \ll 1$, where l is any characteristic length associated with the geometry of Ω. This is the smaller of the minimum radius of curvature of the shell or the minimum spanwise dimension. As before we assume that all length scales have been non-dimensionalized by l *a priori*. We then have $h \ll 1$, where h is now the dimensionless thickness, and we seek an expression for the energy of the elastic shell that is valid to order h^3.

6.2.1 Geometric and Kinematic Formulae

We record a number of formulae pertaining to the geometry of the shell and to the kinematics of the material in a three-dimensional neighborhood of the midsurface. These either coincide with their counterparts in the linear theory—see Chap. 5—or are easily obtained from them by substituting the deformation gradient in place of the displacement gradient of the linear theory.

We seek a model for the shell involving as independent variables the coordinates θ^α that parametrize a curved base surface Ω. To this end we use the standard *normal-coordinate* parametrization of three dimensional space in the vicinity of the midsurface [3, 16, 17, 26]. Thus,

$$\boldsymbol{X}(\theta^\alpha, \varsigma) = \boldsymbol{x}(\theta^\alpha) + \varsigma\, \boldsymbol{N}(\theta^\alpha), \tag{6.108}$$

where $\boldsymbol{x}(\theta^\alpha)$ is the parametrization of Ω with unit-normal field $\boldsymbol{N}(\theta^\alpha)$ and ς is the coordinate in the direction perpendicular to Ω, the latter corresponding to $\varsigma = 0$. The lateral surfaces of the thin three-dimensional body are assumed to correspond to constant values of ς. These are separated by the distance h, the thickness of the shell. For simplicity we assume the thickness to be uniform.

The orientation of Ω is induced by the assumed right-handedness of the coordinate system $(\theta^\alpha, \varsigma)$; thus, $\boldsymbol{A}_1 \times \boldsymbol{A}_2 \cdot \boldsymbol{N} > 0$, where $\boldsymbol{A}_\alpha = \boldsymbol{x}_{,\alpha} \equiv \partial \boldsymbol{x}/\partial\theta^\alpha$ span the tangent plane $T_{\Omega(p)}$ to Ω at the point p with coordinates θ^α. The *curvature* $\boldsymbol{B}$ of the base surface is the symmetric linear map from $T_{\Omega(p)}$ to itself appearing in the Weingarten equation:

$$\mathrm{d}\boldsymbol{N} = -\boldsymbol{B}\,\mathrm{d}\boldsymbol{x}\,, \tag{6.109}$$

where $\mathrm{d}\boldsymbol{x} = \boldsymbol{A}_\alpha \mathrm{d}\theta^\alpha$ and $\mathrm{d}\boldsymbol{N} = \boldsymbol{N}_{,\alpha}\mathrm{d}\theta^\alpha$. Therefore,

$$\mathrm{d}\boldsymbol{X} = \mathrm{d}\boldsymbol{x} + \varsigma\,\mathrm{d}\boldsymbol{N} + \boldsymbol{N}\mathrm{d}\varsigma = \boldsymbol{G}(\mathrm{d}\boldsymbol{x} + \boldsymbol{N}\mathrm{d}\varsigma), \tag{6.110}$$

where

$$\boldsymbol{G} = \boldsymbol{\mu} + \boldsymbol{N}\otimes\boldsymbol{N}, \qquad \boldsymbol{\mu} = \boldsymbol{1} - \varsigma\,\boldsymbol{B}\,, \tag{6.111}$$

and

$$\boldsymbol{1} = \boldsymbol{I} - \boldsymbol{N}\otimes\boldsymbol{N} = \boldsymbol{A}_\alpha \otimes \boldsymbol{A}^\alpha \tag{6.112}$$

is the projection onto—and the identity transformation for – the tangent plane $T_{\Omega(p)}$, on which $\{\boldsymbol{A}^\alpha\}$ is dual to $\{\boldsymbol{A}_\alpha\}$.

In Sect. 6.2.2 we require the volume measure induced by the coordinates. This is (see (5.10)–(5.12)) $\mathrm{d}v = \mu\,\mathrm{d}\varsigma\,\mathrm{d}a$, where $\mathrm{d}a = \boldsymbol{N}\cdot\mathrm{d}\boldsymbol{x}_1 \times \mathrm{d}\boldsymbol{x}_2$ is the area measure on Ω, and

$$\mu = 1 - 2H\varsigma + \varsigma^2 K \tag{6.113}$$

is the (two-dimensional) determinant of $\boldsymbol{\mu}$ in which

$$H = \frac{1}{2}\,\mathrm{tr}\,\boldsymbol{B} \quad \text{and} \quad K = \det \boldsymbol{B} \tag{6.114}$$

are the mean and Gaussian curvatures of Ω, respectively. This may be written

$$\mu = (1 - \varsigma\kappa_1)(1 - \varsigma\kappa_2),$$

where κ_α—the *principal curvatures*—are the eigenvalues of $\boldsymbol{B}$. The transformation from $(\theta^\alpha, \varsigma)$ to $\boldsymbol{X}$ is one-to-one and orientation preserving if and only if $\mu > 0$. Following common practice [7, 19], we refer to the region of space in which this condition holds as *shell space*. This is the region containing the midsurface in which $|\varsigma| < \min\{r_1, r_2\}$, where $r_\alpha = |\kappa_\alpha|^{-1}$ are the principal radii of curvature. Equation (6.110) yields $d\varsigma(\boldsymbol{x}) = \boldsymbol{N} \cdot d\boldsymbol{X}$, which implies that the normal to the base surface is also normal to the lateral surfaces of the shell. We will need this result in Sect. 6.2.3.

Let C^* be the line orthogonal to Ω and intersecting Ω at a point with surface coordinates θ^α. Let $\partial\kappa_C = \partial\Omega \times C$, where C is the collection of such curves, be the ruled generating surface of the thin shell-like region κ obtained by translating the points of $\partial\Omega$ along their associated lines C^*. Let s measure arclength on the curve $\partial\Omega$ with unit tangent $\boldsymbol{\tau}$ and rightward unit normal $\boldsymbol{\nu} = \boldsymbol{\tau} \times \boldsymbol{N}$. The oriented differential surface measure induced by the $(s, \varsigma)-$ parametrization of $\partial\kappa_C$, required in Sect. 6.2.2, is [26] $\boldsymbol{G}^*\boldsymbol{\nu}\, ds\, d\varsigma$, where $\boldsymbol{G}^*$ is the cofactor of $\boldsymbol{G}$ (see Eq. (5.18)). From this it is possible to develop a formula for the oriented area measure on the ruled generators of the shell-like body which, in turn, facilitates the development of an expression for the potential energy of an edge-loaded shell. We do not do so explicitly here, but instead quote the relevant results from [26] as needed in the sequel.

The model to be developed requires expressions for the three-dimensional deformation gradient and its through-thickness derivatives. Let $\tilde{\chi}(\boldsymbol{X})$ be the three-dimensional deformation with gradient $\tilde{\boldsymbol{F}}(\boldsymbol{X})$. We define

$$\hat{\chi}(\theta^\alpha, \varsigma) = \tilde{\chi}\big(\boldsymbol{x}(\theta^\alpha) + \varsigma\boldsymbol{N}(\theta^\alpha)\big) \quad \text{and} \quad \hat{\boldsymbol{F}}(\theta^\alpha, \varsigma) = \tilde{\boldsymbol{F}}\big(\boldsymbol{x}(\theta^\alpha) + \varsigma\boldsymbol{N}(\theta^\alpha)\big). \tag{6.115}$$

Then,

$$\hat{\boldsymbol{F}}(\mu\, d\boldsymbol{x} + \boldsymbol{N} d\varsigma) = d\hat{\chi} = \hat{\chi}_{,\alpha} d\theta^\alpha + \hat{\chi}' d\varsigma = (\nabla\hat{\chi}) d\boldsymbol{x} + \hat{\chi}' d\varsigma\,, \tag{6.116}$$

where $d\theta^\alpha = \boldsymbol{A}^\alpha \cdot d\boldsymbol{x}$, $\hat{\chi}_{,\alpha} = \partial\hat{\chi}/\partial\theta^\alpha$, $\hat{\chi}' = \partial\hat{\chi}/\partial\varsigma$ and

$$\nabla\hat{\chi} = \hat{\chi}_{,\alpha} \otimes \boldsymbol{A}^\alpha \tag{6.117}$$

is the surface deformation gradient. Then,

$$\hat{\boldsymbol{F}}\boldsymbol{1}\mu = \nabla\hat{\chi} \quad \text{and} \quad \hat{\boldsymbol{F}}\boldsymbol{N} = \hat{\chi}', \tag{6.118}$$

in which the orthogonal decomposition $\hat{\boldsymbol{F}} = \hat{\boldsymbol{F}}\boldsymbol{1} + \hat{\boldsymbol{F}}\boldsymbol{N} \otimes \boldsymbol{N}$ has been used. Assuming the configuration κ to be contained in shell space, we conclude that

$$\hat{F} = (\nabla\hat{\chi})\mu^{-1} + \hat{\chi}' \otimes N. \tag{6.119}$$

The mid-surface value of $\hat{F}$, and those of its through-thickness derivatives $\hat{F}'$ and $\hat{F}''$, are needed in Sect. 6.2.2. These, and other variables defined on the midsurface, are identified by the absence of superposed carets. Following the procedure discussed in Chap. 5, they are found to be

$$F = \nabla r + d \otimes N, \qquad F' = \nabla d + (\nabla r)B + g \otimes N \tag{6.120}$$

and

$$F'' = \nabla g + 2(\nabla d)B + 2(\nabla r)B^2 + h \otimes N, \tag{6.121}$$

where

$$r = \chi, \quad d = \chi', \quad g = \chi'' \quad \text{and} \quad h = \chi''' \tag{6.122}$$

are mutually *independent* functions of θ^α and the operation $\nabla(\cdot)$ is defined by (6.117). The first of these is the mid-surface deformation field and the latter three are the *directors*. Together they furnish the coefficient vectors in the expansion

$$\hat{\chi} = r + \varsigma d + \frac{1}{2}\varsigma^2 g + \frac{1}{6}\varsigma^3 h + \cdots \tag{6.123}$$

of the three-dimensional deformation, where $r(\theta^\alpha) = \hat{\chi}(x(\theta^\alpha))$ is the position of a material point on the deformed image ω of the midsurface Ω; its gradient ∇r maps $T_{\Omega(p)}$ to the tangent plane $T_{\omega(p)}$ to ω at the material point p.

6.2.2 Expansion of the Three-Dimensional Energy

The strain energy of the shell is

$$\int_\kappa \mathcal{W}(\hat{F})\,dv = \int_\Omega W\,da, \tag{6.124}$$

where

$$W = \int_C \mu \mathcal{W}(\hat{F})\,d\varsigma, \tag{6.125}$$

with $C = [-h/2, h/2]$, is the areal strain-energy density on Ω. This differs from its counterpart in plate theory by the factor μ appearing in the integrand. As before we use Leibniz' Rule and a Taylor expansion, this time with the result

$$W = h(1 + \frac{1}{12}h^2 K)\mathcal{W} + \frac{1}{24}h^3(\mathcal{W}'' - 4H\mathcal{W}') + o(h^3), \tag{6.126}$$

where, for uniform materials,

$$\mathscr{W} = \mathscr{W}(\boldsymbol{F}), \quad \mathscr{W}' = \boldsymbol{P} \cdot \boldsymbol{F}', \quad \text{and} \quad \mathscr{W}'' = \boldsymbol{P}' \cdot \boldsymbol{F}' + \boldsymbol{P} \cdot \boldsymbol{F}'', \tag{6.127}$$

with

$$\boldsymbol{P}' = \mathscr{M}(\boldsymbol{F})[\boldsymbol{F}']. \tag{6.128}$$

The development is eased considerably by using the decompositions

$$\boldsymbol{P} = \boldsymbol{P1} + \boldsymbol{PN} \otimes \boldsymbol{N} \quad \text{and} \quad \boldsymbol{P}' = \boldsymbol{P}'\boldsymbol{1} + \boldsymbol{P}'\boldsymbol{N} \otimes \boldsymbol{N} \tag{6.129}$$

in (6.127). We obtain

$$\mathscr{W}' = \boldsymbol{P1} \cdot \left[\nabla \boldsymbol{d} + (\nabla \boldsymbol{r})\boldsymbol{B}\right] + \boldsymbol{PN} \cdot \boldsymbol{g} \tag{6.130}$$

and

$$\mathscr{W}'' = \boldsymbol{P}'\boldsymbol{1} \cdot \left[\nabla \boldsymbol{d} + (\nabla \boldsymbol{r})\boldsymbol{B}\right] + \boldsymbol{P}'\boldsymbol{N} \cdot \boldsymbol{g} + \boldsymbol{P1} \cdot \left[\nabla \boldsymbol{g} + 2(\nabla \boldsymbol{d})\boldsymbol{B} + 2(\nabla \boldsymbol{r})\boldsymbol{B}^2\right] + \boldsymbol{PN} \cdot \boldsymbol{h}. \tag{6.131}$$

Assuming the lateral surfaces to be traction free, the load potential in the potential energy (6.5) is found, exactly as in Chapter 5, to be

$$\int_{\partial \kappa_t} \tilde{\boldsymbol{t}} \cdot \tilde{\chi}\, da = \int_{\partial \Omega_t} (\boldsymbol{p}_r \cdot \boldsymbol{r} + \boldsymbol{p}_d \cdot \boldsymbol{d} + \boldsymbol{p}_g \cdot \boldsymbol{g})\, ds + o(h^3), \tag{6.132}$$

where

$$\begin{aligned} \boldsymbol{p}_r &= h\left(1 + \frac{1}{24}h^2\tau^2\right)\boldsymbol{t} + \frac{1}{24}h^3(\boldsymbol{t}'' - 2\kappa_\tau \boldsymbol{t}'), \\ \boldsymbol{p}_d &= \frac{1}{12}h^3(\boldsymbol{t}' - \kappa_\tau \boldsymbol{t}), \qquad \boldsymbol{p}_g = \frac{1}{24}h^3\boldsymbol{t}, \end{aligned} \tag{6.133}$$

and

$$\begin{aligned} \boldsymbol{t} = \boldsymbol{P1\nu}, \qquad \boldsymbol{t}' - \kappa_\tau \boldsymbol{t} &= \boldsymbol{P}'\boldsymbol{1\nu} + \tau \boldsymbol{P1\tau} - \kappa_\tau \boldsymbol{P1\nu}, \\ \boldsymbol{t}'' - 2\kappa_\tau \boldsymbol{t}' &= \boldsymbol{P}''\boldsymbol{1\nu} + 2(\tau \boldsymbol{P}'\boldsymbol{1\tau} - \kappa_\tau \boldsymbol{P}'\boldsymbol{1\nu}) - \tau^2 \boldsymbol{P1\nu}, \end{aligned} \tag{6.134}$$

in which κ_τ is the normal curvature of Ω and τ is the twist. The latter formulas follow by expanding the oriented area measure, alluded to previously, on the ruled generators of the shell (see Chap. 5).

6.2.3 The Optimum Order—h^3 Energy

We seek an order—h^3 truncation of the potential energy that is as accurate as possible by the standard of three-dimensional elasticity theory. To this end we impose restrictions on the director fields $\boldsymbol{d}$, $\boldsymbol{g}$, etc. as required by that theory. Thus we use partial information about three-dimensional equilibria to generate the optimal order—h^3 potential energy functional for the midsurface deformation. Putting it another way, although at this stage $\boldsymbol{r}(\theta^\alpha)$ is unrestricted, the expression for E, to be derived, is not valid for arbitrary kinematically possible three-dimensional states. Instead it is defined on a manifold of configurations defined by equilibrium constraints on the directors.

For example, the exact expressions $\tilde{\boldsymbol{t}}^+ = \tilde{\boldsymbol{P}}^+\boldsymbol{N}$ and $\tilde{\boldsymbol{t}}^+ = -\tilde{\boldsymbol{P}}^-\boldsymbol{N}$ for the tractions at the lateral surfaces with exterior unit normals $\pm\boldsymbol{N}$, together with Taylor expansions of $\tilde{\boldsymbol{P}}^\pm$, furnish

$$\tilde{\boldsymbol{t}}^+ + \tilde{\boldsymbol{t}}^- = h\,\boldsymbol{P}'\boldsymbol{N} + O(h^3) \quad \text{and} \quad \tilde{\boldsymbol{t}}^+ - \tilde{\boldsymbol{t}}^- = 2\,\boldsymbol{P}\boldsymbol{N} + O(h^2). \tag{6.135}$$

Accordingly, for traction-free lateral surfaces,

$$\boldsymbol{P}\boldsymbol{N},\ \boldsymbol{P}'\boldsymbol{N} = O(h^2), \tag{6.136}$$

and with these in hand it follows that the terms in (6.126) involving $\boldsymbol{P}\boldsymbol{N}$ and $\boldsymbol{P}'\boldsymbol{N}$ in the coefficient of h^3 may be dropped. Thus, (6.126) remains valid with the simplifications

$$\mathscr{W}' = \bar{\boldsymbol{P}}\mathbf{1}\cdot\left[\nabla\bar{\boldsymbol{d}} + (\nabla\boldsymbol{r})\boldsymbol{B}\right] \tag{6.137}$$

and

$$\mathscr{W}'' = \bar{\boldsymbol{P}}'\mathbf{1}\cdot\left[\nabla\bar{\boldsymbol{d}} + (\nabla\boldsymbol{r})\boldsymbol{B}\right] + \bar{\boldsymbol{P}}\mathbf{1}\cdot\left[\nabla\bar{\boldsymbol{g}} + 2(\nabla\bar{\boldsymbol{d}})\boldsymbol{B} + 2(\nabla\boldsymbol{r})\boldsymbol{B}^2\right], \tag{6.138}$$

in which $\bar{\boldsymbol{d}}$ and $\bar{\boldsymbol{g}}$ satisfy the constraints $\boldsymbol{P}\boldsymbol{N} = \mathbf{0}$ and $\boldsymbol{P}'\boldsymbol{N} = \mathbf{0}$, respectively, i.e.,

$$\begin{aligned} &\left\{\tilde{\boldsymbol{P}}(\nabla\boldsymbol{r} + \bar{\boldsymbol{d}}\otimes\boldsymbol{N})\right\}\boldsymbol{N} = \mathbf{0} \quad \text{and} \\ &\left\{\mathscr{A}(\nabla\boldsymbol{r} + \bar{\boldsymbol{d}}\otimes\boldsymbol{N})\right\}\bar{\boldsymbol{g}} = -\left\{\mathscr{M}(\nabla\boldsymbol{r} + \bar{\boldsymbol{d}}\otimes\boldsymbol{N})[\nabla\bar{\boldsymbol{d}} + (\nabla\boldsymbol{r})\boldsymbol{B}]\right\}\boldsymbol{N}, \end{aligned} \tag{6.139}$$

and where $\mathscr{A}$ is the acoustic tensor defined by

$$\mathscr{A}(\tilde{\boldsymbol{F}})\boldsymbol{v} = \left\{\mathscr{M}(\tilde{\boldsymbol{F}})[\boldsymbol{v}\otimes\boldsymbol{N}]\right\}\boldsymbol{N}. \tag{6.140}$$

As in the case of plates, strong ellipticity implies that this is positive definite, and $(6.139)_2$ then furnishes $\bar{\boldsymbol{g}}$ uniquely. The existence and uniqueness of solutions to $(6.139)_1$ also follow, as for plates, from strong ellipticity. The notation $\bar{\boldsymbol{P}}$ and $\bar{\boldsymbol{P}}'$ in (6.137) and (6.138) refers to the midsurface values of the stress and its ς—derivative in which $\boldsymbol{d} = \bar{\boldsymbol{d}}$ and $\boldsymbol{g} = \bar{\boldsymbol{g}}$ are imposed. We observe, however, that our argument does not yield $\boldsymbol{d} = \bar{\boldsymbol{d}}$ in the coefficient of h in the expression (6.126) for the strain-energy density. In fact, if this is imposed then an error is incurred which is of the same order as the order—h^3 term exhibited explicitly.

However, as in the case of plates, the solution $\bar{\boldsymbol{d}}$ to $(6.139)_1$ minimizes the midsurface strain energy $\mathscr{W}(\nabla\boldsymbol{r} + \boldsymbol{d}\otimes\boldsymbol{N})$ with respect to $\boldsymbol{d}$ and the solution $\bar{\boldsymbol{g}}$ to $(6.139)_2$

minimizes the quadratic form $\left(\nabla\bar{\boldsymbol{d}} + (\nabla\boldsymbol{r})\boldsymbol{B} + \boldsymbol{g}\otimes\boldsymbol{N}\right)\cdot\mathcal{M}(\boldsymbol{F})[\nabla\bar{\boldsymbol{d}} + (\nabla\boldsymbol{r})\boldsymbol{B} + \boldsymbol{g}\otimes\boldsymbol{N}]$ with respect to $\boldsymbol{g}$. These claims also follow from strong ellipticity. The first result justifies the imposition of $\boldsymbol{d} = \bar{\boldsymbol{d}}$ in *all* terms of the energy, yielding (6.126), with (6.137) and (6.138), in which

$$\mathcal{W} = \mathcal{W}(\nabla\boldsymbol{r} + \bar{\boldsymbol{d}}\otimes\boldsymbol{N}) \quad\text{and}\quad \bar{\boldsymbol{P}}'\boldsymbol{1}\cdot\left[\nabla\bar{\boldsymbol{d}} + (\nabla\boldsymbol{r})\boldsymbol{B}\right] = \boldsymbol{K}\cdot\mathcal{M}(\nabla\boldsymbol{r} + \bar{\boldsymbol{d}}\otimes\boldsymbol{N})[\boldsymbol{K}], \tag{6.141}$$

where

$$\boldsymbol{K} = \nabla\bar{\boldsymbol{d}} + (\nabla\boldsymbol{r})\boldsymbol{B} + \bar{\boldsymbol{g}}\otimes\boldsymbol{N}. \tag{6.142}$$

The term $\boldsymbol{P1}\cdot\nabla\boldsymbol{g}$ appearing in (6.138) may be reduced with the aid of the equilibrium equation (6.3), which can be written in the form $(\tilde{\boldsymbol{P}}_{,i})\boldsymbol{G}^i = \boldsymbol{0}$, where $\{\theta^i\} = \{\theta^\alpha, \varsigma\}$. This is well defined in shell space and hence on the midsurface, where it reduces to $(\boldsymbol{P}_{,\alpha})\boldsymbol{A}^\alpha + \boldsymbol{P}'\boldsymbol{N} = \boldsymbol{0}$. With some effort this may be recast as (see (5.86))

$$\operatorname{div}(\boldsymbol{P1}) + \boldsymbol{P}'\boldsymbol{N} - 2H\boldsymbol{PN} = \boldsymbol{0}, \tag{6.143}$$

where we have

$$\operatorname{div}(\boldsymbol{P1}) = \boldsymbol{P}^\alpha{}_{|\alpha} = A^{-1/2}\left(A^{1/2}\boldsymbol{P}^\alpha\right)_{,\alpha}, \tag{6.144}$$

in which $\boldsymbol{P}^\alpha = \boldsymbol{P}\boldsymbol{A}^\alpha$, $A = \det(A_{\alpha\beta})$ and $A_{\alpha\beta} = \boldsymbol{A}_\alpha\cdot\boldsymbol{A}_\beta$ is the (positive definite) metric on Ω induced by the parametrization $\boldsymbol{x}(\theta^\alpha)$. Integrating over Ω and applying Stokes' theorem, we obtain

$$\int_\Omega \boldsymbol{P1}\cdot\nabla\boldsymbol{g}\,da = \int_{\partial\Omega}\boldsymbol{P1}\boldsymbol{\nu}\cdot\boldsymbol{g}\,ds + \int_\Omega \boldsymbol{g}\cdot(\boldsymbol{P}'\boldsymbol{N} - 2H\boldsymbol{PN})\,da. \tag{6.145}$$

Using (6.136), (6.137) and (6.138) we thus derive

$$\int_\kappa \tilde{\mathcal{W}}\,dv = \int_\Omega W\,da + \frac{1}{24}h^3\int_{\partial\Omega}\bar{\boldsymbol{P}}\boldsymbol{1}\boldsymbol{\nu}\cdot\bar{\boldsymbol{g}}\,ds, \tag{6.146}$$

where W is now given by

$$W = (1 + \frac{1}{12}h^2K)W_1 + W_2 + W_3 + o(h^3), \tag{6.147}$$

with

$$W_1 = h\mathcal{W}(\nabla\boldsymbol{r} + \bar{\boldsymbol{d}}\otimes\boldsymbol{N}),\quad W_2 = \frac{1}{24}h^3\boldsymbol{K}\cdot\mathcal{M}(\nabla\boldsymbol{r} + \bar{\boldsymbol{d}}\otimes\boldsymbol{N})[\boldsymbol{K}]$$
$$\text{and}\quad W_3 = \frac{1}{12}h^3\bar{\boldsymbol{P}}\boldsymbol{1}\cdot\left\{[(\nabla\bar{\boldsymbol{d}})\boldsymbol{B} + (\nabla\boldsymbol{r})\boldsymbol{B}^2] - 2H[\nabla\bar{\boldsymbol{d}} + (\nabla\boldsymbol{r})\boldsymbol{B}]\right\}. \tag{6.148}$$

The term W_3 is non-standard. Its counterpart in Koiter's work [16] was the subject of extensive analysis, the rough conclusion of which is that it may be neglected in comparison to the sum $W_1 + W_2$, provided that the midsurface strain is small and the shell is sufficiently thin. We return to this issue below. Evidently $W = W_1 + W_2$ if the midsurface is flat; that is, if the shell is a plate.

Most treatments of shell theory based on derivations from three-dimensional elasticity are restricted to materials that exhibit reflection symmetry of the material properties with respect to the midsurface. These are exemplified by the case of isotropy, discussed by Koiter, which we will study in detail. Thus we consider strain-energy functions that satisfy $\mathscr{U}(\boldsymbol{E}) = \mathscr{U}(\boldsymbol{Q}^T\boldsymbol{E}\boldsymbol{Q})$ with $\boldsymbol{Q} = \boldsymbol{I} - 2\boldsymbol{N}\otimes\boldsymbol{N}$, which in turn implies that $\mathscr{U}'(E_{ij}) = \mathscr{U}(E_{kl}\boldsymbol{G}^k \otimes \boldsymbol{G}^l)$, where $\boldsymbol{G}^\alpha = \boldsymbol{A}^\alpha$ and $\boldsymbol{G}^3 = \boldsymbol{N}$, is an even function of the transverse shear strain $E_{\alpha 3}\ (= E_{3\alpha})$. Our experience with reflection symmetry in the context of plate theory suggests that we should take the midsurface stress to scale with thickness, to avoid the ill-posedness of the minimization problem of the shell energy that would otherwise occur.

Let $\Gamma(E_{\alpha 3})$ be the function obtained by fixing all components of $\boldsymbol{E}$ except $E_{\alpha 3}$ in the midsurface strain-energy function. Then,

$$\partial\Gamma/\partial E_{\alpha 3} = \boldsymbol{A}^\alpha \cdot (\mathscr{U}_{\boldsymbol{E}})\boldsymbol{N}, \tag{6.149}$$

which vanish by (6.12) and $(6.139)_1$, the latter being equivalent to

$$\boldsymbol{S}\boldsymbol{N} = \boldsymbol{0}. \tag{6.150}$$

In materials that exhibit reflection symmetry these restrictions are satisfied at $E_{\alpha 3} = 0$ because the strain energy is then an even function of the transverse shears [13]. The corresponding strain is

$$\boldsymbol{E} = \boldsymbol{\varepsilon} + E_{33}\boldsymbol{N}\otimes\boldsymbol{N}, \quad \text{where} \quad \boldsymbol{\varepsilon} = E_{\alpha\beta}\boldsymbol{A}^\alpha \otimes \boldsymbol{A}^\beta, \quad E_{33} = \frac{1}{2}(\varphi^2 - 1) \tag{6.151}$$

and φ is the transverse stretch. Comparing with the midsurface strain obtained from $(6.120)_1$, we conclude that

$$\boldsymbol{d} = \varphi\boldsymbol{n}, \tag{6.152}$$

where $\boldsymbol{n}$ is the unit normal to the deformed surface defined by $|\boldsymbol{F}^*\boldsymbol{N}|\boldsymbol{n} = \boldsymbol{F}^*\boldsymbol{N}$, where $\boldsymbol{F}^*$ is the cofactor of $\boldsymbol{F}$, and where $\varphi(>0)$ is obtained in terms of $\boldsymbol{\varepsilon}$ by solving

$$\boldsymbol{N}\cdot\boldsymbol{S}\boldsymbol{N} = 0. \tag{6.153}$$

That (6.152) furnishes a solution to (6.150) in the presence of reflection symmetry was proved in [13, 17]. The energetic optimality of the solution, granted strong ellipticity, was proved in [25, p. 288]. Therefore reflection symmetry and strong ellipticity, combined with (6.150), yield deformations in which the transverse shear strain necessarily vanishes and $\boldsymbol{S}$ is a symmetric 2-tensor that maps the tangent plane $T_{\Omega(p)}$, to the reference surface at the material point p, to itself.

6.2.4 Koiter's Energy as the Leading-Order Model

The restriction (6.153) to positive-definite stress $\boldsymbol{S}$ is associated with the coefficient of h^3 in the strain-energy function. To remove it, and thus to restore the potential well-posedness of the truncated shell energy (6.146) when the pointwise restriction (6.153) is violated, we assume that

$$|\boldsymbol{S}| = O(h) \tag{6.154}$$

after suitable non-dimensionalization. This assumption must be verified *a posteriori*.

We thus have $h^3\boldsymbol{S} = o(h^3)$, and from $\boldsymbol{P1} = (\nabla\boldsymbol{r})\boldsymbol{S}$ we conclude that $W_3 = o(h^3)$. Further, the constitutive hypotheses discussed in the previous section imply that the midsurface value of the strain is sufficiently small—of order $O(h)$—that it may be neglected in the coefficient of h^3 without adverse effect on the accuracy of (6.146). Suppressing, in addition, the small term h^2K in comparison to unity, we arrive at

$$\int_{\kappa} \mathscr{W}(\hat{\boldsymbol{F}})\,dv = \int_{\Omega} W\,da, \quad \text{with} \quad W = W_1 + W_2 + o(h^3) \tag{6.155}$$

in which W_2 is now given by

$$W_2 = \frac{1}{24}h^3\boldsymbol{K}\cdot\mathscr{M}(\boldsymbol{R})[\boldsymbol{K}], \tag{6.156}$$

where $\boldsymbol{R}$ is the rotation factor in the polar decomposition of $\bar{\boldsymbol{F}} = \nabla\boldsymbol{r} + \bar{\boldsymbol{d}}\otimes\boldsymbol{N}$; that is, the replacement of $\bar{\boldsymbol{F}}$ by $\boldsymbol{R}$ entails an error of order $O(h)$, provided that the strain-energy function is sufficiently smooth, and thus an overall error of order $o(h^3)$ which does not affect the order—h^3 truncation. Using (6.11) we then have

$$W_2 = \frac{1}{24}h^3\boldsymbol{R}^T\boldsymbol{K}\cdot\mathscr{C}(\boldsymbol{0})[\boldsymbol{R}^T\boldsymbol{K}], \tag{6.157}$$

with $\boldsymbol{K}$ given by (6.142), where (cf. (6.152))

$$\nabla\bar{\boldsymbol{d}} = \nabla\boldsymbol{n} + \boldsymbol{n}\otimes\nabla\varphi, \tag{6.158}$$

in which $\varphi = 1$ has been imposed for consistency with the suppression of the strain in the coefficient of h^3. For consistency we also impose

$$\nabla\boldsymbol{r} = \bar{\boldsymbol{F}}\boldsymbol{1} = \boldsymbol{R1} \quad \text{and} \quad \boldsymbol{n} = \boldsymbol{RN} \tag{6.159}$$

in this coefficient. The consistent computation of $\bar{\boldsymbol{g}}$ entails the replacement of $(6.139)_2$ by

$$\{\mathscr{A}(\boldsymbol{R})\}\bar{\boldsymbol{g}} = -\{\mathscr{M}(\boldsymbol{R})[\nabla\bar{\boldsymbol{d}} + (\nabla\boldsymbol{r})\boldsymbol{B}]\}\boldsymbol{N}, \tag{6.160}$$

where (cf. (6.140))

$$\mathscr{A}(\boldsymbol{R})\boldsymbol{v} = \{\mathscr{M}(\boldsymbol{R})[\boldsymbol{v} \otimes \boldsymbol{N}]\}\boldsymbol{N}. \tag{6.161}$$

Using (6.11), Eq. (6.160) may then be cast in the form

$$\{\mathscr{C}(\mathbf{0})[\boldsymbol{R}^T\bar{\boldsymbol{g}} \otimes \boldsymbol{N}]\}\boldsymbol{N} = -\{\mathscr{C}(\mathbf{0})[\boldsymbol{R}^T\nabla\bar{\boldsymbol{d}} + \boldsymbol{B}]\}\boldsymbol{N}. \tag{6.162}$$

The strain energy is thus given by (6.155) in which

$$W = h\mathscr{W}(\nabla\boldsymbol{r} + \bar{\boldsymbol{d}} \otimes \boldsymbol{N}) + \frac{1}{24}h^3\boldsymbol{R}^T\boldsymbol{K}\cdot\mathscr{C}(\mathbf{0})[\boldsymbol{R}^T\boldsymbol{K}], \tag{6.163}$$

with $\boldsymbol{K}$ computed as indicated. The load potential (6.132) may also be simplified; we obtain

$$\int_{\partial\kappa_t}\tilde{\boldsymbol{t}}\cdot\tilde{\boldsymbol{\chi}}\,\mathrm{d}a = \int_{\partial\Omega_t}(\boldsymbol{p}_r\cdot\boldsymbol{r} + \boldsymbol{p}_d\cdot\bar{\boldsymbol{d}})\,\mathrm{d}s + o(h^3), \tag{6.164}$$

in which

$$\boldsymbol{p}_r = h\,\boldsymbol{t} + \frac{1}{24}h^3(\boldsymbol{t}'' - 2\kappa_\tau\boldsymbol{t}'), \quad \boldsymbol{p}_d = \frac{1}{12}h^3\boldsymbol{t}' \tag{6.165}$$

and the small term $\frac{1}{24}h^2\tau^2$ has been neglected in comparison to unity.

Here we have assumed $\boldsymbol{d} = \bar{\boldsymbol{d}}$ and $\boldsymbol{g} = \bar{\boldsymbol{g}}$ to obtain on the closure of Ω. Regarding boundary data, we assume that $\tilde{\boldsymbol{\chi}}$ is assigned on $(\partial\Omega \setminus \partial\Omega_t) \times [-h/2, h/2]$. This implies that its midsurface value, $\boldsymbol{r}$, and those of its tangential through-thickness derivatives, $\boldsymbol{d}$ and $\boldsymbol{g}$, are also assigned there. However, the latter two fields cannot be assigned arbitrarily if the foregoing model is to apply on the closure of Ω. The assigned values must agree with the continuous extensions to $\partial\Omega \setminus \partial\Omega_t$ of the functions $\bar{\boldsymbol{d}}$ and $\bar{\boldsymbol{g}}$, respectively. If these extensions are in conflict with the values derived from the data for the three-dimensional problem—this situation being typical—then it is necessary to use three-dimensional theory (or some refinement of the present theory) in a region adjoining the boundary; one then attempts to match its predictions to those of the present model in the interior.

Our constitutive hypotheses furnish the strain

$$\boldsymbol{E} = h\,\bar{\boldsymbol{E}} + o(h), \tag{6.166}$$

where $\bar{\boldsymbol{E}} = O(1)$. Scalings of this kind are sometimes assumed *a priori* in asymptotic treatments (e.g., [4, 29]). The term W_1 in the strain energy (cf. (6.148)) reduces to

$$W_1 = \frac{1}{2}h^3\bar{\boldsymbol{E}}\cdot\mathscr{C}(\mathbf{0})[\bar{\boldsymbol{E}}] + o(h^3), \tag{6.167}$$

yielding

$$W = h^3\bar{W} + o(h^3), \tag{6.168}$$

where

$$\bar{W} = \frac{1}{2}\bar{\boldsymbol{E}} \cdot \mathscr{C}(\mathbf{0})[\bar{\boldsymbol{E}}] + \frac{1}{24}\boldsymbol{R}^T\boldsymbol{K} \cdot \mathscr{C}(\mathbf{0})[\boldsymbol{R}^T\boldsymbol{K}]. \tag{6.169}$$

We show below that this coincides with Koiter's expression for the shell strain-energy density.

Altogether, the potential energy $\mathscr{E}$ is then given by

$$\mathscr{E}/h^3 = \bar{E} + o(h^3)/h^3, \tag{6.170}$$

where

$$\bar{E} = \int_{\Omega} \bar{W}\,da - \int_{\partial\Omega_t} (\bar{\boldsymbol{p}}_r \cdot \boldsymbol{r} + \bar{\boldsymbol{p}}_d \cdot \bar{\boldsymbol{d}})\,ds, \tag{6.171}$$

yielding $E = h^3\bar{E}$ in (6.4). This is the leading order potential energy of the shell under the present hypotheses. In view of (6.170), it would be possible to regard $\bar{E}$ as the rigorous leading-order energy, in the sense of gamma convergence or asymptotic analysis, if the estimate (6.154) on the (plane) stress could be established *a priori* in the interior of Ω. As it stands, the only rationale for (6.154) that we know of is that it yields an expression for the leading-order energy which furnishes a meaningful minimization problem. Examples of such states include, rather obviously, the minimizers (if any) of $\bar{E}$.

As remarked by Koiter [18], the composite energy may then be regarded as furnishing leading-order models in different regions of the shell; the first (order h), wherever the midsurface strain is non-zero, and the second (order h^3), where bending effects dominate.

The associated Euler equations and boundary conditions will be presented in the next subsection. It is customary to express them in terms of covariant derivatives and geometric parameters on the deformed surface [3, 17]. To connect the boundary conditions to the load potential in (6.171), we show that the variation of the latter may be expressed in the form

$$\int_{\partial\Omega_t} \left[\bar{\boldsymbol{p}}_r \cdot \boldsymbol{u} + \bar{\boldsymbol{p}}_d \cdot (\bar{\boldsymbol{d}})^{\cdot}\right] ds = \int_{\partial\omega_t} (\boldsymbol{f} \cdot \boldsymbol{u} - M\bar{\boldsymbol{\tau}} \cdot \boldsymbol{\omega})\,d\bar{s}, \tag{6.172}$$

where $\boldsymbol{u} = \dot{\boldsymbol{r}}$, superposed dots are used to denote variational (Gateaux) derivatives, $\boldsymbol{f}$ is the net force density on the deformed edge with unit tangent $\bar{\boldsymbol{\tau}}$, the vector $\boldsymbol{\omega}$ is the virtual rotation of the shell midsurface defined by $\dot{\boldsymbol{n}} = \boldsymbol{\omega} \times \boldsymbol{n}$ and M is the density of bending couple along the edge.

To this end we observe that to within negligible errors of order h, arclength on $\partial\Omega$ coincides with that on $\partial\omega$, and $\bar{\boldsymbol{d}} = \boldsymbol{n}$ (see (6.152)), whereas [20]

$$\dot{\boldsymbol{n}} = (\bar{\boldsymbol{\tau}} \cdot \boldsymbol{\omega})\bar{\boldsymbol{\nu}} - (\boldsymbol{n} \cdot \boldsymbol{u}_{,\bar{s}})\bar{\boldsymbol{\tau}}\,, \tag{6.173}$$

where $\bar{\boldsymbol{\nu}} = \bar{\boldsymbol{\tau}} \times \boldsymbol{n}$ and $(\cdot)_{,\bar{s}} = d(\cdot)/d\bar{s}$. Thus, to consistent order,

$$\bar{\boldsymbol{p}}_r \cdot \boldsymbol{u} + \bar{\boldsymbol{p}}_d \cdot (\bar{\boldsymbol{d}})^{\cdot} = \left\{ \bar{\boldsymbol{p}}_r - [(\bar{\boldsymbol{\tau}} \cdot \bar{\boldsymbol{p}}_d)\boldsymbol{n}]_{,\bar{s}} \right\} \cdot \boldsymbol{u} + (\bar{\boldsymbol{\nu}} \cdot \bar{\boldsymbol{p}}_d)\bar{\boldsymbol{\tau}} \cdot \boldsymbol{\omega} + [(\bar{\boldsymbol{\tau}} \cdot \bar{\boldsymbol{p}}_d)\boldsymbol{n} \cdot \boldsymbol{u}]_{,\bar{s}} \,. \tag{6.174}$$

Because $\boldsymbol{u}$ vanishes identically on $\partial\omega \setminus \partial\omega_t$ the integral of the last term on the right may be extended to $\partial\omega$ and is thereby seen to contribute nil to the net working if $\partial\omega$ is smooth, i.e., if $\bar{\boldsymbol{\tau}}$ is continuous. The same term generates corner forces in the case of a piecewise smooth edge with a finite number of jumps in $\bar{\boldsymbol{\tau}}$. We thus derive (6.172) with

$$\boldsymbol{f} = \bar{\boldsymbol{p}}_r - \left[(\bar{\boldsymbol{\tau}} \cdot \bar{\boldsymbol{p}}_d)\boldsymbol{n}\right]_{,\bar{s}} \quad \text{and} \quad M = -\bar{\boldsymbol{\nu}} \cdot \bar{\boldsymbol{p}}_d \,. \tag{6.175}$$

It is interesting that whereas $\bar{\boldsymbol{p}}_r$ and $\bar{\boldsymbol{p}}_d$ are fixed in a formulation based on the three-dimensional dead-load problem (see (6.165)), the fields $\boldsymbol{f}$ and M are not fixed because they involve the unknown orientation of the deformed surface ω at the edge $\partial\omega_t$. Said differently, $\dot{\boldsymbol{f}}$ and $\dot{M}$ do not vanish on $\partial\omega_t$ although the variations of $\bar{\boldsymbol{p}}_r$ and $\bar{\boldsymbol{p}}_d$ vanish there.

6.2.5 The Explicit Energy for Isotropic Materials

For isotropic materials we have the well-known representation

$$\mathscr{C}(\boldsymbol{0})[\boldsymbol{A}] = \lambda(\operatorname{tr} \boldsymbol{A})\boldsymbol{I} + 2\mu \operatorname{sym} \boldsymbol{A} \,, \tag{6.176}$$

where λ and μ are the classical Lamé moduli, satisfying the inequalities $3\lambda + 2\mu > 0$ and $\mu > 0$ associated with the positivity of $\mathscr{C}(\boldsymbol{0})$. This furnishes

$$\left\{\mathscr{C}(\boldsymbol{0})[\boldsymbol{v} \otimes \boldsymbol{N}]\right\}\boldsymbol{N} = (\lambda + 2\mu)\, v\, \boldsymbol{N} + 2\mu \boldsymbol{1}\boldsymbol{v} \tag{6.177}$$

for any vector $\boldsymbol{v}$, where $v = \boldsymbol{v} \cdot \boldsymbol{N}$. Using this in (6.162), we derive

$$\begin{aligned} \boldsymbol{N} \cdot \boldsymbol{R}^T \bar{\boldsymbol{g}} &= -(\lambda + 2\mu)^{-1}\left\{\lambda \operatorname{tr}[\boldsymbol{R}^T \nabla \bar{\boldsymbol{d}} + \boldsymbol{B}] + 2\mu \boldsymbol{N} \cdot \left(\operatorname{sym}[\boldsymbol{R}^T \nabla \bar{\boldsymbol{d}} + \boldsymbol{B}]\right)\boldsymbol{N}\right\} \\ \text{and} \quad \boldsymbol{1}(\boldsymbol{R}^T \bar{\boldsymbol{g}}) &= -2\,\boldsymbol{1}\left(\operatorname{sym}[\boldsymbol{R}^T \nabla \bar{\boldsymbol{d}} + \boldsymbol{B}]\right)\boldsymbol{N}, \end{aligned} \tag{6.178}$$

which in turn generate $\boldsymbol{R}^T \bar{\boldsymbol{g}} = \boldsymbol{1}(\boldsymbol{R}^T \bar{\boldsymbol{g}}) + (\boldsymbol{N} \cdot \boldsymbol{R}^T \bar{\boldsymbol{g}})\boldsymbol{N}$ for use in (6.142) and (6.169).

From (6.158) the term $\boldsymbol{R}^T \nabla \bar{\boldsymbol{d}}$ is seen to involve

$$\boldsymbol{R}^T \nabla \boldsymbol{n} = \boldsymbol{\kappa}, \quad \text{with} \quad \boldsymbol{\kappa} = -(\nabla \boldsymbol{r})^T \boldsymbol{b}(\nabla \boldsymbol{r}), \tag{6.179}$$

where $\boldsymbol{b}$ is the curvature tensor on the deformed surface. In terms of surface coordinates,

$$\boldsymbol{\kappa} = -b_{\alpha\beta} \boldsymbol{A}^{\alpha} \otimes \boldsymbol{A}^{\beta}; \qquad b_{\alpha\beta} = \boldsymbol{n} \cdot \boldsymbol{r}_{,\alpha\beta} \,. \tag{6.180}$$

From $\boldsymbol{R}^T \boldsymbol{n} = \boldsymbol{N}$ it then follows that

$$\boldsymbol{R}^T\nabla\bar{\boldsymbol{d}} + \boldsymbol{B} = -\boldsymbol{\rho} + \boldsymbol{N}\otimes\nabla\varphi \quad \text{and} \quad \boldsymbol{R}^T\bar{\boldsymbol{g}} = -\frac{\lambda}{\lambda+2\mu}(\operatorname{tr}\boldsymbol{\rho})\boldsymbol{N} - \nabla\varphi, \tag{6.181}$$

where

$$\boldsymbol{\rho} = -(\boldsymbol{\kappa} + \boldsymbol{B}), \tag{6.182}$$

which is given in terms of components by Koiter's expression [3, 17]

$$\boldsymbol{\rho} = \rho_{\alpha\beta}\boldsymbol{A}^\alpha \otimes \boldsymbol{A}^\beta \quad \text{with} \quad \rho_{\alpha\beta} = b_{\alpha\beta} - B_{\alpha\beta}\,. \tag{6.183}$$

This combines with (6.142) and (6.178) to give

$$\boldsymbol{R}^T\boldsymbol{K} = \boldsymbol{\rho} - \frac{\lambda}{\lambda+2\mu}(\operatorname{tr}\boldsymbol{\rho})\boldsymbol{N}\otimes\boldsymbol{N} + \boldsymbol{N}\otimes\nabla\varphi - \nabla\varphi\otimes\boldsymbol{N}, \tag{6.184}$$

and (6.169) then gives the classical bending energy

$$\mathscr{C}(\boldsymbol{0})[\boldsymbol{R}^T\boldsymbol{K}]\cdot\boldsymbol{R}^T\boldsymbol{K} = \frac{2\lambda\mu}{\lambda+2\mu}(\operatorname{tr}\boldsymbol{\rho})^2 + 2\mu\,|\boldsymbol{\rho}|^2, \tag{6.185}$$

in precise agreement with the result obtained by the method of gamma convergence [10] or asymptotic analysis [4].

To reduce the stretching term we invoke $(6.151)_1$ and (6.153), obtaining

$$\boldsymbol{E} = \boldsymbol{\varepsilon} - \frac{\lambda}{\lambda+2\mu}(\operatorname{tr}\boldsymbol{\varepsilon})\boldsymbol{N}\otimes\boldsymbol{N} \quad \text{with} \quad \boldsymbol{\varepsilon} = \frac{1}{2}\big(a_{\alpha\beta} - A_{\alpha\beta}\big)\boldsymbol{A}^\alpha\otimes\boldsymbol{A}^\beta,$$
$$\text{and} \quad \mathscr{C}(\boldsymbol{0})[\boldsymbol{E}]\cdot\boldsymbol{E} = \frac{2\lambda\mu}{\lambda+2\mu}(\operatorname{tr}\boldsymbol{\varepsilon})^2 + 2\mu\,|\boldsymbol{\varepsilon}|^2, \tag{6.186}$$

apart from errors of order $o(h)$ and $o(h^2)$, respectively, arriving finally at Koiter's energy (6.107), in which $\boldsymbol{\varepsilon} = O(h)$ and $\mathscr{D} = \mathscr{D}^{\alpha\beta\gamma\delta}\boldsymbol{A}_\alpha\otimes\boldsymbol{A}_\beta\otimes\boldsymbol{A}_\gamma\otimes\boldsymbol{A}_\delta$, with

$$\mathscr{D}^{\alpha\beta\gamma\delta} = \frac{2\lambda\mu}{\lambda+2\mu}A^{\alpha\beta}A^{\gamma\delta} + \mu\big(A^{\alpha\gamma}A^{\beta\delta} + A^{\alpha\delta}A^{\beta\gamma}\big). \tag{6.187}$$

6.3 Equilibrium Equations

We characterize equilibria of the shell as states that satisfy the virtual-power equality

$$\dot{E} = P, \tag{6.188}$$

where the superposed dot is again used to denote a variational derivative and P is the virtual power of the loads acting on the shell, the form of which is made explicit below. Here E is a functional of the midsurface deformation $\boldsymbol{r}$.

As the shell deforms the strain-energy function (6.107) evolves in response to variations $\dot{a}_{\alpha\beta}$ and $\dot{b}_{\alpha\beta}$ of the metric and curvature of the deforming surface ω. This follows from the structure of (6.107) and the definitions $(6.186)_2$ of the surface strain $\boldsymbol{\varepsilon}$ and (6.183) of the bending strain $\boldsymbol{\rho}$. Thus, $\dot{W}$ is given by

$$\dot{W} = \frac{1}{2} N^{\alpha\beta} \dot{a}_{\alpha\beta} + M^{\alpha\beta} \dot{b}_{\alpha\beta}, \tag{6.189}$$

where, by abuse of notation,

$$N^{\alpha\beta} = \frac{\partial W}{\partial a_{\alpha\beta}} + \frac{\partial W}{\partial a_{\beta\alpha}}, \qquad M^{\alpha\beta} = \frac{1}{2}\Big(\frac{\partial W}{\partial b_{\alpha\beta}} + \frac{\partial W}{\partial b_{\beta\alpha}}\Big), \tag{6.190}$$

in which we have exploited the symmetries $\dot{a}_{\alpha\beta} = \dot{a}_{\beta\alpha}$ and $\dot{b}_{\alpha\beta} = \dot{b}_{\beta\alpha}$ and the factor $\frac{1}{2}$ is included in (6.189) rather than $(6.190)_1$ for the sake of later convenience.

Explicit expressions for the response functions $N^{\alpha\beta}$ and $M^{\alpha\beta}$ may be derived by computing $\dot{W}$ from (6.107) and comparing the result with (6.189). Using

$$\dot{\boldsymbol{\varepsilon}} = \frac{1}{2} \dot{a}_{\alpha\beta}\, \boldsymbol{A}^{\alpha} \otimes \boldsymbol{A}^{\beta} \quad \text{and} \quad \dot{\boldsymbol{\rho}} = -\dot{b}_{\alpha\beta}\, \boldsymbol{A}^{\alpha} \otimes \boldsymbol{A}^{\beta}, \tag{6.191}$$

together with the minor symmetries of $\mathscr{D}$, we obtain

$$N^{\alpha\beta} = h\, \mathscr{D}^{\alpha\beta\gamma\delta}(\boldsymbol{x})\, \varepsilon_{\gamma\delta} \tag{6.192}$$

and

$$M^{\alpha\beta} = \frac{1}{12} h^3 \mathscr{D}^{\alpha\beta\gamma\delta}(\boldsymbol{x})\, \rho_{\gamma\delta}\,. \tag{6.193}$$

We proceed to express (6.189) in terms of the variation $\boldsymbol{u}(\boldsymbol{x}) = \dot{\boldsymbol{r}}$ of the midsurface position field. To this end we use

$$\dot{a}_{\alpha\beta} = \boldsymbol{a}_{\alpha} \cdot \boldsymbol{u}_{,\beta} + \boldsymbol{a}_{\beta} \cdot \boldsymbol{u}_{,\alpha} \tag{6.194}$$

together with

$$\boldsymbol{u}_{;\alpha\beta} = \boldsymbol{u}_{,\alpha\beta} - \Gamma^{\lambda}_{\alpha\beta} \boldsymbol{u}_{,\lambda} \tag{6.195}$$

in which $\Gamma^{\lambda}_{\alpha\beta}$ are the Christoffel symbols on the equilibrium surface ω and subscripts preceded by semi-colons identify covariant derivatives on ω. By varying the Gauss equations

$$\boldsymbol{r}_{,\alpha\beta} = \Gamma^{\lambda}_{\alpha\beta} \boldsymbol{a}_{\lambda} + b_{\alpha\beta} \boldsymbol{n}\,, \tag{6.196}$$

and using $\dot{\boldsymbol{a}}_{\lambda} = \boldsymbol{u}_{,\lambda}$, we reduce (6.195) to

$$\boldsymbol{u}_{;\alpha\beta} = \dot{\Gamma}^{\lambda}_{\alpha\beta} \boldsymbol{a}_{\lambda} + b_{\alpha\beta} \dot{\boldsymbol{n}} + \dot{b}_{\alpha\beta} \boldsymbol{n}\,, \tag{6.197}$$

and conclude that

$$\dot{b}_{\alpha\beta} = \boldsymbol{n} \cdot \boldsymbol{u}_{;\alpha\beta} \,. \tag{6.198}$$

However, for our present purposes it is more appropriate to express $\dot{b}_{\alpha\beta}$ in terms of $\boldsymbol{u}_{,\alpha}$ and $\boldsymbol{u}_{|\alpha\beta}$, where

$$\boldsymbol{u}_{|\alpha\beta} = \boldsymbol{u}_{,\alpha\beta} - \bar{\Gamma}^{\lambda}_{\alpha\beta}\boldsymbol{u}_{,\lambda} \,, \tag{6.199}$$

in which $\bar{\Gamma}^{\lambda}_{\alpha\beta}$ are the Christoffel symbols on Ω, is the referential covariant derivative. To achieve this we solve for $\boldsymbol{u}_{,\alpha\beta}$ and substitute into (6.195), obtaining

$$\boldsymbol{u}_{;\alpha\beta} = \boldsymbol{u}_{|\alpha\beta} - S^{\lambda}_{\alpha\beta}\boldsymbol{u}_{,\lambda} \,, \tag{6.200}$$

where

$$S^{\lambda}_{\alpha\beta} = \Gamma^{\lambda}_{\alpha\beta} - \bar{\Gamma}^{\lambda}_{\alpha\beta} \,. \tag{6.201}$$

Thus,

$$\dot{b}_{\alpha\beta} = \boldsymbol{n} \cdot \left(\boldsymbol{u}_{|\alpha\beta} - S^{\lambda}_{\alpha\beta}\boldsymbol{u}_{,\lambda}\right). \tag{6.202}$$

It then follows from (6.189), (6.194) and (6.202) that

$$\dot{W} = \boldsymbol{N}^{\alpha} \cdot \boldsymbol{u}_{,\alpha} + \boldsymbol{M}^{\alpha\beta} \cdot \boldsymbol{u}_{|\alpha\beta} \,, \tag{6.203}$$

with

$$\boldsymbol{N}^{\alpha} = N^{\alpha\beta}\boldsymbol{a}_{\beta} + N^{\alpha}\boldsymbol{n} \,, \qquad \boldsymbol{M}^{\alpha\beta} = M^{\alpha\beta}\boldsymbol{n} \,, \tag{6.204}$$

in which

$$N^{\alpha} = -M^{\beta\lambda}S^{\alpha}_{\beta\lambda} \,. \tag{6.205}$$

The non-standard terms N^{α} arise from the use of covariant derivatives on Ω, whereas standard shell theory is typically developed in terms of covariant derivatives on ω [3, 17].

Proceeding with the reduction of (6.188), we define

$$\varphi^{\alpha} = \boldsymbol{T}^{\alpha} \cdot \boldsymbol{u} + \boldsymbol{M}^{\alpha\beta} \cdot \boldsymbol{u}_{,\beta} \,, \tag{6.206}$$

with

$$\boldsymbol{T}^{\alpha} = \boldsymbol{N}^{\alpha} - \boldsymbol{M}^{\alpha\beta}{}_{|\beta} \,, \tag{6.207}$$

where

$$\boldsymbol{M}^{\alpha\beta}{}_{|\beta} = M^{\alpha\beta}{}_{|\beta}\,\boldsymbol{n} - M^{\alpha\lambda}b^{\beta}_{\lambda}\boldsymbol{a}_{\beta} \,, \tag{6.208}$$

in which use has been made of the Weingarten equation

$$\boldsymbol{n}_{,\beta} = -b^{\lambda}_{\beta}\boldsymbol{a}_{\lambda} \,, \tag{6.209}$$

where $b^{\lambda}_{\beta} = a^{\lambda\alpha} b_{\alpha\beta}$ are the mixed curvature components and $(a^{\alpha\beta}) = (a_{\alpha\beta})^{-1}$ is the reciprocal metric. With this we have

$$\dot{W} = \varphi^{\alpha}{}_{|\alpha} - \boldsymbol{u} \cdot \boldsymbol{T}^{\alpha}{}_{|\alpha} \tag{6.210}$$

and Stokes' theorem may then be used to write

$$\dot{E} = \int_{\Omega} \dot{W}\, da = \int_{\partial\Omega} \varphi^{\alpha} \nu_{\alpha}\, ds - \int_{\Omega} \boldsymbol{u} \cdot \boldsymbol{T}^{\alpha}{}_{|\alpha}\, da, \tag{6.211}$$

wherein $\boldsymbol{\nu} = \nu_{\alpha} \boldsymbol{A}^{\alpha}$ is the rightward unit normal to $\partial\Omega$.

The virtual power statement (6.188) thus accommodates a distributed load $\boldsymbol{g}$, per unit area of Ω, which contributes $\int_{\Omega} \boldsymbol{g} \cdot \boldsymbol{u}\, da$ to the virtual power. It follows immediately from the Fundamental Lemma of the Calculus of Variations that the equilibrium equation, holding in Ω, is

$$\boldsymbol{T}^{\alpha}{}_{|\alpha} + \boldsymbol{g} = \boldsymbol{0}. \tag{6.212}$$

Turning to the boundary terms, a standard integration-by-parts procedure [30] may be used to recast the first integral in the right-hand side of (6.211) as

$$\begin{aligned} \int_{\partial\Omega} \varphi^{\alpha} \nu_{\alpha}\, ds = & \int_{\partial\Omega} \{[\boldsymbol{T}^{\alpha} \nu_{\alpha} - (\boldsymbol{M}^{\alpha\beta} \nu_{\alpha} \tau_{\beta})'] \cdot \boldsymbol{u} + \boldsymbol{M}^{\alpha\beta} \nu_{\alpha} \nu_{\beta} \cdot \boldsymbol{u}_{,\nu}\}\, ds \\ & - \sum \left[\boldsymbol{M}^{\alpha\beta} \nu_{\alpha} \tau_{\beta}\right]_i \cdot \boldsymbol{u}_i\,, \end{aligned} \tag{6.213}$$

where $\boldsymbol{\tau} = \tau_{\alpha} \boldsymbol{A}^{\alpha} = \boldsymbol{N} \times \boldsymbol{\nu}$ is the unit tangent to $\partial\Omega$, $\boldsymbol{u}_{,\nu} = \nu^{\alpha} \boldsymbol{u}_{,\alpha}$ is the normal derivative of $\boldsymbol{u}$, $(\cdot)' = d(\cdot)/ds$ and the square bracket refers to the forward jump as a corner of the boundary is traversed. Thus, $[\,\cdot\,] = (\cdot)_{+} - (\cdot)_{-}$, where the subscripts $\pm$ identify the limits as a corner located at arclength station s is approached through larger and smaller values of arclength, respectively. The sum accounts for the contributions from all corners. Here we admit piecewise smooth boundaries having a finite number of jumps in the unit tangent $\boldsymbol{\tau}$.

From the previous considerations it follows that the virtual power is of the form

$$P = \int_{\Omega} \boldsymbol{g} \cdot \boldsymbol{u}\, da + \int_{\partial\Omega_t} \boldsymbol{t} \cdot \boldsymbol{u}\, ds + \int_{\partial\Omega_m} \boldsymbol{m} \cdot \boldsymbol{u}_{,\nu}\, ds + \sum_{*} \boldsymbol{f}_i \cdot \boldsymbol{u}_i\,, \tag{6.214}$$

and the Fundamental Lemma furnish

$$\boldsymbol{t} = \boldsymbol{T}^{\alpha} \nu_{\alpha} - (Q\boldsymbol{n})', \qquad \boldsymbol{m} = M\boldsymbol{n} \qquad \text{and} \qquad \boldsymbol{f}_i = -[Q\boldsymbol{n}]_i\,, \tag{6.215}$$

with

$$Q = M^{\alpha\beta} \nu_{\alpha} \tau_{\beta} \qquad \text{and} \qquad M = M^{\alpha\beta} \nu_{\alpha} \nu_{\beta}, \tag{6.216}$$

where $\boldsymbol{t}$, $\boldsymbol{m}$ and $\boldsymbol{f}_i$ are the edge traction, edge double force density [32] and the corner force at the ith corner, respectively. Here, $\partial\Omega_t$ and $\partial\Omega_m$ respectively are parts of $\partial\Omega$ where $\boldsymbol{r}$ and $\boldsymbol{r}_{,\nu}$ are not assigned, and the starred sum includes only the corners where position is not assigned. We suppose that $\boldsymbol{r}$ and $\boldsymbol{r}_{,\nu}$ are assigned on $\partial\Omega \setminus \partial\Omega_t$ and $\partial\Omega \setminus \partial\Omega_m$, respectively, and that position is assigned at corners not included in the starred sum.

To interpret the double force in conventional terms we consider the case in which no kinematical data are assigned anywhere on $\partial\Omega$, so that $\partial\Omega_t = \partial\Omega_m = \partial\Omega$ and rigid-body variations of the deformation are kinematically admissible. The variational derivative of such a deformation is expressible as $\boldsymbol{u} = \boldsymbol{\omega} \times \boldsymbol{r} + \boldsymbol{a}$, where $\boldsymbol{a}$ and $\boldsymbol{\omega}$ are arbitrary spatially uniform vectors. Because the strain-energy function is invariant under such variations, we have $\dot{E} = 0$ and (6.188) reduces to $P = 0$, i.e.,

$$\begin{aligned} &\boldsymbol{a} \cdot \Big(\int_\Omega \boldsymbol{g}\, da + \int_{\partial\Omega} \boldsymbol{t}\, ds + \sum \boldsymbol{f}_i \Big) \\ &+ \boldsymbol{\omega} \cdot \Big[\int_\Omega \boldsymbol{r} \times \boldsymbol{g}\, da + \int_{\partial\Omega} (\boldsymbol{r} \times \boldsymbol{t} + \boldsymbol{c})\, ds + \sum \boldsymbol{r}_i \times \boldsymbol{f}_i \Big] = 0, \end{aligned} \tag{6.217}$$

where

$$\boldsymbol{c} = \boldsymbol{r}_{,\nu} \times \boldsymbol{m}\,. \tag{6.218}$$

The arbitrariness of $\boldsymbol{a}$ and $\boldsymbol{\omega}$ then yield the overall force and moment balances

$$\begin{aligned} &\int_\Omega \boldsymbol{g}\, da + \int_{\partial\Omega} \boldsymbol{t}\, ds + \sum \boldsymbol{f}_i = \boldsymbol{0} \quad \text{and} \\ &\int_\Omega \boldsymbol{r} \times \boldsymbol{g}\, da + \int_{\partial\Omega} (\boldsymbol{r} \times \boldsymbol{t} + \boldsymbol{c})\, ds + \sum \boldsymbol{r}_i \times \boldsymbol{f}_i = \boldsymbol{0}, \end{aligned} \tag{6.219}$$

yielding the interpretation of $\boldsymbol{c}$ as the couple traction.

From (6.215) and (6.219), with a little effort we can show that

$$\boldsymbol{c} = M\nu^\alpha \boldsymbol{a}_\alpha \times \boldsymbol{n} = -JM\tau_\alpha \boldsymbol{a}^\alpha, \tag{6.220}$$

where $J = \sqrt{a/A}$, with $a = \det(a_{\alpha\beta})$ and $A = \det(A_{\alpha\beta})$. Using $d\theta^\alpha = \tau^\alpha ds = \tau^{*\alpha} ds^*$, where s^* measures arclength on $\partial\omega$ with unit tangent $\boldsymbol{\tau}^* = \tau^{*\alpha}\boldsymbol{a}_\alpha$, together with $ds \simeq ds^*$ at leading order in strain, we infer that $\tau_\alpha = A_{\alpha\beta}\tau^\beta \simeq a_{\alpha\beta}\tau^\beta \simeq a_{\alpha\beta}\tau^{*\beta} = \tau^*_\alpha$, and hence that

$$\boldsymbol{c} \simeq -M\boldsymbol{\tau}^*, \tag{6.221}$$

a pure bending couple on $\partial\omega$. However, from the variational point of view it is not consistent to specify the couple traction in a boundary value problem. Rather, it is the double force that may be specified, and the couple traction can then be computed from (6.218), if desired, once the deformation is known.

6.4 Exercises

6.1 With reference to Exercise 4.2 (see Chap. 4) evaluate the corner forces at the corners of a square plate with lateral deformation given by

$$w(x, y) = Wxy\,,$$

where W is a constant and the x, y–axes are aligned with the edges of the plate. Show that the plate is in equilibrium with zero lateral load and that the tractions and bending couples vanish on the edges. Sketch the deformed plate with the corner forces acting.

6.2 Obtain a system of equations for a hemispherical shell under uniform lateral pressure, clamped along the equatorial edge. Assume an axisymmetric deformation mode without twist. What are the appropriate conditions to impose at the apex of the shell?

6.3 Consider an isotropic plate with an isotropic state of pre-stress, i.e., $S_{\alpha\beta} = T\delta_{\alpha\beta}$ Assume that $T = O(1)$ and derive the leading-order linear (membrane) problem for the transverse deflection w under a lateral distributed load g of order h. Show that this is given by

$$T\Delta w + g = 0.$$

Solve this on the interior of an ellipse, assuming that g is uniform and $w = 0$ on the elliptical boundary.

References

1. Cerda, E., Mahadevan, L.: Geometry and physics of wrinkling. Phys. Rev. Lett. **90**, 1–4 (2003)
2. Ciarlet, P.G.: Mathematical Elasticity, vol. III. Theory of Shells, North-Holland, Amsterdam (2000)
3. Ciarlet, P.G.: An Introduction to Differential Geometry with Applications to Elasticity. Springer, Dordrecht (2005)
4. Ciarlet, P.G., Roquefort, A.: Justification d'un modèle bi-dimensionnel non linéaire de coque analogue à celui de W.T. Koiter. C.R. Acad. Sci. I **331**, 411–416 (2000)
5. Ciarlet, P.G., Mardare, C.: An existence theorem for a two-dimensional nonlinear shell model of Koiter's type. Math. Models Methods Appl. Sci. **28**(14), 2833–2861 (2018)
6. Dacarogna, B.: Direct Methods in the Calculus of Variations. Springer, Berlin (1989)
7. Dikmen, M.: Theory of Thin Elastic Shells. Pitman Advanced Pub. Program (1982)
8. Fox, D.D., Raoult, A., Simo, J.C.: A justification of nonlinear properly invariant plate theories. Arch. Ration. Mech. Anal. **124**, 157–199 (1993)
9. Friesecke, G., James, R.D., Müller, S.: A hierarchy of plate models derived from nonlinear elasticity by Gamma-convergence. Arch. Ration. Mech. Anal. **180**, 183–236 (2006)
10. Friesecke, G., James R.D., Mora, M.G., Müller, S.: Derivation of nonlinear bending theory for shells from three-dimensional elasticity by Gamma-convergence. C.R. Acad. Sci. I **336**, 697–702 (2003)
11. Healey, T.J., Li, Q., Cheng, R.-B.: Wrinkling behavior of highly stretched rectangular elastic films via parametric global bifurcation. J. Nonlin. Sci. **23**, 777–805 (2013)

12. Hilgers, M.G., Pipkin, A.C.: Elastic sheets with bending stiffness. Q. Jl. Mech. Appl. Math. **45**, 57–75 (1992)
13. Hilgers, M.G., Pipkin, A.C.: Bending energy of highly elastic membranes. Quart. Appl. Math. **50**, 389–400 (1992)
14. Hilgers, M.G., Pipkin, A.C.: The Graves condition for variational problems of arbitrary order. IMA J. Appl. Math. **48**, 265–269 (1992)
15. Hilgers, M.G., Pipkin, A.C.: Bending energy of highly elastic membranes II. Quart. Appl. Math. **54**, 307–316 (1996)
16. Koiter, W.T.: A consistent first approximation in the general theory of thin elastic shells. In: Koiter, W.T. (ed.) Proceeding of the 1st IUTAM Symposium on the Theory of Thin Elastic Shells (Delft 1959), pp. 12–33. North-Holland, Amsterdam (1960)
17. Koiter, W.T.: On the nonlinear theory of thin elastic shells. Proc. Knonklijke Nederlandse Akademie van Wetenschappen **B69**, 1–54 (1966)
18. Koiter, W.T.: Foundations and basic equations of shell theory: a survey of recent progress. In: Niordson, F.I. (ed.) Theory of Thin Shells, IUTAM Symposium Copenhagen 1967, pp. 93–105. Springer, Berlin (1969)
19. Naghdi, P.M.: The theory of shells and plates. In: Flügge, W. (ed.), Handbuch der Physik, vol. VIa/2, pp. 425–640. Springer, Berlin (1972)
20. Pietraszkiewicz, W.: Geometrically nonlinear theories of thin elastic shells. Adv. Mech. **12**, 52–130 (1989)
21. Song, Z.L., Dai, H.-H.: On a consistent dynamic finite-strain plate theory and its linearization. J. Elast. **125**, 149–183 (2016)
22. Paroni, P.: Theory of linearly elastic residually stressed plates. Math. Mech. Solids **11**, 137–159 (2006)
23. Steigmann, D.: Tension-field theory. Proc. R. Soc. Lond. **A429**, 141–173 (1990)
24. Steigmann, D.: Thin-plate theory for large elastic deformations. Int. J. Non-Linear Mech. **42**, 233–240 (2007)
25. Steigmann, D.: Applications of polyconvexity and strong ellipticity to nonlinear elasticity and elastic plate theory. In: Schröder, J., Neff, P. (eds.) Poly-, Quasi-, and Rank-One Convexity in Applied Mechanics, Ser. CISM Courses and Lectures, vol. 516, pp. 265–299. Springer, Wien and New York (2010)
26. Steigmann, D.J.: Extension of Koiter's linear shell theory to materials exhibiting arbitrary symmetry. Int. J. Eng. Sci. **51**, 216–232 (2012)
27. Steigmann, D.J.: Koiter's shell theory from the perspective of three-dimensional nonlinear elasticity. J. Elast. **111**, 91–107 (2013)
28. Taylor, M., Bertoldi, K., Steigmann, D.J.: Spatial resolution of wrinkle patterns in thin elastic sheets at finite strain. J. Mech. Phys. Solids **62**, 163–180 (2014)
29. Steigmann, D.J.: Mechanics of materially-uniform thin films. Math. Mech. Solids **20**, 309–326 (2015)
30. Steigmann, D.J.: Equilibrium of elastic lattice shells. J. Eng. Math. **109**, 47–61 (2018)
31. Taylor, M., Shirani, M., Dabiri, Y., Guccione, J., Steigmann, D.J.: Finite elastic wrinkling deformations of incompressible fiber-reinforced plates. Int. J. Eng. Sci. **144**, 103138 (2019)
32. Toupin, R.A.: Theories of elasticity with couple-stress. Arch. Rational Mech. Anal. **17**, 85–112 (1964)

Chapter 7
Buckling of Elastic Plates

Abstract The classical theory of plate buckling is shown here to emerge from our dimension reduction procedure applied to incremental elasticity theory, concerned with the linearized theory or small deformations superposed upon large. Plate buckling theory emerges as the leading-order-in-thickness model when the underlying pre-stress scales appropriately with respect to thickness.

Introductory Remarks

We discuss the relationship between classical plate-buckling theory and the linear three-dimensional theory of incremental elasticity following the approach presented in the paper [18]. We show that the plate-buckling theory, despite the seemingly ad hoc assumptions underpinning its foundations [1, 5], emerges naturally as the limit of the three-dimensional linear theory of incremental elasticity if the thickness is sufficiently small, in the sense that the classical theory furnishes the leading order model in the small-thickness limit. Using elementary methods we derive precisely the same results as would be obtained by using the method of gamma convergence, which has become a popular tool for dimension reduction procedure in static elasticity theory.

The classical plate-buckling model is derived in the literature on the basis of several ad hoc assumptions, such as (see [1, pp. 137–138]):

- The state of stress is approximately plane and parallel to the middle surface.
- Normals to the undeformed middle surface remain normal to the deformed middle surface.
- The fundamental state is linear, i.e. the underlying deformation of the plate prior to buckling may be described using linear elasticity theory.

The derivation is based on the idea of minimizing the second variation of the energy at equilibrium, yielding linear incremental elasticity with null incremental

D. J. Steigmann et al., *Lecture Notes on the Theory of Plates and Shells*,
Solid Mechanics and Its Applications 274,
https://doi.org/10.1007/978-3-031-25674-5_7

data, similar to the developments of [5]. The novelty in the present approach is the relaxation of the a priori assumptions, which means a strengthening of the theoretical underpinnings of the theory. Thus, the classical assumptions are derived here as consequences of the three-dimensional theory in the small-thickness limit. We only make two important hypotheses, which concern the regularity of solutions of the three-dimensional incremental elasticity problem in the presence of strong ellipticity, together with the stipulation that the pre-stress vanishes with plate thickness. The assumption on regularity is discussed in Sect. 7.3, while the condition on pre-stress is introduced to ensure that the considered model furnishes a meaningful minimization problem.

In Sect. 7.1 we summarize some prerequisite material on nonlinear elasticity theory, including a review of the connection between stability and uniqueness of equilibria in the dead-load problem. In Sect. 7.2 we outline the linear incremental theory, which provides the basis for the study of bifurcation of equilibria. In Sect. 7.3 we develop in detail the small-thickness limit of the theory and present the main results of this chapter.

7.1 Nonlinear Elasticity and Stability

Consider the equilibrium boundary-value problem in the absence of body forces, which has the standard form

$$\operatorname{Div} \boldsymbol{P} = \boldsymbol{0} \quad \text{in } \kappa \quad \text{with} \qquad \boldsymbol{P}\boldsymbol{n} = \boldsymbol{t} \ \text{ on } \partial\kappa_t \quad \text{and} \qquad \boldsymbol{\chi} = \boldsymbol{\phi} \ \text{ on } \partial\kappa_\phi\,, \tag{7.1}$$

where κ is a fixed reference configuration with piecewise smooth boundary $\partial\kappa = \partial\kappa_t \cup \partial\kappa_\phi$, and $\boldsymbol{n}$ is the unit outward normal to the boundary. Here, Div is the divergence operation based on position $\boldsymbol{x} \in \kappa$, while $\boldsymbol{\chi}(\boldsymbol{x})$ is the deformation function, yielding the position $\boldsymbol{y} = \boldsymbol{\chi}(\boldsymbol{x})$ of the material point $\boldsymbol{x}$ after deformation. The vector fields $\boldsymbol{t}$ and $\boldsymbol{\phi}$ are specified functions. We use the convention that the divergence operates on the second index when applied to a second-order tensor, e.g. $(\operatorname{Div} \boldsymbol{P})_i = \partial P_{i\alpha}/\partial x_\alpha$ in Cartesian components. Also, we denote by $\boldsymbol{P}$ the first Piola-Kirchhoff stress tensor, which is given by

$$\boldsymbol{P} = \mathscr{W}_{\boldsymbol{F}}\,, \tag{7.2}$$

where $\mathscr{W}(\boldsymbol{F})$ is the strain-energy function (per unit volume). The tensor $\boldsymbol{F}$ is the deformation gradient given by $\boldsymbol{F} = \mathrm{D}\boldsymbol{\chi}$, where D is the gradient operation based on $\boldsymbol{x}$, and we impose that $\det \boldsymbol{F} > 0$. The second Piola-Kirchhoff stress $\boldsymbol{S}$ is defined in terms of the first Piola-Kirchhoff stress $\boldsymbol{P}$ by

$$\boldsymbol{P} = \boldsymbol{F}\boldsymbol{S}\,. \tag{7.3}$$

As usual, bold subscripts are used to denote the gradients of the strain energy with respect to the indicated tensor, as in (7.2). Another example is the fourth-order tensor of elastic moduli, denoted $\mathcal{M}$, which is defined as

$$\mathcal{M} = \mathcal{W}_{\boldsymbol{FF}} . \tag{7.4}$$

For any fourth-order tensor $\mathcal{L}$ and second-order tensor $\boldsymbol{A}$ we use the notation $\mathcal{L}[\boldsymbol{A}]$ to represent the second-order tensor with Cartesian components $L_{ijkl} A_{kl}$. The transpose $\mathcal{L}^T$ is defined by $\boldsymbol{B} \cdot \mathcal{L}^T[\boldsymbol{A}] = \boldsymbol{A} \cdot \mathcal{L}[\boldsymbol{B}]$. The tensor $\mathcal{L}$ is said to possess major symmetry if $\mathcal{L}^T = \mathcal{L}$. Also, $\mathcal{L}$ is said to possess minor symmetry if $\boldsymbol{A} \cdot \mathcal{L}[\boldsymbol{B}] = \boldsymbol{A}^T \cdot \mathcal{L}[\boldsymbol{B}]$ and $\boldsymbol{A} \cdot \mathcal{L}[\boldsymbol{B}] = \boldsymbol{A} \cdot \mathcal{L}[\boldsymbol{B}^T]$. For example, $\mathcal{M}$ possesses major symmetry but not minor symmetry.

The moment-of-momentum balance equation is satisfied identically provided the strain-energy function is invariant under superposed rigid motions. This ensures the symmetry of the second Piola-Kirchhoff stress tensor $\boldsymbol{S}$, which is then given by

$$\boldsymbol{S} = \mathcal{U}_{\boldsymbol{E}} , \tag{7.5}$$

where $\boldsymbol{E} = \frac{1}{2}(\boldsymbol{F}^T \boldsymbol{F} - \boldsymbol{I})$ is the Green-Lagrange strain tensor and we denote by

$$\mathcal{U}(\boldsymbol{E}) = \mathcal{W}(\boldsymbol{F}). \tag{7.6}$$

In what follows we consider only smooth deformations and limit our attention to deformations which satisfy the strong-ellipticity condition

$$(\boldsymbol{a} \otimes \boldsymbol{b}) \cdot \mathcal{M}(\boldsymbol{F})[\boldsymbol{a} \otimes \boldsymbol{b}] > 0 \quad \text{for all} \quad \boldsymbol{a} \otimes \boldsymbol{b} \neq \boldsymbol{0}. \tag{7.7}$$

We consider boundary-value problems is of mixed type, involving position and traction data on complementary parts of the boundary, and confine our attention to *dead*, i.e. configuration-independent, tractions. Thus, the vector fields $\boldsymbol{t}$ and $\boldsymbol{\phi}$ in (7.1) are to be regarded as assigned functions of $\boldsymbol{x}$.

The potential energy of the body consists of the strain energy and the load potential; it is given as a functional of $\chi(\boldsymbol{x})$ by the following expression

$$\mathcal{E}[\chi] = \int_\kappa \mathcal{W}(\boldsymbol{F})\, \mathrm{d}v - \int_{\partial\kappa_t} \boldsymbol{t} \cdot \chi \, \mathrm{d}a, \tag{7.8}$$

apart from an unimportant constant. It is well known (see, for example, Sect. 3.3 or [10]) that equilibrium deformations satisfying the boundary-value problem (7.1) render $\mathcal{E}$ stationary, and conversely. This is the so-called virtual-work principle, in the case of the position/dead-load problem. The energy criterion of elastic stability [9] affirms that the deformation χ is stable if and only if it minimizes the energy relative to kinematically admissible alternatives, that is if and only if

$$\mathcal{E}[\chi] < \mathcal{E}[\chi + \Delta\chi] \tag{7.9}$$

for all $\Delta\chi(x)$ vanishing on $\partial\kappa_\phi$ but not vanishing identically in κ. The inequality (7.9) implies that $\mathscr{E}[\,\cdot\,]$ is stationary at χ. Hence, the deformation χ is equilibrated, i.e. it solves the boundary-value problem (7.1).

Moreover, the relation (7.9) also implies that solutions of (7.1) are unique. Indeed, let us prove this property by adapting an argument of Hill [7]: suppose that $\chi_1(x)$ and $\chi_2(x)$ are two solutions of the boundary-value problem. If both are strict minimizers of the energy, then, by selecting χ and $\Delta\chi$ appropriately it follows that

$$\int_\kappa \big[\mathscr{W}(F_2) - \mathscr{W}(F_1) - P_1 \cdot (F_2 - F_1)\big]\,\mathrm{d}v > 0 \tag{7.10}$$

provided that $F_2 \neq F_1$, and

$$\int_\kappa \big[\mathscr{W}(F_1) - \mathscr{W}(F_2) - P_2 \cdot (F_1 - F_2)\big]\,\mathrm{d}v > 0 \tag{7.11}$$

where $F_\gamma = \mathrm{D}\chi_\gamma$, while $P_\gamma = \mathscr{W}_F(F_\gamma)$ are the associated stresses ($\gamma = 1, 2$). These inequalities follow from the fact that both fields χ_1 and χ_2 satisfy the Eqs. (7.1). Summation of the inequalities (7.10) and (7.11) implies

$$\int_\kappa (P_2 - P_1) \cdot (F_2 - F_1)\,\mathrm{d}v > 0, \qquad F_2 \neq F_1\,. \tag{7.12}$$

But in view of (7.1) we have $\mathrm{Div}(P_2 - P_1) = \mathbf{0}$. If we take the inner product of this with $\chi_2 - \chi_1$, integrate over the domain κ and use the boundary conditions, then we deduce

$$\int_\kappa (P_2 - P_1) \cdot (F_2 - F_1)\,\mathrm{d}v = 0, \tag{7.13}$$

which is reconciled with (7.12) if and only if $F_2 = F_1$. Integrating the latter with respect to x, we arrive at $\chi_2 = \chi_1$ apart from a rigid-body translation. In view of the position boundary data, this rigid-body translation vanishes. Thus, we have proved the claim that stability, in the sense of the energy criterion, implies uniqueness of solution of the mixed position/dead-load problem. Conversely, non-uniqueness implies that equilibria are not stable in the sense of strict inequality in (7.9).

If the inequality in (7.9) is semi-strict, with equality holding for some admissible $\Delta\chi$, then the deformation χ is called neutrally stable. This semi-strict inequality again implies that $\mathscr{E}[\,\cdot\,]$ is stationary at χ and hence that the latter is equilibrated. In this case, non-uniqueness is possible because strict inequality in (7.9) is sufficient for uniqueness of solution of the present equilibrium boundary-value problem. Non-uniqueness of solutions then implies that the strict inequality does not obtain. Accordingly, the non-uniqueness feature signals a failure of stability and thus a potential instability. This property is Euler's well-known adjacent-equilibrium criterion of elastic stability, adapted to nonlinear elasticity. Note that the energy criterion of elastic stability as stated is heuristic in the sense that no rigorous connection to stability in the dynamical sense is known in the case of continuous systems [9]. However, Koiter has given convincing arguments in support of this criterion in [8].

7.2 Linear Incremental Elasticity and Bifurcation of Equilibria

Let us summarize below the basic theory of incremental elasticity as it relates to bifurcation of equilibria. We mention that this theory is presented in detail in [10], together with extensions to configuration-dependent loading. Consider a one-parameter family of kinematically possible deformations $\boldsymbol{\chi}(\boldsymbol{x}\,;\epsilon)$ satisfying the fixed position data

$$\boldsymbol{\chi}(\boldsymbol{x}\,;\epsilon) = \boldsymbol{\phi}(\boldsymbol{x}) \quad \text{on} \quad \partial\kappa_\phi\,. \tag{7.14}$$

We regard the potential energy (7.8) as a function of ϵ and write it in the form

$$F(\epsilon) = \mathscr{E}\big[\boldsymbol{\chi}(\boldsymbol{x}\,;\epsilon)\big]. \tag{7.15}$$

If we assume the deformation corresponding to $\epsilon = 0$ to be equilibrated, and expand F for small ϵ (i.e., for small displacements from the equilibrium configuration), then we obtain

$$F(\epsilon) = F(0) + \epsilon^2 \mathscr{G}[\boldsymbol{\chi}, \dot{\boldsymbol{\chi}}] + o(\epsilon^2) \tag{7.16}$$

where

$$\mathscr{G}[\boldsymbol{\chi}, \dot{\boldsymbol{\chi}}] = \frac{1}{2}\ddot{F}, \tag{7.17}$$

in which the superposed dots stand for derivatives with respect to ϵ, evaluated at $\epsilon = 0$, and $\boldsymbol{\chi}(\boldsymbol{x}) = \boldsymbol{\chi}(\boldsymbol{x}\,;0)$ is the underlying finite equilibrium deformation. Here, $\mathscr{G}$ stands for the second variation of the potential energy at the considered equilibrium state. Notice that the first variation does not appear in (7.16), since it vanishes by the virtual-work principle.

Let us compute the first an second variations of the potential energy. Using (7.8), (7.15) and the fact that $\boldsymbol{\chi}'$ vanishes on $\partial\kappa_\phi$, we deduce

$$F'(\epsilon) = \int_\kappa \mathscr{W}_{\boldsymbol{F}} \cdot \mathrm{D}\boldsymbol{\chi}'\,\mathrm{d}v - \int_{\partial\kappa} \boldsymbol{t} \cdot \boldsymbol{\chi}'\,\mathrm{d}a, \tag{7.18}$$

where primes are used to denote derivatives at any value of ϵ. In view of (7.2), the last relation yields

$$F'(\epsilon) = \int_{\partial\kappa_t} (\boldsymbol{P}\boldsymbol{n} - \boldsymbol{t}) \cdot \boldsymbol{\chi}'\,\mathrm{d}a - \int_\kappa \boldsymbol{\chi}' \cdot \mathrm{Div}\,\boldsymbol{P}\,\mathrm{d}v. \tag{7.19}$$

As already mentioned, the first variation (7.19) vanishes at $\epsilon = 0$, by virtue of (7.1). To obtain the second variation, we differentiate the relation (7.18) an find

$$F''(\epsilon) = \int_{\kappa} \mathscr{W}_F \cdot \mathrm{D}\chi'' \, dv - \int_{\partial\kappa} \boldsymbol{t} \cdot \chi'' \, da + \int_{\kappa} \mathrm{D}\chi' \cdot \mathscr{M}(\boldsymbol{F})[\mathrm{D}\chi'] \, dv. \tag{7.20}$$

The first two terms cancel at $\epsilon = 0$, by virtue of (7.18) and the relation $F'(0) = 0$. Evaluating (7.20) at $\epsilon = 0$ and using (7.17) we obtain

$$\mathscr{G}[\chi, \dot{\chi}] = \frac{1}{2} \int_{\kappa} \dot{\boldsymbol{F}} \cdot \mathscr{M}(\boldsymbol{F})[\dot{\boldsymbol{F}}] \, dv, \tag{7.21}$$

in which $\dot{\boldsymbol{F}} = \mathrm{D}\dot{\chi}$ is the incremental deformation gradient and $\boldsymbol{F} = \mathrm{D}\chi$ is now the gradient of the underlying finite equilibrium deformation. If the equilibrium state is stable, then from (7.9) and (7.16) we derive that

$$0 < [F(\epsilon) - F(0)]/\epsilon = \mathscr{G}[\chi, \dot{\chi}] + o(\epsilon^2)/\epsilon^2 , \tag{7.22}$$

and in the limit we obtain

$$\mathscr{G}[\chi, \dot{\chi}] > 0. \tag{7.23}$$

Let us derive now an expression for the second variation based on the strain-energy function $\mathscr{U}(\boldsymbol{E})$, which will be required later. To this aim, we start from the incremental form of (7.3), which can be written

$$\dot{\boldsymbol{P}} = \dot{\boldsymbol{F}}\boldsymbol{S} + \boldsymbol{F}\dot{\boldsymbol{S}} , \tag{7.24}$$

where

$$\dot{\boldsymbol{S}} = \mathscr{C}(\boldsymbol{E})[\dot{\boldsymbol{E}}] , \qquad \mathscr{C} = \mathscr{U}_{\boldsymbol{EE}} , \tag{7.25}$$

and

$$\dot{\boldsymbol{E}} = \frac{1}{2}(\dot{\boldsymbol{F}}^T \boldsymbol{F} + \boldsymbol{F}^T \dot{\boldsymbol{F}}) . \tag{7.26}$$

Note that the fourth-order tensor $\mathscr{C}$ introduced in (7.25) possesses both major and minor symmetry. We can combine this with (7.21) to obtain an expression for the second variation that shows explicit the role played by the pre-stress $\boldsymbol{S}(\boldsymbol{x})$ induced by $\chi(\boldsymbol{x})$. Using the relation

$$\dot{\boldsymbol{P}} = \mathscr{M}(\boldsymbol{F})[\dot{\boldsymbol{F}}] , \tag{7.27}$$

we find in particular that

$$\mathscr{M}(\boldsymbol{F})[\boldsymbol{A}] = \boldsymbol{A}\boldsymbol{S} + \frac{1}{2} \boldsymbol{F}\mathscr{C}(\boldsymbol{E})[\boldsymbol{A}^T \boldsymbol{F} + \boldsymbol{F}^T \boldsymbol{A}] \tag{7.28}$$

for any (second-order) tensor $\boldsymbol{A}$.

According to the discussion at the end of Sect. 7.1, we know that stability is lost in the strict sense if χ satisfies

$$\mathscr{G}[\chi, \dot{\chi}] = 0 \tag{7.29}$$

for some non-trivial admissible $\dot{\chi}$. To see the relationship to bifurcation of equilibrium, suppose that $\chi(\boldsymbol{x}\,; \mu)$ is a one-parameter family of equilibrium deformations which satisfies the boundary-value problem (7.1) at all values of μ in some open neighborhood of 0. If we differentiate the equations in (7.1) with respect to μ, evaluate them at $\mu = 0$ and identify the associated value of the μ-derivative of $\chi(\boldsymbol{x}\,; \mu)$ with $\dot{\chi}$ above, we deduce

$$\operatorname{Div} \dot{\boldsymbol{P}} = \boldsymbol{0} \quad \text{in } \kappa\,, \qquad \dot{\boldsymbol{P}}\boldsymbol{n} = \boldsymbol{0} \quad \text{on } \partial\kappa_t \qquad \text{and} \qquad \dot{\chi} = \boldsymbol{0} \quad \text{on } \partial\kappa_\phi\,. \tag{7.30}$$

Thus, we obtain a homogeneous linear boundary-value problem for the incremental bifurcation $\dot{\chi}$. This problem is also the linearization of the equilibrium boundary-value problem for the deformation $\chi + \dot{\chi}$ in which χ is equilibrated.

In view of the relations (7.21), (7.27) and the major symmetry of $\mathcal{M}$, we can write the variation of the functional $\mathscr{G}$ with respect to $\dot{\chi}$ as

$$\delta\mathscr{G}[\chi, \dot{\chi}, \delta\dot{\chi}] = \frac{1}{2}\int_\kappa \{\mathrm{D}(\delta\dot{\chi}) \cdot \dot{\boldsymbol{P}} + \dot{\boldsymbol{F}} \cdot \mathcal{M}(\boldsymbol{F})[\mathrm{D}(\delta\dot{\chi})]\}\mathrm{d}v = \int_\kappa \mathrm{D}(\delta\dot{\chi}) \cdot \dot{\boldsymbol{P}}\,\mathrm{d}v \tag{7.31}$$

This equation may be re-written in the form

$$\delta\mathscr{G}[\chi, \dot{\chi}, \delta\dot{\chi}] = \int_{\partial\kappa_t} \delta\dot{\chi} \cdot \dot{\boldsymbol{P}}\boldsymbol{n}\,\mathrm{d}a - \int_\kappa \delta\dot{\chi} \cdot \operatorname{Div} \dot{\boldsymbol{P}}\,\mathrm{d}v, \tag{7.32}$$

Note that the variation (7.32) vanishes for any admissible $\delta\dot{\chi}$ if and only the equations in (7.30) hold. Accordingly, a deformation $\chi(\boldsymbol{x})$ is stable only if the inequality

$$\mathscr{G}[\chi, \dot{\chi}] \geq 0 \tag{7.33}$$

holds in the *strict* sense for all kinematically admissible $\dot{\chi}$. The deformation $\chi(\boldsymbol{x})$ is *potentially* unstable if the equality is satisfied for nonzero $\dot{\chi}$, i.e. if the linearized bifurcation problem (7.30) has a non-trivial solution. We can justify the latter claim from the fact $\mathscr{G}$ vanishes for such $\dot{\chi}$. Indeed, to show this we employ (7.30) and (7.32) to write

$$\mathscr{G}[\chi, \dot{\chi}] = \frac{1}{2}\delta\mathscr{G}[\chi, \dot{\chi}, \dot{\chi}],$$

which vanishes for all solutions $\dot{\chi}$ to the bifurcation problem. This criterion is due to Hill [7] in the present framework. If this criterion is verified, then the energy comparison is dominated by the higher-order variations. For instance, instability of the underlying finite equilibrium deformation is guaranteed if the third variation can be made non-zero. In this sense, we can say that a non-trivial bifurcation signals a potential instability.

We refer to such solutions as buckling modes or eigenmodes, as it is usual in the literature, see e.g. [7, 10]. Remark that it is not possible to distinguish between neutral stability and instability of the equilibrium deformation $\boldsymbol{\chi}$ on the basis of linearized bifurcation theory. To make such distinction it is necessary to consider nonlinear terms. The influence of such terms is investigated by the post-bifurcation theory [3, 5], but they will not be taken into account here. In what follows, we denote by

$$\tilde{\boldsymbol{u}}(\boldsymbol{x}) = \dot{\boldsymbol{\chi}}(\boldsymbol{x}) \quad \text{and} \quad \tilde{\boldsymbol{H}}(\boldsymbol{x}) = \dot{\boldsymbol{F}}(\boldsymbol{x}), \tag{7.34}$$

respectively, the incremental displacement and displacement gradient fields, such that $\tilde{\boldsymbol{H}} = \mathrm{D}\tilde{\boldsymbol{u}}$.

7.3 Two–dimensional Model for Plate-Buckling

7.3.1 *Description of Plates as Thin Prismatic Bodies*

As usual for plate models, we parametrize the reference configuration κ in the form

$$\boldsymbol{x} = \boldsymbol{r} + \varsigma\, \boldsymbol{k}, \qquad \text{with} \;\; \boldsymbol{r} \in \omega, \tag{7.35}$$

where ω is the plate midsurface, $\boldsymbol{k}$ is the fixed unit normal to the plate and $\varsigma \in [-h/2, h/2]$. The plate thickness h is assumed to be uniform and small, such that $h/l \ll 1$, where l is any other length scale in the problem such as a typical spanwise dimension of ω. The set ω is assumed to be simply connected, so that Green's integral theorem may be applied. For the sake of simplicity, we adopt l as the unit of length, i.e. $l = 1, h \ll 1$. In this way, we presume all length scales to be non-dimensionalized by l and henceforth regard h as a small dimensionless parameter.

Let us denote

$$\hat{\boldsymbol{u}}(\boldsymbol{r}, \varsigma) = \tilde{\boldsymbol{u}}(\boldsymbol{r} + \varsigma\, \boldsymbol{k}), \qquad \hat{\boldsymbol{H}}(\boldsymbol{r}, \varsigma) = \tilde{\boldsymbol{H}}(\boldsymbol{r} + \varsigma\, \boldsymbol{k}). \tag{7.36}$$

We designate by $\nabla(\cdot)$ the two-dimensional gradient with respect to $\boldsymbol{r}$ at fixed ς and by $(\cdot)'$ the derivative $\partial(\cdot)/\partial\varsigma$ at fixed $\boldsymbol{r}$. We consider the projection onto the translation (vector) space ω' of ω, given by

$$\boldsymbol{1} = \boldsymbol{I} - \boldsymbol{k} \otimes \boldsymbol{k}, \tag{7.37}$$

where $\boldsymbol{I}$ is the identity for three-space. With these notations we can write (see Eq. (4.14) or [14])

$$\tilde{\boldsymbol{H}}\boldsymbol{1} = \nabla\hat{\boldsymbol{u}}, \qquad \tilde{\boldsymbol{H}}\boldsymbol{k} = \hat{\boldsymbol{u}}', \tag{7.38}$$

as well as the orthogonal decomposition

$$\hat{\boldsymbol{H}} = \nabla\hat{\boldsymbol{u}} + \hat{\boldsymbol{u}}' \otimes \boldsymbol{k}. \tag{7.39}$$

By differentiating (7.39) with respect to ς, we obtain

$$\boldsymbol{H}_0 = \nabla\boldsymbol{u} + \boldsymbol{a} \otimes \boldsymbol{k}, \qquad \boldsymbol{H}_0' = \nabla\boldsymbol{a} + \boldsymbol{b} \otimes \boldsymbol{k}, \qquad \boldsymbol{H}_0'' = \nabla\boldsymbol{b} + \boldsymbol{c} \otimes \boldsymbol{k}, \tag{7.40}$$

where the zero subscript indicates evaluation at $\varsigma = 0$. Consider the thickness-wise power expansion of the three-dimensional incremental displacement field

$$\hat{\boldsymbol{u}} = \boldsymbol{u} + \varsigma\,\boldsymbol{a} + \frac{1}{2}\,\varsigma^2\boldsymbol{b} + \frac{1}{6}\,\varsigma^3\boldsymbol{c} + \dots\,, \tag{7.41}$$

where the coefficient vectors are

$$\boldsymbol{u} = \hat{\boldsymbol{u}}_0\,, \qquad \boldsymbol{a} = \hat{\boldsymbol{u}}_0'\,, \qquad \boldsymbol{b} = \hat{\boldsymbol{u}}_0''\,, \qquad \boldsymbol{c} = \hat{\boldsymbol{u}}_0'''\,. \tag{7.42}$$

Notice that we have assumed more regularity than has been proved, particularly for the underlying finite-deformation equilibrium problem. It is in this sense that our analysis is formal. Henceforth we will not discuss the degree of regularity required by the analysis. Classical regularity results relying on strong ellipticity are available in the linear theory [2] and it seems likely that much of these results can be extended to the incremental theory. Since we consider equilibrium states, we construct an expression for the plate energy that presumes C^3 continuity at the outset, which is the class of admissible functions, subject to additional boundary conditions.

We designate by c^* be the line orthogonal to ω and intersecting $\partial\omega$ at a point with position $\boldsymbol{r}$, and by $\partial\kappa_c = \partial\omega \times c$, where c is the collection of such lines, the cylindrical boundary surface of the plate-like region κ obtained by translating the points of $\partial\omega$ along their associated lines c^*. Denote by s the arclength on the boundary curve $\partial\omega$, which has unit tangent $\boldsymbol{\tau}$ and rightward unit normal $\boldsymbol{\nu} = \boldsymbol{\tau} \times \boldsymbol{k}$.

The set $\partial\kappa_t$ consists of the major surfaces of the plate together with a part $\partial\omega_t \times c$ of $\partial\kappa_c$, where $\partial\omega_t$ is a subset of $\partial\omega$. From the relation $(7.30)_2$ we derive $\dot{\boldsymbol{P}}\boldsymbol{1}\boldsymbol{\nu} = \boldsymbol{0}$ on $\partial\omega_t \times c$, so in particular it holds

$$\dot{\boldsymbol{P}}_0\boldsymbol{1}\boldsymbol{\nu} = \boldsymbol{0} \quad \text{on} \quad \partial\omega_t\,, \tag{7.43}$$

where $\dot{\boldsymbol{P}}_0$ is the restriction of $\dot{\boldsymbol{P}}$ to ω. Let $\partial\omega_\phi = \partial\omega \setminus \partial\omega_t$. On the boundary part $\partial\omega_\phi \times c$ position is then assigned, so the incremental displacement vanishes. Hence, $\hat{\boldsymbol{u}}$ vanishes identically on $\partial\omega_\phi \times c$ and the expansion (7.41) infers

$$\boldsymbol{u} = \boldsymbol{a} = \boldsymbol{b} = \dots = \boldsymbol{0} \quad \text{on} \quad \partial\omega_\phi\,. \tag{7.44}$$

7.3.2 Thickness-Wise Power Expansion of the Second Variation

We consider the second variation of the potential energy $\mathscr{G}$ and seek the optimal expression for the functional G in the expansion

$$\mathscr{G} = G + o(h^3). \tag{7.45}$$

Let us write the functional $\mathscr{G}$ as the multiple integral

$$\mathscr{G} = \frac{1}{2} \int_\omega \int_{-h/2}^{h/2} \dot{\boldsymbol{P}} \cdot \tilde{\boldsymbol{H}} \, \mathrm{d}\varsigma \, \mathrm{d}a \tag{7.46}$$

Then, using the Leibniz rule we obtain the Taylor expansion

$$\int_{-h/2}^{h/2} \dot{\boldsymbol{P}} \cdot \tilde{\boldsymbol{H}} \, \mathrm{d}\varsigma = h(\dot{\boldsymbol{P}} \cdot \hat{\boldsymbol{H}})_0 + \frac{1}{24} h^3 (\dot{\boldsymbol{P}} \cdot \hat{\boldsymbol{H}})_0'' + o(h^3), \tag{7.47}$$

where we have

$$\begin{aligned} (\dot{\boldsymbol{P}} \cdot \hat{\boldsymbol{H}})_0 &= \dot{\boldsymbol{P}}_0 \cdot \boldsymbol{H}_0, \\ (\dot{\boldsymbol{P}} \cdot \hat{\boldsymbol{H}})_0'' &= 4\dot{\boldsymbol{P}}_0' \cdot \boldsymbol{H}_0' + 2\dot{\boldsymbol{P}}_0 \cdot \boldsymbol{H}_0'' + \boldsymbol{H}_0 \cdot \mathscr{M}_0''[\boldsymbol{H}_0] - 2\boldsymbol{H}_0' \cdot \mathscr{M}_0[\boldsymbol{H}_0'], \end{aligned} \tag{7.48}$$

since it holds (cf. (7.27))

$$\dot{\boldsymbol{P}}_0' = \mathscr{M}_0'[\boldsymbol{H}_0] + \mathscr{M}_0[\boldsymbol{H}_0']. \tag{7.49}$$

The thickness-wise derivatives $\mathscr{M}'$ and $\mathscr{M}''$ are calculated in the case of uniform materials on the basis of through-thickness variations of the gradient $\boldsymbol{F}$. However, their computation requires higher-order elastic moduli, e.g. $\mathscr{M}_{ijkl}' = \mathscr{A}_{ijklmn} F_{mn}'$, where $\mathscr{A}$ is the tensor of second-order moduli [10]. Thus, the finite deformation χ may be used to generate a through-thickness *functional gradient* in the incremental response.

The above expansion is terminated at order h^3 since this is the lowest order at which the derived two-dimensional model remains well posed while accommodating buckling in the presence of compressive pre-stress. Further discussions on this reason is presented in Sect. 7.3.5 below. In what follows, we suppose the initial finite deformation to be such that $\boldsymbol{F}$ does not vary through the thickness of the plate. This assumption allows us to keep the treatment simple and to have direct contact with classical plate buckling theory. Notice that higher-order moduli are not needed in this case, so we get a simplification of the model while making allowance for in-plane functional gradients. Then, in view of (7.40) we derive

$$
\begin{aligned}
(\dot{\boldsymbol{P}} \cdot \hat{\boldsymbol{H}})_0 &= \dot{\boldsymbol{P}}_0 \mathbf{1} \cdot \nabla \boldsymbol{u} + \dot{\boldsymbol{P}}_0 \boldsymbol{k} \cdot \boldsymbol{a}\,, \\
(\dot{\boldsymbol{P}} \cdot \hat{\boldsymbol{H}})''_0 &= 2\big(\dot{\boldsymbol{P}}'_0 \mathbf{1} \cdot \nabla \boldsymbol{a} + \dot{\boldsymbol{P}}'_0 \boldsymbol{k} \cdot \boldsymbol{b} + \dot{\boldsymbol{P}}_0 \mathbf{1} \cdot \nabla \boldsymbol{b} + \dot{\boldsymbol{P}}_0 \boldsymbol{k} \cdot \boldsymbol{c}\big).
\end{aligned}
\tag{7.50}
$$

7.3.3 The Refined Model

According to the above relations, we can obtain the functional G in (7.45) by substituting the Eqs. (7.49) and (7.50) into (7.47). Then, the resulting expression of G is found to depend on the vector fields $\boldsymbol{u}$, $\boldsymbol{a}$, $\boldsymbol{b}$, and $\boldsymbol{c}$. Since these fields are independent at this level of the development, each would generate an associated Euler equation and natural boundary condition. This system of equations would then constitute the linearized bifurcation problem for the thin plate.

Nevertheless, one can optimize the functional G in respect of the three-dimensional theory by imposing certain a priori constraints among the vector fields. We follow this line and derive below the expression of G, which involves the single field $\boldsymbol{u}$ and furnishes the optimal approximation of order h^3 to the second variation for a given mid-surface incremental displacement field. This model automatically encodes restrictions arising in the three-dimensional parent theory.

Consider the lateral surfaces of the plate characterized by thickness coordinates $\varsigma = \pm h/2$ and denote by $\dot{\boldsymbol{P}}^{\pm} = \dot{\boldsymbol{P}}_{|\varsigma=\pm h/2}$. If the lateral surfaces of the plate are traction-free, then from $(7.30)_2$ it follows that $\dot{\boldsymbol{P}}^{\pm}\boldsymbol{k} = \mathbf{0}$. Consider the two Taylor expansions corresponding to the major surfaces

$$
(\mathbf{0} =)\; \dot{\boldsymbol{P}}^{\pm}\boldsymbol{k} = \dot{\boldsymbol{P}}_0\boldsymbol{k} \pm \frac{h}{2}\,\dot{\boldsymbol{P}}'_0\boldsymbol{k} + \frac{h^2}{8}\,\dot{\boldsymbol{P}}''_0\boldsymbol{k} + O(h^3).
$$

By addition and subtraction of these two equations we deduce

$$
\dot{\boldsymbol{P}}_0\boldsymbol{k} = O(h^2), \qquad \dot{\boldsymbol{P}}'_0\boldsymbol{k} = O(h^2). \tag{7.51}
$$

Hence, $\dot{\boldsymbol{P}}_0\boldsymbol{k}$ and $\dot{\boldsymbol{P}}'_0\boldsymbol{k}$ may be suppressed *in the coefficient of* h^3 without loss of accuracy in the expansion of the second variation of order h^3. Accordingly, we impose in relation $(7.50)_2$ that

$$
\dot{\boldsymbol{P}}_0\boldsymbol{k} = \mathbf{0} = \dot{\boldsymbol{P}}'_0\boldsymbol{k}\,. \tag{7.52}
$$

Remark that the system of equations (7.52) may be solved explicitly for $\boldsymbol{a}$ and $\boldsymbol{b}$. Let us prove this claim: in view of $(7.40)_1$ we have

$$
\dot{\boldsymbol{P}}_0\boldsymbol{k} = \big(\mathscr{M}[\nabla \boldsymbol{u}]\big)\boldsymbol{k} + \boldsymbol{A}_{(\boldsymbol{k})}\boldsymbol{a}\,, \tag{7.53}
$$

where $\boldsymbol{A}_{(\boldsymbol{k})}$ stands for the the tensor defined by

$$A_{(k)}\boldsymbol{v} = \big(\mathscr{M}[\boldsymbol{v}\otimes\boldsymbol{k}]\big)\boldsymbol{k} \qquad \text{for all vectors } \boldsymbol{v}. \tag{7.54}$$

Note that $\boldsymbol{A}_{(k)}$ is referred to as the *acoustic tensor* in other contexts. By virtue of the strong-ellipticity inequality (7.7) it follows that this tensor is positive definite and hence invertible. Thus, the Eq. $(7.52)_1$ determines $\boldsymbol{a}$ uniquely in terms of $\nabla\boldsymbol{u}$, i.e.

$$\boldsymbol{a} = \boldsymbol{g}(\nabla\boldsymbol{u}), \tag{7.55}$$

where the function $\boldsymbol{g}$ is determined by material properties and the underlying finite deformation. In view of $(7.52)_1$ and (7.53), the function $\boldsymbol{g}$ has the explicit form

$$\boldsymbol{g}(\nabla\boldsymbol{u}) = -\boldsymbol{A}_{(k)}^{-1}\big(\mathscr{M}[\nabla\boldsymbol{u}]\big)\boldsymbol{k}\,. \tag{7.56}$$

Similarly, the Eqs. (7.49) and $(7.52)_2$ determine the vector $\boldsymbol{b}$. If we do not account for the through-thickness functional gradient, we obtain the relation

$$\boldsymbol{b} = \boldsymbol{g}(\nabla\boldsymbol{a}). \tag{7.57}$$

This relation is modified in the presence of a through-thickness functional gradient, but we can anyway use this procedure to determine the vector $\boldsymbol{b}$ explicitly. In this way, the Eq. $(7.50)_2$ becomes

$$(\dot{\boldsymbol{P}}\cdot\hat{\boldsymbol{H}})''_0 = 2\big(\dot{\boldsymbol{P}}'_0\mathbf{1}\cdot\nabla\boldsymbol{a} + \dot{\boldsymbol{P}}_0\mathbf{1}\cdot\nabla\boldsymbol{b}\big) + O(h^2), \tag{7.58}$$

where the vectors $\boldsymbol{a}$ and $\boldsymbol{b}$ are expressed by (7.55) and (7.57), respectively.

However, notice that we are *not* justified to neglect $\dot{\boldsymbol{P}}_0\boldsymbol{k}$ in the order-h term, since the relation $(7.51)_1$ implies that $\dot{\boldsymbol{P}}_0\boldsymbol{k}$ makes a net contribution at order h^3 and is thus comparable to other terms that have been retained. We discuss this issue later on.

We can further simplify the model by writing

$$\dot{\boldsymbol{P}}_0\mathbf{1}\cdot\nabla\boldsymbol{b} = \mathrm{div}\big[(\dot{\boldsymbol{P}}_0\mathbf{1})^T\boldsymbol{b}\big] - \boldsymbol{b}\cdot\mathrm{div}(\dot{\boldsymbol{P}}_0\mathbf{1}), \tag{7.59}$$

where div is the (two-dimensional) divergence with respect to position $\boldsymbol{r}$ on the mid-plane ω. Then, we can use the decomposition $\dot{\boldsymbol{P}} = \dot{\boldsymbol{P}}\mathbf{1} + \dot{\boldsymbol{P}}\boldsymbol{k}\otimes\boldsymbol{k}$ of the incremental stress to write the Eq. $(7.30)_1$ in the form

$$\mathrm{div}(\dot{\boldsymbol{P}}\mathbf{1}) + \dot{\boldsymbol{P}}'\boldsymbol{k} = \mathbf{0}$$

and evaluating this on ω we deduce

$$\mathrm{div}(\dot{\boldsymbol{P}}_0\mathbf{1}) + \dot{\boldsymbol{P}}'_0\boldsymbol{k} = \mathbf{0}. \tag{7.60}$$

Using (7.59), (7.60) and the estimate $(7.51)_2$ we find

$$\dot{\boldsymbol{P}}_0\mathbf{1}\cdot\nabla\boldsymbol{b} = \mathrm{div}\big[(\dot{\boldsymbol{P}}_0\mathbf{1})^T\boldsymbol{b}\big] + O(h^2). \tag{7.61}$$

Then, in view of of Green's theorem and the data (7.43) and (7.44), we infer that the term $\dot{\boldsymbol{P}}_0\mathbf{1}\cdot\nabla\boldsymbol{b}$ may be neglected in the order-h^3 expansion of the second variation without loss of accuracy. Thus, we arrive at

$$G = \int_\omega \bar{W}\,da, \qquad \text{with} \tag{7.62}$$

$$\bar{W} = \frac{1}{2}h(\dot{\boldsymbol{P}}_0\mathbf{1}\cdot\nabla\boldsymbol{u} + \dot{\boldsymbol{P}}_0\boldsymbol{k}\cdot\boldsymbol{a}) + \frac{1}{24}h^3\dot{\boldsymbol{P}}_0'\mathbf{1}\cdot\nabla\boldsymbol{a}\,, \tag{7.63}$$

where we impose (7.55) and (7.57) in the coefficient of h^3.

Since the present model is assumed to hold on the closure of ω, we see that the vectors $\boldsymbol{a}$ and $\boldsymbol{b}$ obtained from (7.55) and (7.57) must fulfill the boundary conditions (7.44). This represents a restriction on admissible data. Thus, using the decomposition (see relation Eq $(4.76)_2$)

$$\nabla\boldsymbol{u} = \boldsymbol{u}_{,s}\otimes\boldsymbol{\tau} + \boldsymbol{u}_{,\nu}\otimes\boldsymbol{\nu}\,, \tag{7.64}$$

and the Eq. (7.55) we deduce that the condition $\boldsymbol{a} = \mathbf{0}$ on $\partial\omega_\phi$ is satisfied provided that $\boldsymbol{u}_{,s}$ and $\boldsymbol{u}_{,\nu}$ vanish on $\partial\omega_\phi$. We see that the data for $\boldsymbol{u}$ and $\boldsymbol{a}$ reduce in this case to the clamping conditions

$$\boldsymbol{u} = \boldsymbol{u}_{,\nu} = \mathbf{0} \quad \text{on} \quad \partial\omega_\phi\,. \tag{7.65}$$

However, if the conditions (7.65) are imposed together with $\boldsymbol{b} = \mathbf{0}$ on $\partial\omega_\phi$, then the derived model is overspecified. For this reason, we shall relax the condition on $\boldsymbol{b}$ in what follows.

In view of (7.63), we remark that the energy $\bar{W}$ involves the vector $\boldsymbol{a}$ only in the coefficient of h. It appears namely in the combination

$$H(\nabla\boldsymbol{u},\boldsymbol{a}) = \frac{1}{2}\boldsymbol{H}_0\cdot\mathcal{M}[\boldsymbol{H}_0]\,. \tag{7.66}$$

if we compute the variation of this term with respect to $\boldsymbol{a}$ we get

$$\delta H = \delta\boldsymbol{a}\cdot\big(\mathcal{M}[\boldsymbol{H}_0]\big)\boldsymbol{k}\,. \tag{7.67}$$

Thus, the Euler equation for the functional G associated with the variable $\boldsymbol{a}$ is given by $(7.52)_1$, which is solved by (7.55). By virtue of the strong ellipticity, we see that the solution (7.55) represents the minimum of H with respect to $\boldsymbol{a}$ and hence also the pointwise minimum of $\bar{W}$ with respect to $\boldsymbol{a}$. To prove this assertion one can adapt the argument discussed in [17, Eqs. (22)–(28)]. Since one obtains a pointwise minimizer of the energy function G, the adoption of (7.55) in the coefficient of h yields the optimal criterion for buckling. This property follows because a field $\{\boldsymbol{u},\boldsymbol{a}\}$ that satisfies the bifurcation criterion in which (7.55) is imposed will, in the alternative case, satisfy the semi-strict inequality (7.33). In other words, the adoption of (7.55) promotes bifurcation, whereas its onset is delayed in the alternative case. This furnishes justification for the assumption of plane stress (on which classical treatments are based), mentioned in the Introduction of this chapter.

Consequently, we obtain the energy function G given by (7.62) in which $\bar{W}$ has the expression

$$W(\nabla \boldsymbol{u}, \nabla\nabla \boldsymbol{u}) = \frac{1}{2} h\, \dot{\boldsymbol{P}}_0 \mathbf{1} \cdot \nabla \boldsymbol{u} + \frac{1}{24} h^3 \dot{\boldsymbol{P}}'_0 \mathbf{1} \cdot \nabla \boldsymbol{a}\,, \tag{7.68}$$

with (7.55) and (7.57) incorporated in all terms.

Now, the buckling equations are simply the Euler equations and natural boundary conditions associated with the functional G. One can write these equations using Cartesian tensor notation in the form (see [14] or Sect. 4.3)

$$T_{i\alpha,\alpha} = 0 \quad \text{in} \quad \omega \tag{7.69}$$

and

$$T_{i\alpha}\nu_\alpha - (M_{i\alpha\beta}\nu_\alpha\tau_\beta)_{,s} = 0, \qquad M_{i\alpha\beta}\nu_\alpha\nu_\beta = 0 \quad \text{on} \quad \partial\omega_t\,, \tag{7.70}$$

where

$$T_{i\alpha} = N_{i\alpha} - M_{i\beta\alpha,\beta}\,, \quad \text{with} \quad N_{i\alpha} = \frac{\partial W}{\partial u_{i,\alpha}} \quad \text{and} \quad M_{i\alpha\beta} = \frac{\partial W}{\partial u_{i,\alpha\beta}}\,. \tag{7.71}$$

We have denoted here as usual by $u_i = \boldsymbol{u} \cdot \boldsymbol{e}_i$ (with $\boldsymbol{e}_3 = \boldsymbol{k}$) the orthogonal components of $\boldsymbol{u}$ The subscript s stands for the arclength derivative along $\partial\omega$ (traversed counterclockwise), with unit tangent $\boldsymbol{\tau} = \boldsymbol{k} \times \boldsymbol{\nu}$ to $\partial\omega$ and rightward unit normal $\boldsymbol{\nu}$. The boundary conditions $(7.70)_1$ and $(7.70)_2$, respectively, are to be interpreted as the vanishing of the incremental force and moment (per unit length) on $\partial\omega_t$.

7.3.4 Conditions Pertaining to Reflection Symmetry and the Pre-stress

In what follows, we limit attention to strain-energy functions that exhibit reflection symmetry with respect to the midplane ω, i.e.

$$\mathscr{U}(\boldsymbol{E}) = \mathscr{U}(\boldsymbol{Q}^T \boldsymbol{E}\, \boldsymbol{Q}) \qquad \text{with} \quad \boldsymbol{Q} = \boldsymbol{I} - 2\boldsymbol{k} \otimes \boldsymbol{k}.$$

This is in accordance with the restrictions imposed in the classical theory [5] and implies that the function $\mathscr{U}'(E_{ij}) := \mathscr{U}(E_{kl}\boldsymbol{e}_k \otimes \boldsymbol{e}_l)$ depends on $E_{3\alpha}$ $(= E_{\alpha 3})$ through their squares and the product $E_{31}E_{32}$ (see [4, Sect. 5.4(a)]). For instance, any material that is isotropic relative to κ satisfies this condition. As in the classical theory, we suppose that the second Piola-Kirchhoff stress associated to the underlying finite deformation is a function only of the in-plane coordinates and subject to null-traction conditions on the major surfaces, which means

$$S^{\pm}k = 0, \qquad \text{where} \quad S^{\pm} = S_{|\varsigma=\pm h/2}\,.$$

These conditions imply the pointwise restriction

$$Sk = 0 \tag{7.72}$$

on the pre-stress throughout the plate, which shows that the underlying finite deformation is associated with a state of *plane stress*.

We denote with $\gamma_\alpha = E_{\alpha 3} = E_{3\alpha}$ the transverse shear strains associated with the finite pre-strain, and with $\Gamma(\gamma_\alpha)$ the function obtained by holding fixed all components of E other than the γ_α in the strain-energy function. The derivatives of $\Gamma(\gamma_\alpha)$ vanish, since we have

$$\frac{\partial \Gamma}{\partial \gamma_\alpha} = e_\alpha \cdot (\mathscr{U}_E)k = 0, \tag{7.73}$$

by virtue of (7.5) and (7.72). Any material with reflection symmetry satisfies these restrictions automatically at $\gamma_\alpha = 0$ because the strain energy is then an even function of the transverse shears. The corresponding strain tensor has the form

$$E = \epsilon + \frac{1}{2}(\varphi^2 - 1)k \otimes k\,, \tag{7.74}$$

where $\epsilon = E_{\alpha\beta} e_\alpha \otimes e_\beta$ and φ is the transverse stretch. At any point (x_α, ς), the corresponding deformation gradient has the form

$$F = f + \varphi\, n \otimes k\,, \tag{7.75}$$

where the vector n is a local unit normal to the material surface $\varsigma =$ constant after deformation. The tensor f maps ω' to the local tangent plane of this surface. As it was proved in [16, p. 288], this is the only mode of deformation that is consistent with the restriction (7.72) in the presence of the strong ellipticity condition (7.7). Thus, reflection symmetry and strong ellipticity, combined with restriction (7.72), furnish deformations in which the transverse shear strain necessarily vanishes.

The orientations of the planes $\varsigma =$ constant are also assumed to remain unchanged by the finite deformation. This assumption is consistent with reflection symmetry and restriction (7.72), and means that we can put $n = k$. In this case, f is an invertible map from ω' to itself.

On the basis of (7.24) and (7.72), the condition $\dot{P}_0 k = 0$ implies that $\dot{S}_0 k = 0$, which can be rewritten

$$\{\mathscr{C}(E)[\dot{E}_0]\}k = 0. \tag{7.76}$$

Here, all orthogonal components of the tensor $\mathscr{C}$ of elastic moduli with an odd number of subscripts equal to 3 vanish, by virtue of reflection symmetry [4]. Using the orthogonal decompositions

$$\boldsymbol{u} = \boldsymbol{v} + w\,\boldsymbol{k}, \qquad \boldsymbol{a} = \boldsymbol{\alpha} + a\,\boldsymbol{k}, \qquad \text{with} \;\; \boldsymbol{v} = \mathbf{1}\boldsymbol{u} \;\; \text{and} \;\; \boldsymbol{\alpha} = \mathbf{1}\boldsymbol{a}, \tag{7.77}$$

we have

$$2\dot{\boldsymbol{E}}_0 = \boldsymbol{f}^T(\nabla\boldsymbol{v}) + (\nabla\boldsymbol{v})^T\boldsymbol{f} + \boldsymbol{k}\otimes(\boldsymbol{f}^T\boldsymbol{\alpha} + \varphi\,\nabla w) + (\boldsymbol{f}^T\boldsymbol{\alpha} + \varphi\,\nabla w)\otimes\boldsymbol{k} + 2\varphi\,a\,\boldsymbol{k}\otimes\boldsymbol{k}. \tag{7.78}$$

In the decomposition $(7.77)_1$ we have denoted by $\boldsymbol{v}$ the in-plane displacement of points on ω and by w the transverse displacement, which is the variable of principal interest in plate-buckling theory. We see that (7.76) is equivalent to the Eq. $(7.52)_1$, provided that the condition (7.72) is satisfied. If we designate by $C = \mathscr{C}_{3333}$, then we can write (7.76) with help of components in the following form

$$\mathscr{C}_{\alpha 3\beta 3}\dot{E}_{0\beta 3} = 0, \qquad \mathscr{C}_{33\alpha\beta}\dot{E}_{0\alpha\beta} + C\,\dot{E}_{033} = 0. \tag{7.79}$$

Taking into account the relations (7.28) and (7.54), as well as the stated restrictions on the material and the underlying finite deformation, we find that the acoustic tensor reduces to

$$\boldsymbol{A}_{(\boldsymbol{k})} = A_{\alpha\beta}\boldsymbol{e}_\alpha\otimes\boldsymbol{e}_\beta + A\,\boldsymbol{k}\otimes\boldsymbol{k}, \quad \text{where} \quad A_{\alpha\beta} = f_{\alpha\gamma}f_{\beta\delta}\mathscr{C}_{\gamma 3\delta 3} \;\; \text{and} \;\; A = \varphi^2 C. \tag{7.80}$$

Then, the assumed strong ellipticity implies that $A > 0$ and that $(A_{\alpha\beta})$ is positive definite. In view of relation $(7.80)_2$ we deduce that the matrix $(\mathscr{C}_{\alpha 3\beta 3})$ is also positive definite, since $\det\boldsymbol{f} > 0$ (cf. (7.75)). Hence, the Eq. $(7.79)_1$ yields $\dot{E}_{0\alpha 3} = 0$, while $(7.79)_2$ delivers $\dot{E}_{033}$ in terms of $\dot{E}_{0\alpha\beta}$. Thus,

$$\boldsymbol{f}^T\boldsymbol{\alpha} = -\varphi\,\nabla w \qquad \text{and} \qquad \varphi\,a = -C^{-1}\mathscr{C}_{33\alpha\beta}\dot{E}_{0\alpha\beta}\,, \tag{7.81}$$

in which

$$2\dot{E}_{0\alpha\beta} = f_{\lambda\alpha}v_{\lambda,\beta} + f_{\lambda\beta}v_{\lambda,\alpha}\,. \tag{7.82}$$

These equations together with $(7.77)_2$ then furnish the function $\boldsymbol{g}$ in (7.56).

Remark that Eq. $(7.81)_{1,2}$ comprise the classical Kirchhoff-Love hypothesis with thickness distension. These conditions correspond to the kinematic assumption listed in the Introduction of this chapter concerning the preservation of the surface normal and adopted a priori in classical approaches. In contrast, these conditions are here derived rather than postulated.

Consider the decomposition

$$\boldsymbol{b} = \boldsymbol{\beta} + b\,\boldsymbol{k} \qquad \text{with} \qquad \boldsymbol{\beta} = \mathbf{1}\boldsymbol{b}\,. \tag{7.83}$$

Then, from (7.57) we get

$$f^T \boldsymbol{\beta} = -\varphi \nabla a \quad \text{and} \quad \varphi b = -C^{-1} \mathscr{C}_{33\alpha\beta} \dot{E}'_{0\alpha\beta} , \tag{7.84}$$

where

$$2\dot{E}'_{0\gamma\beta} = f_{\lambda\gamma} \alpha_{\lambda,\beta} + f_{\lambda\beta} \alpha_{\lambda,\gamma} . \tag{7.85}$$

These relations also follow directly by imposing $\dot{S}'_0 \boldsymbol{k} = \boldsymbol{0}$, which is equivalent, provided that (7.72) holds, to $\dot{\boldsymbol{P}}'_0 \boldsymbol{k} = \boldsymbol{0}$ (cf. (7.24)).

In view of the symmetries of the inner product, we can use these results to deduce

$$\dot{\boldsymbol{P}}'_0 \mathbf{1} \cdot \nabla \boldsymbol{a} = \nabla a \cdot \boldsymbol{S}(\nabla a) + (\nabla \boldsymbol{\alpha}) \boldsymbol{S} \cdot \nabla \boldsymbol{\alpha} + \dot{S}'_0 \cdot \boldsymbol{f}^T \nabla \boldsymbol{\alpha} , \tag{7.86}$$

where

$$\dot{S}'_{0\alpha\beta} = \mathscr{D}_{\alpha\beta\gamma\delta} \dot{E}'_{0\gamma\delta} , \tag{7.87}$$

and

$$\mathscr{D}_{\alpha\beta\gamma\delta} = \mathscr{C}_{\alpha\beta\gamma\delta} - C^{-1} \mathscr{C}_{\alpha\beta 33} \mathscr{C}_{\gamma\delta 33} \tag{7.88}$$

are the plane-stress elastic moduli.

Any deformation satisfying the Euler equations (7.69)–(7.71) should represent a minimum of the quadratic functional G. But for the existence of a minimum it is necessary that the operative Legendre-Hadamard condition be satisfied (see Sect. 3.4). In our framework this condition yields the requirement that the term in the integrand W in (7.68) involving the highest-order derivative $u_{i,\alpha\beta}$ (the components of $\nabla\nabla \boldsymbol{u}$) be positive definite when $u_{i,\alpha\beta}$ is replaced by $y_i\, z_\alpha z_\beta$ for any three-vector $\boldsymbol{y}$ and any two-vector $\boldsymbol{z}$ [6]. Making the choice $y_3 = 0$ we reduce this requirement to the restriction

$$\nabla a \cdot \boldsymbol{S}(\nabla a) > 0 \tag{7.89}$$

in which ∇a can be assigned arbitrary values by choice of $y_\gamma\, z_\alpha z_\beta$. The requirement thus limits the present model to the case of a positive-definite pre-stress. However, this in turn precludes its application to precisely the kinds of problems for which it was designed.

In order to remove the above restriction, we notice that it arises from the coefficient of h^3 in the energy function for the plate. Hence, to restore the applicability of the theory to plate buckling, we assume that

$$|\boldsymbol{S}| = o(1) \tag{7.90}$$

for small h, that is $\boldsymbol{S}$ tends to zero as $h \to 0$. This condition is a restriction on solutions of the underlying finite-deformation problem. If (7.90) is satisfied, then all terms involving $\boldsymbol{S}$ may be suppressed in the coefficient of h^3 without affecting the accuracy of the order-h^3 expansion of the energy. Thus, we obtain the following simplification of the order-h^3 plate energy

$$W(\nabla \boldsymbol{u}, \nabla\nabla \boldsymbol{u}) = \frac{1}{2} h \, \dot{\boldsymbol{P}}_0 \mathbf{1} \cdot \nabla \boldsymbol{u} + \frac{1}{24} h^3 \dot{\boldsymbol{S}}_0' \cdot \boldsymbol{f}^T \nabla \boldsymbol{\alpha} \,, \tag{7.91}$$

where

$$\dot{\boldsymbol{P}}_0 \mathbf{1} \cdot \nabla \boldsymbol{u} = \nabla w \cdot \boldsymbol{S}(\nabla w) + (\nabla \boldsymbol{v}) \boldsymbol{S} \cdot \nabla \boldsymbol{v} + \dot{\boldsymbol{S}}_0 \cdot \boldsymbol{f}^T \nabla \boldsymbol{v} \tag{7.92}$$

and

$$\dot{S}_{0\alpha\beta} = \mathscr{D}_{\alpha\beta\gamma\delta} \dot{E}_{0\gamma\delta} \,. \tag{7.93}$$

Remark Note that the necessary condition (7.89) does not apply in the three-dimensional theory. Here, it arises from the fact that the order-h^3 truncation of the energy does not account fully for the energy of the expansion (7.41). This fact requires that the pre-stress be limited by the condition (7.90), such that the truncation of order h^3 furnishes a well-posed minimization problem. Otherwise, the ill-posedness may be alternatively eliminated by retaining all terms in the energy associated with a finite truncation of the power expansion (7.41), but this procedure would lead to a more complicated model. Such a model would then be limited by the approximations inherent in truncations of this kind. In this respect, we will show that the condition (7.90) is in fact not restrictive, since it is satisfied by the pre-stress associated with a non-trivial bifurcation mode.

In view of the above relations we see that the energy decouples additively into the pure stretching part W_s and pure bending part W_b, i.e.

$$W = W_s + W_b \,, \tag{7.94}$$

where, by virtue of (7.87) and (7.93),

$$\begin{aligned} W_s &= \frac{1}{2} h \big\{ (\nabla \boldsymbol{v}) \boldsymbol{S} \cdot \nabla \boldsymbol{v} + \boldsymbol{f}^T \nabla \boldsymbol{v} \cdot \mathscr{D}[\boldsymbol{f}^T \nabla \boldsymbol{v}] \big\}, \\ W_b &= \frac{1}{2} h \, \nabla w \cdot \boldsymbol{S}(\nabla w) + \frac{1}{24} h^3 \boldsymbol{f}^T \nabla \boldsymbol{\alpha} \cdot \mathscr{D}[\boldsymbol{f}^T \nabla \boldsymbol{\alpha}], \end{aligned} \tag{7.95}$$

with $\boldsymbol{\alpha}$ given by (7.81) and $\mathscr{D}$ defined by (7.88). Taking into account (7.90), we obtain the *leading-order* (order-h) stretching energy by suppressing the term involving $\boldsymbol{S}$ explicitly in $(7.95)_1$, so we get

$$W_s = \frac{1}{2} h \, \boldsymbol{f}^T \nabla \boldsymbol{v} \cdot \mathscr{D}[\boldsymbol{f}^T \nabla \boldsymbol{v}]. \tag{7.96}$$

7.3.5 Derivation of the Classical Model

If we exploit the full implications of the restriction (7.90) on the pre-stress, we can further simplify the problem. Assume that the reference configuration κ is stress free, the thickness h is sufficiently small and the strain energy $\mathscr{U}(\boldsymbol{E})$ is convex in a neighborhood of the origin in strain space. Then, the initial strain is also small and vanishes in the zero-thickness limit, i.e. $|\boldsymbol{E}| = o(1)$. Thus, we can neglect the pre-strain in the coefficient of h^3 with *no effect* on the order-h^3 accuracy of the strain-energy function. Further, we can replace the fourth-order tensor $\mathscr{D}$ (which in principle is evaluated at the underlying finite strain) with the tensor of linear-elastic moduli relative to κ, denoted by

$$\mathscr{D}^{(\kappa)} = \mathscr{D}_{|\boldsymbol{E}=\boldsymbol{0}} \,. \tag{7.97}$$

Since we restrict our attention to uniform materials, the tensor field $\mathscr{D}^{(\kappa)}$ is *spatially uniform*. This leads to a further simplification of the model which is fully consistent with order-h^3 accuracy.

Let Orth$^+$ denote the group of proper-orthogonal tensors, or rotations. If we impose that $\boldsymbol{F} \in \text{Orth}^+$, then the same degree of accuracy is preserved in the coefficient of h^3. Indeed, to show this we write the polar decomposition of $\boldsymbol{F}$ as the product of a rotation $\boldsymbol{R}$ and the right stretch tensor, which is of order unity by virtue of the restriction on $|\boldsymbol{E}|$. Thus, we have $\boldsymbol{F} = \boldsymbol{R} + o(1)$ with $\boldsymbol{R} \in \text{Orth}^+$, which in turn implies that order-h^3 accuracy is preserved by substituting $\boldsymbol{R}^T\nabla\boldsymbol{\alpha}$ in place of $\boldsymbol{f}^T\nabla\boldsymbol{\alpha}(= \boldsymbol{F}^T\nabla\boldsymbol{\alpha})$ in the coefficient of h^3. Here, $\boldsymbol{f}^T\boldsymbol{\alpha} = -\varphi\,\nabla w$ (see $(7.81)_1$) is replaced by $\boldsymbol{R}^T\boldsymbol{\alpha} = -\nabla w$.

According to the equations (8) and (9) of [13], we deduce that the gradient of $\boldsymbol{R}$ is small if the gradient of the strain is small. This leads to the following important simplification: we restrict our further attention to the practically important case in which the gradient of $\boldsymbol{E}$ is of order $o(1)$. This confinement yields the conclusion that, to within an error of order $o(1)$, $\boldsymbol{R}$ is *spatially uniform*. The error contributes at order $o(h^3)$ and may be suppressed in the present model with no adverse effect on accuracy. Thus, we obtain the consistent-order estimate $\boldsymbol{f}^T\nabla\boldsymbol{\alpha} = -\nabla\nabla w$, so we can replace the bending energy by

$$W_b = \frac{1}{2}\,h\,\nabla w \cdot \boldsymbol{S}(\nabla w) + \frac{1}{24}\,h^3\nabla\nabla w \cdot \mathscr{D}^{(\kappa)}[\nabla\nabla w], \tag{7.98}$$

with *no effect* on the accuracy of the truncation $(7.95)_2$.

Remark The relation (7.98) represents the reason for truncating the thickness-wise expansion of the energy at order h^3. In particular, this is the lowest order at which the bending energy remains well posed in the presence of a compressive pre-stress. Hence, the order h^3 is the lowest order at which a two-dimensional model, derived from the three-dimensional theory, can furnish a meaningful basis on which classical plate buckling may be analyzed.

The stress $\boldsymbol{S}$ associated with the underlying pre-buckling deformation satisfies the equation Div $\boldsymbol{P} = \boldsymbol{0}$, where $\boldsymbol{P} = \boldsymbol{F}\boldsymbol{S}$. As before, the assumption $|\boldsymbol{S}| = o(1)$, taken together with the stated restriction on the gradient of the pre-strain and our constitutive hypotheses, imply that the *leading order* restriction on this stress is obtained by replacing the factor $\boldsymbol{F}$ in the relation $\boldsymbol{P} = \boldsymbol{F}\boldsymbol{S}$ with the rotation tensor $\boldsymbol{R}$. Since the rotation $\boldsymbol{R}$ is *uniform*, the equilibrium equation infers

$$\operatorname{Div} \boldsymbol{S} = \boldsymbol{0} \quad \text{in} \quad \kappa\,, \tag{7.99}$$

to leading order in thickness, where we have

$$\boldsymbol{S} = \mathscr{D}^{(\kappa)}[\boldsymbol{\epsilon}]\,, \tag{7.100}$$

in which $|\boldsymbol{\epsilon}| = o(1)$ (see (7.74)).

In a similar way, we can replace the leading order stretching energy (without affecting order-h accuracy) by

$$W_s = \frac{1}{2} h\, \nabla\bar{\boldsymbol{v}} \cdot \mathscr{D}^{(\kappa)}[\nabla\bar{\boldsymbol{v}}]\,, \tag{7.101}$$

where $\bar{\boldsymbol{v}} = \boldsymbol{R}^T \boldsymbol{v}$ is a rigidly-rotated displacement field. In view of our assumptions, the tensor $\mathscr{D}^{(\kappa)}$ is positive definite. This implies that $\boldsymbol{v} = \boldsymbol{0}$ when the boundary part $\partial\omega_\phi$ is non-empty, like in the classical theory of generalized plane stress.

Thus, the third main assumption of the classical treatments has been justified; namely, that the underlying pre-buckling deformation associated with the order-h^3 model may be described using classical linear elasticity theory. This follows from the fact that the model calls for the ground-state moduli $\mathscr{D}^{(\kappa)}$ while the pre-stress $\boldsymbol{S}$ satisfies the equilibrium and constitutive equations of the linear theory.

Remark In contrast to the incremental theory in which *pre-stress* is induced by the pre-buckling deformation, in plate models based a priori on conventional three-dimensional linear elasticity theory with *initial* stress [12, 15] both the initial stress and the elastic moduli are constrained by the nature of the material symmetry in the configuration κ. For instance, if the material in κ is isotropic, the initial stress is a uniform pure pressure that vanishes by virtue of the traction data on the lateral surfaces of the plate. Thus, we see that this approach does not yield a plate buckling model in the case of isotropy. On the contrary, the pre-stress in our approach is induced by the underlying strain of order $o(1)$. The symmetry of the material in κ is described only through the moduli in relation (7.100), whereas the pre-stress is delivered by the linearly elastic boundary-value problem. So it is important to distinguish between initial stress and pre-stress, see the detailed discussion in [11].

We note that classical plate-buckling theory [1, 5] is associated with the special case of relation (7.90) in which

$$\boldsymbol{S} = h^2 \bar{\boldsymbol{S}} + o(h^2), \quad \text{with} \quad |\bar{\boldsymbol{S}}| = O(1). \tag{7.102}$$

This yields the bending energy of the form

$$W_b = \frac{1}{2} h^3 \Big\{ \nabla w \cdot \bar{\mathbf{S}}(\nabla w) + \frac{1}{12} \nabla\nabla w \cdot \mathscr{D}^{(\kappa)}[\nabla\nabla w] \Big\}. \tag{7.103}$$

Writing the corresponding boundary-value problem (7.69)–(7.71) for $i = 3$, we get the equations

$$\frac{1}{12} \mathscr{D}^{(\kappa)}_{\alpha\beta\gamma\delta}\, w_{,\alpha\beta\gamma\delta} = \bar{S}_{\alpha\beta}\, w_{,\alpha\beta} \quad \text{in} \quad \omega \tag{7.104}$$

and boundary conditions

$$w = 0, \qquad \boldsymbol{\nu} \cdot \nabla w = 0 \quad \text{on} \quad \partial\omega_\phi\,, \tag{7.105}$$

with

$$\begin{aligned} &\Big(\bar{S}_{\alpha\beta}\, w_{,\beta} - \frac{1}{12} \mathscr{D}^{(\kappa)}_{\alpha\beta\gamma\delta}\, w_{,\beta\gamma\delta}\Big)\nu_\alpha - \frac{1}{12}\big(\mathscr{D}^{(\kappa)}_{\alpha\beta\gamma\delta}\, w_{,\gamma\delta}\nu_\alpha\tau_\beta\big)_{,s} = 0, \\ &\mathscr{D}^{(\kappa)}_{\alpha\beta\gamma\delta}\, w_{,\gamma\delta}\nu_\alpha\nu_\beta = 0 \qquad \text{on} \qquad \partial\omega_t\,. \end{aligned} \tag{7.106}$$

These equations represent the classical plate-buckling problem for anisotropic materials, which also incorporate the classical plate-buckling problem for isotropic materials found in the literature.

Since the above problem contains no small parameters, it does not exhibit localized boundary-layer effects. Thus, the bifurcation modes are global. This property is well known in the technical literature [1, 5]. We remark that in this case the bending energy W_b furnishes the *rigorous leading-order energy* of the thin plate. Indeed, one can show this by dividing the exact energy $\mathscr{G}$ given in (7.45) by h^3 and passing to the limit.

Note that the scaling (7.102) assumed by the classical plate-buckling theory represents the smallest pre-stress for which a non-trivial bifurcation mode can exist in the order-h^3 model. Indeed, smaller initial stresses contribute at order $o(h^3)$ and therefore play no role in the model, whereas the positivity of $\mathscr{D}^{(\kappa)}$ then allows only the trivial solution $w = 0$ to the boundary-value problem.

The relevance of the classical model (7.104)–(7.106) is reflected by the large number of plate-buckling problems that have been solved using the classical theory. All these solutions exhibit eigenvalues $\bar{\mathbf{S}}$ that satisfy the relation $(7.102)_2$. This fact furnishes a *post-facto* justification for the imposition of the restriction (7.90) on the pre-stress, which we have earlier otherwise motivated.

7.4 Exercises

7.1 Assuming the classical scaling for the pre-stress in connection with plate buckling, solve the following buckling problem for isotropic materials: Assume a rectangular plate having edges of length a and b aligned with the x and y axes, respectively, and a (scaled) pre-stress given by

$$\overline{\boldsymbol{S}} = -P\boldsymbol{e}_1 \otimes \boldsymbol{e}_1 \ ,$$

with P constant and positive. This corresponds to a state of uniaxial compression. Suppose the plate is "simply-supported" in the sense that w and the bending moment vanish along the entire boundary. Assume a solution of the form

$$w\,(x, y) = C \sin\left(\frac{m\pi x}{a}\right) \sin\left(\frac{n\pi y}{b}\right),$$

in which C is a constant and m, n are positive integers. (Note: the buckling equations do not determine C because the same equations and boundary conditions are satisfied for all values of C; this is typical of eigenvalue problems, exemplified by plate buckling).

(a) Find the buckling stress P_{crit} in terms of the mode numbers m and n.
(b) Show, graphically if necessary, that the lowest value of P_{crit} is given approximately by $4\pi^2 D/b^2$, where D is the usual flexural rigidity of the plate. (*Hint*: minimize P_{crit} with respect to n and plot the result against the ratio a/b for several values of m.)

References

1. Cox, H.L.: The Buckling of Plates and Shells. Pergamon-McMillan, New York (1963)
2. Fichera, G.: Existence Theorems in Elasticity. In: Flügge, W. (ed.) Handbuch der Physik, vol. VIa/2, pp. 347–389. Springer, Berlin (1972)
3. Fu, Y.B.: Perturbation Methods and Nonlinear Stability Analysis. In: Fu, Y.B., Ogden, R.W. (ed.) Nonlinear Elasticity, Theory and Applications. London Mathematical Society Lecture Note Series, vol. 283. Cambridge University Press, Cambridge (2001)
4. Green, A.E., Zerna, W.: Theoretical Elasticity, 2nd edn. Oxford University Press, Oxford (1968)
5. van der Heijden, A.M.A.: W. T. Koiter's Elastic Stability of Solids and Structures. Cambridge University Press, Cambridge (2009)
6. Hilgers, M.G., Pipkin, A.C.: The Graves condition for variational problems of arbitrary order. IMA. J. Appl. Math. **48**, 265–269 (1992)
7. Hill, R.: On uniqueness and stability in the theory of finite elastic strain. J. Mech. Phys. Solids **5**, 229–241 (1957)
8. Koiter, W.T.: A Basic Open Problem in the Theory of Elastic Stability. In: Joint IUTAM/IMU Symposium on Applications of Methods of Functional Analysis to Problems in Mechanics, 1–6, Part XXIX, Marseille, 1975. Springer, Berlin, Heidelberg (1975)

9. Knops, R.J., Wilkes, E.W.: Theory of Elastic Stability. In: Flügge, W. (ed.) Handbuch der Physik, vol. VIa/3, pp. 125–302. Springer, Berlin (1973)
10. Ogden, R.W.: Non-linear Elastic Deformations. Dover, New York (1997)
11. Ogden, R.W., Singh, B.: Propagation of waves in an incompressible transversely isotropic solid with initial stress: Biot revisited. J. Mech. Mater. Struct. **6**, 453–477 (2011)
12. Paroni, P.: Theory of linearly elastic residually stressed plates. Math. Mech. Solids **11**, 137–159 (2006)
13. Shield, R.T.: The rotation associated with large strain. SIAM J. Appl. Math. **25**, 483–491 (1973)
14. Steigmann, D.J.: Two-dimensional models for the combined bending and stretching of plates and shells based on three-dimensional linear elasticity. Int. J. Eng. Sci. **46**, 654–676 (2008)
15. Steigmann, D.J.: Linear theory for the bending and extension of a thin, residually stressed, fiber-reinforced lamina. Int. J. Eng. Sci. **47**, 1367–1378 (2009)
16. Steigmann, D.: Applications of polyconvexity and strong ellipticity to nonlinear elasticity and elastic plate theory. In: Schröder, J., Neff, P. (eds.) Poly-, Quasi-, and Rank-One Convexity in Applied Mechanics. CISM Courses and Lectures, vol. 516, pp. 265–299. Springer, Wien and New York (2010)
17. Steigmann, D.J.: Refined theory for linearly elastic plates: laminae and laminates. Math. Mech. Solids **17**, 351–363 (2012)
18. Steigmann, D.J., Ogden, R.W.: Classical plate buckling theory as the small-thickness limit of three-dimensional incremental elasticity. Z. Angew. Math. Mech. (ZAMM) **94**, 7–20 (2014)

Chapter 8
Saint-Venant Problem for General Cylindrical Shells

Abstract In this chapter we investigate the deformation of cylindrical linearly elastic shells using the Koiter model. We formulate and solve the relaxed Saint-Venant's problem for thin cylindrical tubes made of isotropic and homogeneous elastic materials. We present a general solution procedure to determine closed-form solutions for the extension, bending, torsion and flexure problems. To this aim, we employ a method established in the three-dimensional theory of elasticity and determine the corresponding Saint-Venant's solutions for shells [7]. Finally, the special case of circular cylindrical shells is discussed in details.

8.1 Relaxed Saint-Venant's Problem for Cylindrical Shells

In this chapter we apply the linear Koiter model for shells to solve the relaxed Saint-Venant's problem for thin cylindrical tubes. The derivation of the governing equations for linearly elastic shells has been presented in Chap. 5. Then, in Chaps. 6 and 7 we have continued the theoretical development by extending the derivation procedure to the case of nonlinear elastic shells and, respectively, buckling of plates. In what follows, we return to the linear shell theory and present a significant application of the Koiter approach.

The Saint-Venant's problem consists in determining the equilibrium deformation of a cylindrical shell subjected to mechanical loads distributed over its end edges. This problem has been intensively studied both in the three-dimensional linear elasticity (i.e., for solid cylinders), as well as in the shell theory [1, 10, 13]. Barré de Saint-Venant [14] have found a set of solutions to the relaxed formulation of the problem, which represent a classical result in the three-dimensional elasticity and are widely used in engineering. In the sequel, we follow the approach presented in the paper [7] and derive the analogues of the classical Saint-Venant's solutions for the deformation of cylindrical shells (thin elastic tubes) by adapting the method established by Ieşan [8, 9] in the context of three-dimensional elasticity.

Let us recapitulate briefly the Koiter's equations for linearly elastic shells (which were derived in Chap. 5) and formulate the relaxed Saint-Venant's problem for cylindrical shells. The reference midsurface $\mathscr{S}$ of the shell is described by the position

D. J. Steigmann et al., *Lecture Notes on the Theory of Plates and Shells*,
Solid Mechanics and Its Applications 274,
https://doi.org/10.1007/978-3-031-25674-5_8

vector $\boldsymbol{r}(\theta^1, \theta^2)$. The metric tensor is $a_{\alpha\beta}$ and the curvature tensor $\kappa_{\alpha\beta}$ are then given by (see Eqs. (5.2), (5.3))

$$a_{\alpha\beta} = \boldsymbol{a}_\alpha \cdot \boldsymbol{a}_\beta\,, \qquad \kappa_{\alpha\beta} = -\boldsymbol{n}_{,\alpha} \cdot \boldsymbol{a}_\beta\,. \tag{8.1}$$

If $\boldsymbol{u}(\theta^1, \theta^2)$ denotes the displacement field, then the linearized strain tensors are the change of metric tensor $\epsilon_{\alpha\beta}$ and change of curvature tensor $\rho_{\alpha\beta}$, given by

$$\epsilon_{\alpha\beta} = \frac{1}{2}(\boldsymbol{a}_\alpha \cdot \boldsymbol{u}_{,\beta} + \boldsymbol{a}_\beta \cdot \boldsymbol{u}_{,\alpha}), \qquad \rho_{\alpha\beta} = \boldsymbol{n} \cdot \boldsymbol{u}_{;\alpha\beta} = \boldsymbol{n} \cdot (\boldsymbol{u}_{,\alpha\beta} - \Gamma^\lambda_{\alpha\beta}\boldsymbol{u}_{,\lambda}), \tag{8.2}$$

see relations (5.168) and (5.170).

According to (5.235) and (5.252), the equilibrium equations can be written in the form

$$\boldsymbol{T}^\alpha{}_{;\alpha} + \boldsymbol{g} = \boldsymbol{0}, \tag{8.3}$$

or equivalently,

$$P^{\beta\alpha}{}_{;\alpha} - \kappa^\beta_\alpha S^\alpha + g^\beta = 0, \qquad S^\alpha{}_{;\alpha} + \kappa_{\beta\alpha}P^{\beta\alpha} + g = 0, \tag{8.4}$$

where $\boldsymbol{g} = g^\alpha \boldsymbol{a}_\alpha + g\,\boldsymbol{n}$ is the assigned force per unit area and

$$\boldsymbol{T}^\alpha = P^{\beta\alpha}\boldsymbol{a}_\beta + S^\alpha \boldsymbol{n}\,. \tag{8.5}$$

The tangential tractions $P^{\beta\alpha}$ and transverse shear tractions S^α are expressed in terms of the stress and couple stress tensors $N^{\alpha\beta}$ and $M^{\alpha\beta}$ by the relations

$$P^{\alpha\beta} = N^{\alpha\beta} + \kappa^\alpha_\gamma M^{\gamma\beta}, \qquad S^\alpha = -M^{\beta\alpha}{}_{;\beta} = -M^{\beta\alpha}{}_{,\beta} - \Gamma^\beta_{\gamma\beta}M^{\gamma\alpha} - \Gamma^\alpha_{\gamma\beta}M^{\beta\gamma}. \tag{8.6}$$

For isotropic materials the stress-strain relations are

$$N^{\alpha\beta} = C\left[\nu a^{\alpha\beta}\epsilon^\gamma_\gamma + (1-\nu)\epsilon^{\alpha\beta}\right], \qquad M^{\alpha\beta} = D\left[\nu a^{\alpha\beta}\rho^\gamma_\gamma + (1-\nu)\rho^{\alpha\beta}\right], \tag{8.7}$$

where $C = \frac{Eh}{1-\nu^2}$ is the stretching stiffness and $D = \frac{Eh^3}{12(1-\nu^2)}$ the bending stiffness of the shell.

Let $\mathscr{C}$ be the boundary curve of the surface $\mathscr{S}$, see Fig. 8.1. The boundary conditions in the Koiter model are (cf. (5.244))

$$\boldsymbol{T}^\alpha \nu_\alpha - \left(M^{\alpha\beta}\nu_\alpha \tau_\beta\, \boldsymbol{n}\right)_{,s} = \boldsymbol{f} \quad \text{and} \quad M^{\alpha\beta}\nu_\alpha\nu_\beta\,\boldsymbol{n} = \boldsymbol{c}\,, \tag{8.8}$$

where $\boldsymbol{f}$ and $\boldsymbol{c}$ are the assigned force and moment per unit length of $\mathscr{C}$. Here, $\boldsymbol{\tau} = \tau_\alpha \boldsymbol{a}^\alpha$ is the unit tangent vector along $\mathscr{C}$ and $\boldsymbol{\nu} = \nu_\alpha \boldsymbol{a}^\alpha$ is the unit outer normal to $\mathscr{C}$ lying in the tangent plane, while s is the arclength parameter along $\mathscr{C}$.

On the basis of the Eq. (8.4) and boundary conditions (8.8) we can show that the following relations hold in case of equilibrium

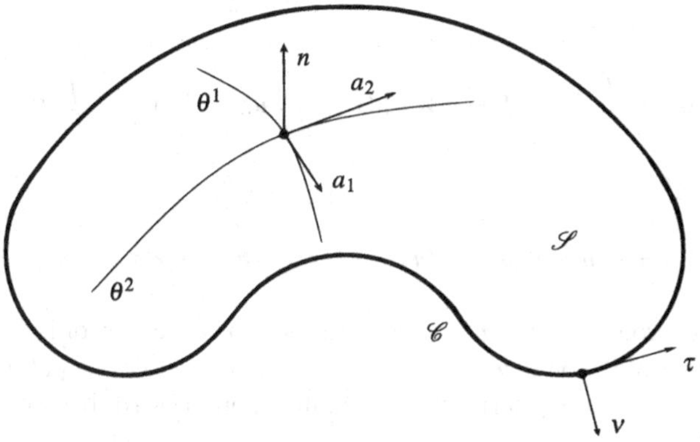

Fig. 8.1 The reference midsurface $\mathscr{S}$ of the shell

$$\int_{\mathscr{S}} \boldsymbol{g}\, \mathrm{d}a + \int_{\mathscr{C}} \boldsymbol{f}\, \mathrm{d}s = \boldsymbol{0}, \qquad \int_{\mathscr{S}} \boldsymbol{r} \times \boldsymbol{g}\, \mathrm{d}a + \int_{\mathscr{C}} (\boldsymbol{r} \times \boldsymbol{f} + \boldsymbol{\nu} \times \boldsymbol{c}) \mathrm{d}s = \boldsymbol{0}. \quad (8.9)$$

These equations express that the resultant assigned force and the resultant assigned moment are zero for equilibrated shells. Indeed, let us prove the relations (8.9): using the equilibrium equations (8.4) and the boundary conditions (8.8) we can write

$$\int_{\mathscr{C}} \boldsymbol{f}\, \mathrm{d}s = \int_{\mathscr{C}} \big[\boldsymbol{T}^{\alpha} \nu_{\alpha} - \big(M^{\alpha\beta} \nu_{\alpha} \tau_{\beta}\, \boldsymbol{n}\big)_{,s}\big] \mathrm{d}s = \int_{\mathscr{C}} \boldsymbol{T}^{\alpha} \nu_{\alpha}\, \mathrm{d}s = \int_{\mathscr{S}} \boldsymbol{T}^{\alpha}{}_{;\alpha}\, \mathrm{d}s\,,$$

in view of the divergence formula (2.148) and the fact that $\mathscr{C}$ is a closed curve. So, the relation $(8.9)_1$ reduces to

$$\int_{\mathscr{S}} \big(\boldsymbol{g} + \boldsymbol{T}^{\alpha}{}_{;\alpha}\big)\, \mathrm{d}s = \boldsymbol{0}\,,$$

which holds true by virtue of the Eq. (8.3). To prove the second equilibrium condition $(8.9)_2$, we transform the left-hand side in the following way

$$\begin{aligned}
\int_{\mathscr{S}} \boldsymbol{r} \times \boldsymbol{g}\, \mathrm{d}a + \int_{\mathscr{C}} (\boldsymbol{r} \times \boldsymbol{f} + \boldsymbol{\nu} \times \boldsymbol{c}) \mathrm{d}s &= \int_{\mathscr{S}} \boldsymbol{r} \times \boldsymbol{g}\, \mathrm{d}a \\
&+ \int_{\mathscr{C}} \big[\boldsymbol{r} \times \boldsymbol{T}^{\alpha} \nu_{\alpha} - \boldsymbol{r} \times \big(M^{\alpha\beta} \nu_{\alpha} \tau_{\beta}\, \boldsymbol{n}\big)_{,s} + \boldsymbol{\nu} \times \big(M^{\alpha\beta} \nu_{\alpha} \nu_{\beta}\, \boldsymbol{n}\big)\big] \mathrm{d}s \\
&= \int_{\mathscr{S}} \big[\boldsymbol{r} \times \boldsymbol{g} + (\boldsymbol{r} \times \boldsymbol{T}^{\alpha})_{;\alpha}\big] \mathrm{d}a + \int_{\mathscr{C}} \big[-\big(\boldsymbol{r} \times M^{\alpha\beta} \nu_{\alpha} \tau_{\beta}\, \boldsymbol{n}\big)_{,s} \\
&\qquad + \boldsymbol{\tau} \times \big(M^{\alpha\beta} \nu_{\alpha} \tau_{\beta}\, \boldsymbol{n}\big) - M^{\alpha\beta} \nu_{\alpha} \nu_{\beta}\, \boldsymbol{\tau}\big] \mathrm{d}s \\
&= \int_{\mathscr{S}} \big[\boldsymbol{r} \times \underbrace{\big(\boldsymbol{g} + \boldsymbol{T}^{\alpha}{}_{;\alpha}\big)}_{=0} + \boldsymbol{a}_{\alpha} \times \boldsymbol{T}^{\alpha}\big] \mathrm{d}a + \int_{\mathscr{C}} \big(M^{\alpha\beta} \nu_{\alpha} \tau_{\beta}\, \boldsymbol{\nu} - M^{\alpha\beta} \nu_{\alpha} \nu_{\beta}\, \boldsymbol{\tau}\big) \mathrm{d}s \\
&= \int_{\mathscr{S}} \boldsymbol{a}_{\alpha} \times \boldsymbol{T}^{\alpha}\, \mathrm{d}a + \int_{\mathscr{C}} (\boldsymbol{\nu} \otimes \boldsymbol{\tau} - \boldsymbol{\tau} \otimes \boldsymbol{\nu}) \boldsymbol{M} \boldsymbol{\nu}\, \mathrm{d}s\,,
\end{aligned}$$

so we have

$$\int_{\mathscr{S}} \boldsymbol{r} \times \boldsymbol{g}\,\mathrm{d}a + \int_{\mathscr{C}} (\boldsymbol{r} \times \boldsymbol{f} + \boldsymbol{\nu} \times \boldsymbol{c})\mathrm{d}s = \int_{\mathscr{S}} \boldsymbol{a}_\alpha \times \boldsymbol{T}^\alpha\,\mathrm{d}a - \int_{\mathscr{C}} \boldsymbol{n} \times \boldsymbol{M}\boldsymbol{\nu}\,\mathrm{d}s\,, \tag{8.10}$$

since it holds

$$\boldsymbol{\nu} \otimes \boldsymbol{\tau} - \boldsymbol{\tau} \otimes \boldsymbol{\nu} = -(\boldsymbol{n} \times \boldsymbol{\tau}) \otimes \boldsymbol{\tau} - (\boldsymbol{n} \times \boldsymbol{\nu}) \otimes \boldsymbol{\nu} = -\boldsymbol{n} \times (\boldsymbol{\tau} \otimes \boldsymbol{\tau} + \boldsymbol{\nu} \otimes \boldsymbol{\nu}) = -\boldsymbol{n} \times \mathbf{1}.$$

Here, the cross product between a vector and a second order tensor is defined on the basis of relations of the type $\boldsymbol{u} \times (\boldsymbol{v} \otimes \boldsymbol{w}) = (\boldsymbol{u} \times \boldsymbol{v}) \otimes \boldsymbol{w}$, see e.g. [11, p. 44]. Using the divergence theorem for surfaces (2.159), the relation (8.10) becomes

$$\int_{\mathscr{S}} \boldsymbol{r} \times \boldsymbol{g}\,\mathrm{d}a + \int_{\mathscr{C}} (\boldsymbol{r} \times \boldsymbol{f} + \boldsymbol{\nu} \times \boldsymbol{c})\mathrm{d}s = \int_{\mathscr{S}} \big[\boldsymbol{a}_\alpha \times \boldsymbol{T}^\alpha - \mathrm{div}_s(\boldsymbol{n} \times \boldsymbol{M})\big]\mathrm{d}a\,. \tag{8.11}$$

Further, we compute the two terms in the last integral: in view of (8.5), (8.6), we can write

$$\begin{aligned} \boldsymbol{a}_\alpha \times \boldsymbol{T}^\alpha &= \boldsymbol{a}_\alpha \times \big[(N^{\beta\alpha} + \kappa^\beta_\gamma M^{\gamma\alpha})\boldsymbol{a}_\beta + (-\,M^{\beta\alpha}{}_{;\beta})\boldsymbol{n}\big] \\ &= (N^{\beta\alpha} + \kappa^\beta_\gamma M^{\gamma\alpha})\varepsilon_{\alpha\beta}\,\boldsymbol{n} \;+\; \boldsymbol{n} \times (M^{\beta\alpha}{}_{;\beta}\,\boldsymbol{a}_\alpha) \\ &= \varepsilon_{\alpha\beta}\kappa^\beta_\gamma M^{\gamma\alpha}\,\boldsymbol{n} \;+\; \boldsymbol{n} \times \mathrm{div}_s\,\boldsymbol{M}\,, \end{aligned} \tag{8.12}$$

since $\varepsilon_{\alpha\beta}N^{\beta\alpha} = 0$. Also, from the definition (2.153) and the relation $\boldsymbol{n}_{,\alpha} = -\kappa^\beta_\alpha\,\boldsymbol{a}_\beta$ we have

$$\begin{aligned} \mathrm{div}_s(\boldsymbol{n} \times \boldsymbol{M}) = \;(\boldsymbol{n} \times \boldsymbol{M})_{,\alpha}\,\boldsymbol{a}^\alpha \;&=\; \boldsymbol{n}_{,\alpha} \times \boldsymbol{M}\,\boldsymbol{a}^\alpha + (\boldsymbol{n} \times \boldsymbol{M}_{,\alpha})\,\boldsymbol{a}^\alpha \\ &= -\kappa^\beta_\alpha\,\boldsymbol{a}_\beta \times (M^{\gamma\alpha}\boldsymbol{a}_\gamma) + \boldsymbol{n} \times (\boldsymbol{M}_{,\alpha}\,\boldsymbol{a}^\alpha) \\ &= -\varepsilon_{\beta\gamma}\,\kappa^\beta_\alpha M^{\gamma\alpha}\,\boldsymbol{n} \;+\; \boldsymbol{n} \times \mathrm{div}_s\,\boldsymbol{M}\,. \end{aligned} \tag{8.13}$$

By comparing (8.12) and (8.13) we remark that they are equal (since $M^{\alpha\gamma} = M^{\gamma\alpha}$), so we deduce that

$$\boldsymbol{a}_\alpha \times \boldsymbol{T}^\alpha = \mathrm{div}_s(\boldsymbol{n} \times \boldsymbol{M}). \tag{8.14}$$

Using (8.14) into (8.11) we obtain that the relation $(8.9)_2$ holds true. □

We consider general cylindrical shells, i.e. cylindrical thin tubes of arbitrary cross section. The cylindrical midsurface $\mathscr{S}$ of the shell is referred to an orthogonal Cartesian coordinate frame $Ox_1x_2x_3$ with unit vectors $\boldsymbol{e}_i$ along the axes Ox_i, see Fig. 8.2. The curvilinear coordinates on $\mathscr{S}$ are denoted by $\theta^1 = s,\ \theta^2 = z$ and the parametrization of $\mathscr{S}$ is given by

$$\boldsymbol{r}(s,z) = x_\alpha(s)\,\boldsymbol{e}_\alpha + z\,\boldsymbol{e}_3\,, \qquad s \in [0,\bar{s}],\;\; z \in [0,\ell], \tag{8.15}$$

where $x_\alpha(s)$ are given functions of class $C^2[0,\bar{s}]$, which describe the form of the cross-section curve. Here, s is the arclength parameter along the cross-section curves

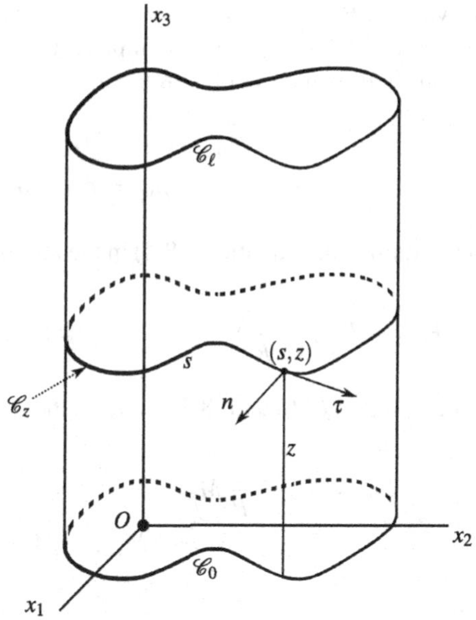

Fig. 8.2 The reference configuration of the cylindrical shell

and $z = x_3$ is the distance to the coordinate plane Ox_1x_2. Thus, $\bar{s}$ stands for the length of the cross-section curve.

We denote by $\mathscr{C}_z$ the cross-section curve at the distance $x_3 = z$; accordingly, $\mathscr{C}_0$ and $\mathscr{C}_\ell$ are the end edge curves described by $x_3 = 0$ and $x_3 = \ell$, respectively. We remark that the curvilinear coordinate system $\theta^1 = s, \theta^2 = z$ is orthonormal and we have

$$
\begin{gathered}
\boldsymbol{a}_1 = \boldsymbol{a}^1 = \frac{\partial \boldsymbol{r}}{\partial s} = \boldsymbol{\tau} = x'_\alpha(s)\boldsymbol{e}_\alpha\,, \qquad \boldsymbol{a}_2 = \boldsymbol{a}^2 = \frac{\partial \boldsymbol{r}}{\partial z} = \boldsymbol{e}_3\,, \\
\boldsymbol{n} = \frac{\boldsymbol{a}_1 \times \boldsymbol{a}_2}{|\boldsymbol{a}_1 \times \boldsymbol{a}_2|} = e_{\alpha\beta}x'_\beta(s)\boldsymbol{e}_\alpha\,,
\end{gathered}
\tag{8.16}
$$

where $\boldsymbol{\tau}$ and $\boldsymbol{n}$ are the unit tangent and normal vectors to the cross-section curve $\mathscr{C}_z$ and $e_{\alpha\beta}$ is the two-dimensional alternator (i.e., $e_{12} = -e_{21} = 1$, $e_{11} = e_{22} = 0$). Let $R(s)$ be the curvature radius of the planar curve $\mathscr{C}_z$. In view of the Frenet–Serret equations, it holds

$$
\boldsymbol{\tau}'(s) = -\frac{1}{R(s)}\boldsymbol{n}(s), \quad \boldsymbol{n}'(s) = \frac{1}{R(s)}\boldsymbol{\tau}(s) \quad \text{and} \quad \frac{1}{R(s)} = e_{\alpha\beta}x'_\alpha(s)x''_\beta(s). \tag{8.17}
$$

The covariant and contravariant components of any tensor are equal, since the covariant basis $\{\boldsymbol{a}_1, \boldsymbol{a}_2\}$ is orthonormal. Thus, we have $a_{\alpha\beta} = \delta_{\alpha\beta}$ (the Kronecker symbol) and

$$
\kappa_{11} = -\frac{1}{R(s)}\,, \qquad \kappa_{12} = \kappa_{21} = \kappa_{22} = 0, \qquad \Gamma^\gamma_{\alpha\beta} = 0. \tag{8.18}
$$

In view of $\theta^1 = s$, $\theta^2 = z$, we shall use the subscripts s and z instead of 1 and 2, respectively, for any tensor components. Thus, the strain measures (8.2) for cylindrical shells can be written in the form

$$\begin{aligned} \epsilon_{ss} &= \boldsymbol{u}_{,s} \cdot \boldsymbol{\tau}\,, \qquad \epsilon_{sz} = \epsilon_{zs} = \tfrac{1}{2}(\boldsymbol{u}_{,z} \cdot \boldsymbol{\tau} + \boldsymbol{u}_{,s} \cdot \boldsymbol{e}_3), \qquad \epsilon_{zz} = \boldsymbol{u}_{,z} \cdot \boldsymbol{e}_3\,, \\ \rho_{ss} &= \boldsymbol{u}_{,ss} \cdot \boldsymbol{n}\,, \qquad \rho_{sz} = \rho_{zs} = \boldsymbol{u}_{,sz} \cdot \boldsymbol{n}, \qquad \rho_{zz} = \boldsymbol{u}_{,zz} \cdot \boldsymbol{n}\,. \end{aligned} \tag{8.19}$$

The equilibrium equations (8.4) in the absence of body forces (i.e., $\boldsymbol{g} = \boldsymbol{0}$) reduce to

$$P_{ss,s} + P_{sz,z} + \frac{1}{R}\, S_s = 0, \qquad P_{zs,s} + P_{zz,z} = 0, \qquad S_{s,s} + S_{z,z} - \frac{1}{R}\, P_{ss} = 0. \tag{8.20}$$

According to (8.6) and (8.7) we have the constitutive relations

$$\begin{aligned} P_{ss} &= N_{ss} - \frac{1}{R}\, M_{ss}\,, \qquad P_{sz} = N_{sz} - \frac{1}{R}\, M_{sz}\,, \qquad P_{zs} = N_{zs}\,, \\ P_{zz} &= N_{zz}\,, \qquad S_s = -M_{ss,s} - M_{zs,z}\,, \qquad S_z = -M_{sz,s} - M_{zz,z} \end{aligned} \tag{8.21}$$

and

$$\begin{aligned} N_{ss} &= C(\epsilon_{ss} + \nu\epsilon_{zz}), \quad N_{sz} = N_{zs} = C(1-\nu)\epsilon_{sz}\,, \quad N_{zz} = C(\nu\epsilon_{ss} + \epsilon_{zz}), \\ M_{ss} &= D(\rho_{ss} + \nu\rho_{zz}), \quad M_{sz} = M_{zs} = D(1-\nu)\rho_{sz}\,, \quad M_{zz} = D(\nu\rho_{ss} + \rho_{zz}). \end{aligned} \tag{8.22}$$

The boundary of $\mathscr{S}$ comprises of the end edge curves $\mathscr{C}_0$ and $\mathscr{C}_\ell$. Thus, the boundary conditions (8.8) become

$$\begin{aligned} -\big(P_{sz}\boldsymbol{\tau} + P_{zz}\boldsymbol{e}_3 + S_z\boldsymbol{n}\big) + \big(M_{zs}\boldsymbol{n}\big)_{,s} &= \boldsymbol{f} \quad \text{and} \quad M_{zz}\boldsymbol{n} = \boldsymbol{c} \;\; \text{on } \mathscr{C}_0\,, \\ \big(P_{sz}\boldsymbol{\tau} + P_{zz}\boldsymbol{e}_3 + S_z\boldsymbol{n}\big) - \big(M_{zs}\boldsymbol{n}\big)_{,s} &= \boldsymbol{f} \quad \text{and} \quad M_{zz}\boldsymbol{n} = \boldsymbol{c} \;\; \text{on } \mathscr{C}_\ell\,. \end{aligned} \tag{8.23}$$

The problem of Saint-Venant consists in determining the equilibrium of cylindrical shells subjected to given loads $\boldsymbol{f}, \boldsymbol{c}$ on the end edges $\mathscr{C}_0$ and $\mathscr{C}_\ell$, i.e. solving the Eqs. (8.19)–(8.22) with boundary conditions (8.23).

In the relaxed formulation of Saint-Venant's problem we replace the pointwise assignment of loads on the end edges by prescribing the resultant force and resultant moment of the given loads. If we denote by $\mathscr{R}^0 = \mathscr{R}_i^0\boldsymbol{e}_i$ the resultant force and by $\mathscr{M}^0 = \mathscr{M}_i^0\boldsymbol{e}_i$ the resultant moment about O on the end edge $\mathscr{C}_0$, then we have

$$\int_{\mathscr{C}_0} \boldsymbol{f}\,\mathrm{d}s = \mathscr{R}^0, \qquad \int_{\mathscr{C}_0} (\boldsymbol{r} \times \boldsymbol{f} + \boldsymbol{\nu} \times \boldsymbol{c})\,\mathrm{d}s = \mathscr{M}^0. \tag{8.24}$$

Inserting the expressions $(8.23)_1$ into (8.24) and making some simplifications we obtain the following form of the resultant boundary conditions on $\mathscr{C}_0$

$$
\begin{aligned}
&\int_{\mathscr{C}_0} \left(P_{sz}\boldsymbol{\tau} + P_{zz}\boldsymbol{e}_3 + S_z\boldsymbol{n}\right) \mathrm{d}s = -\mathscr{R}^0,\\
&\int_{\mathscr{C}_0} \left[\boldsymbol{r} \times \left(P_{sz}\boldsymbol{\tau} + P_{zz}\boldsymbol{e}_3 + S_z\boldsymbol{n}\right) - M_{sz}\boldsymbol{e}_3 + M_{zz}\boldsymbol{\tau}\right] \mathrm{d}s = -\mathscr{M}^0.
\end{aligned}
\tag{8.25}
$$

Since the whole boundary is $\partial\mathscr{S} = \mathscr{C}_0 \cup \mathscr{C}_\ell$, by virtue of the equilibrium conditions (8.9) and the relations (8.24) we deduce that the corresponding boundary conditions on the end edge $\mathscr{C}_\ell$ are automatically satisfied and do not represent an additional condition to be imposed on the solution. For the sake of completeness, we record the end edge conditions on $\mathscr{C}_\ell$ in the form

$$
\int_{\mathscr{C}_\ell} \boldsymbol{f}\, \mathrm{d}s = -\mathscr{R}^0, \qquad \int_{\mathscr{C}_\ell} (\boldsymbol{r} \times \boldsymbol{f} + \boldsymbol{\nu} \times \boldsymbol{c})\, \mathrm{d}s = -\mathscr{M}^0. \tag{8.26}
$$

To summarize, the relaxed Saint-Venant's problem is: find the equilibrium displacement field $\boldsymbol{u}$ which satisfies the governing equations (8.19)–(8.22) and the resultant boundary conditions (8.25) on the end edge $\mathscr{C}_0$.

8.2 Closed–form Exact Solutions

We determine an exact solution to the relaxed Saint-Venant's problem. To this aim, we employ the method elaborated by Ieşan [8, 9] in the classical elasticity.

For any displacement field $\boldsymbol{u}$ we define the linear functionals $\mathscr{R}(\boldsymbol{u}) = \mathscr{R}_i(\boldsymbol{u})\boldsymbol{e}_i$ and $\mathscr{M}(\boldsymbol{u}) = \mathscr{M}_i(\boldsymbol{u})\boldsymbol{e}_i$ by the relations

$$
\begin{aligned}
\mathscr{R}(\boldsymbol{u}) &= -\int_{\mathscr{C}_0} \left[P_{sz}(\boldsymbol{u})\boldsymbol{\tau} + P_{zz}(\boldsymbol{u})\boldsymbol{e}_3 + S_z(\boldsymbol{u})\boldsymbol{n}\right]\mathrm{d}s,\\
\mathscr{M}(\boldsymbol{u}) &= -\int_{\mathscr{C}_0} \left[\boldsymbol{r} \times \left(P_{sz}(\boldsymbol{u})\boldsymbol{\tau} + P_{zz}(\boldsymbol{u})\boldsymbol{e}_3 + S_z(\boldsymbol{u})\boldsymbol{n}\right) - M_{sz}(\boldsymbol{u})\boldsymbol{e}_3 + M_{zz}(\boldsymbol{u})\boldsymbol{\tau}\right]\mathrm{d}s,
\end{aligned}
\tag{8.27}
$$

which represent the corresponding resultant force and moment on the end edge $\mathscr{C}_0$. Then, the end edge conditions (8.25) can be written in the compact form

$$
\mathscr{R}(\boldsymbol{u}) = \mathscr{R}^0, \qquad \mathscr{M}(\boldsymbol{u}) = \mathscr{M}^0. \tag{8.28}
$$

The following result will be useful in the sequel.

Theorem 8.1 *Let $\boldsymbol{u}$ be a displacement field which satisfies the governing equations (8.19)–(8.22). If the field $\frac{\partial \boldsymbol{u}}{\partial z}$ is of class $C^1(\bar{\mathscr{S}}) \cap C^2(\mathscr{S})$, then $\frac{\partial \boldsymbol{u}}{\partial z}$ is also a solution of the governing equations (8.19)–(8.22) and it satisfies*

$$
\mathscr{R}\left(\frac{\partial \boldsymbol{u}}{\partial z}\right) = \boldsymbol{0}, \qquad \mathscr{M}\left(\frac{\partial \boldsymbol{u}}{\partial z}\right) = e_{\alpha\beta}\mathscr{R}_\beta(\boldsymbol{u})\, \boldsymbol{e}_\alpha\,. \tag{8.29}
$$

Proof In view of the linearity of the equations and the fact that all the coefficients in the relations (8.19)–(8.22) are independent of z, we deduce by differentiation that $\frac{\partial \boldsymbol{u}}{\partial z}$ also satisfies the field Eqs. (8.19)–(8.22).

Using the equilibrium equations (8.20) and the relations (8.17) we get

$$\begin{aligned}\mathscr{R}\left(\frac{\partial \boldsymbol{u}}{\partial z}\right) &= -\int_{\mathscr{C}_0} \frac{\partial}{\partial z}\big[P_{sz}(\boldsymbol{u})\boldsymbol{\tau} + P_{zz}(\boldsymbol{u})\boldsymbol{e}_3 + S_z(\boldsymbol{u})\boldsymbol{n}\big]\,\mathrm{d}s \\ &= \int_{\mathscr{C}_0} \Big[\big(P_{ss,s} + \frac{1}{R}\,S_s\big)\boldsymbol{\tau} + P_{zs,s}\,\boldsymbol{e}_3 + \big(S_{s,s} - \frac{1}{R}\,P_{ss}\big)\boldsymbol{n}\Big]\,\mathrm{d}s \\ &= \int_{\mathscr{C}_0} \frac{\partial}{\partial s}\big(P_{ss}\boldsymbol{\tau} + P_{zs}\boldsymbol{e}_3 + S_s\boldsymbol{n}\big)\,\mathrm{d}s = \boldsymbol{0}.\end{aligned}$$

Analogously, we obtain

$$\begin{aligned}\mathscr{M}_3\left(\frac{\partial \boldsymbol{u}}{\partial z}\right) &= \int_{\mathscr{C}_0} \big[e_{\alpha\beta}x'_\alpha x_\beta P_{sz,z}(\boldsymbol{u}) + x_\alpha x'_\alpha S_{z,z}(\boldsymbol{u}) + M_{sz,z}(\boldsymbol{u})\big]\,\mathrm{d}s \\ &= -\int_{\mathscr{C}_0} \Big[e_{\alpha\beta}x'_\alpha x_\beta\big(P_{ss,s} + \frac{1}{R}S_s\big) + x_\alpha x'_\alpha\big(S_{s,s} - \frac{1}{R}P_{ss}\big) + \big(S_s + M_{ss,s}\big)\Big]\mathrm{d}s \\ &= -\int_{\mathscr{C}_0} \frac{\partial}{\partial s}\big(e_{\alpha\beta}x'_\alpha x_\beta P_{ss} + x_\alpha x'_\alpha S_s + M_{ss}\big)\,\mathrm{d}s = 0\end{aligned}$$

and

$$\begin{aligned}\mathscr{M}_\alpha\left(\frac{\partial \boldsymbol{u}}{\partial z}\right) &= -\int_{\mathscr{C}_0} \frac{\partial}{\partial z}\big[e_{\alpha\beta}x_\beta P_{zz}(\boldsymbol{u}) + x'_\alpha M_{zz}(\boldsymbol{u})\big]\,\mathrm{d}s \\ &= \int_{\mathscr{C}_0} \big[e_{\alpha\beta}x_\beta P_{zs,s} + x'_\alpha\big(S_z + M_{sz,s}\big)\big]\,\mathrm{d}s \\ &= -\int_{\mathscr{C}_0} \Big[e_{\alpha\beta}\big(x'_\beta P_{sz} + e_{\beta\gamma}x'_\gamma S_z\big) + \big(\frac{e_{\alpha\beta}x'_\beta}{R} + x''_\alpha\big)M_{sz}\Big]\mathrm{d}s = e_{\alpha\beta}\mathscr{R}_\beta(\boldsymbol{u}),\end{aligned}$$

since $x''_\alpha = e_{\beta\alpha}x'_\beta R^{-1}$. □

In virtue of the linearity of the theory, we can decompose the relaxed Saint-Venant's problem into two problems:

- the extension-bending-torsion problem (characterized by the resultant loads $\mathscr{R}^0 = \mathscr{R}_3^0\boldsymbol{e}_3$ and $\mathscr{M}^0 = \mathscr{M}_i^0\boldsymbol{e}_i$) ;
- the flexure problem (characterized by the resultants of the form $\mathscr{R}^0 = \mathscr{R}_\alpha^0\boldsymbol{e}_\alpha$ and $\mathscr{M}^0 = \boldsymbol{0}$).

8.2.1 Extension-Bending-Torsion Problem

We can determine an exact closed-form solution to this problem, which is given in the next theorem.

Theorem 8.2 *Consider the extension-bending-torsion problem for cylindrical shells: solve the field Eqs. (8.19)–(8.22) under the end edge conditions*

$$\mathscr{R}(\boldsymbol{u}) = \mathscr{R}_3^0\, \boldsymbol{e}_3\,, \qquad \mathscr{M}(\boldsymbol{u}) = \mathscr{M}_i^0\, \boldsymbol{e}_i\,, \tag{8.30}$$

for given resultants $\mathscr{R}_3^0$, $\mathscr{M}_i^0$ $(i = 1, 2, 3)$. *This problem admits the following solution*

$$\begin{aligned}
\boldsymbol{u} = &-\big[\tfrac{1}{2}A_\alpha(z^2 - \nu x_\beta x_\beta) + \nu(A_\beta x_\beta + \bar{A}_3)x_\alpha + K\, z\, e_{\alpha\beta}x_\beta\big]\boldsymbol{e}_\alpha \\
&+\big[z(A_\alpha x_\alpha + \bar{A}_3) + K\varphi(s)\big]\boldsymbol{e}_3 + \frac{1}{C}\Big(s\,\boldsymbol{B}\times\boldsymbol{e}_3 - \int_0^s \frac{1}{R}(\boldsymbol{B}\cdot\hat{\boldsymbol{r}})\boldsymbol{\tau}\,\mathrm{d}s\Big) \\
&-\int_0^s \boldsymbol{n}\int_0^s \Big[\frac{\nu}{R}\boldsymbol{A} + \Big(\frac{1}{D} + \frac{1}{CR^2}\Big)\boldsymbol{B}\Big]\cdot\hat{\boldsymbol{r}}\,\mathrm{d}s\,\mathrm{d}s,
\end{aligned} \tag{8.31}$$

where we denote by

$$\hat{\boldsymbol{r}} = x_\alpha\boldsymbol{e}_\alpha + \bar{s}\,\boldsymbol{e}_3\,, \qquad \boldsymbol{A} = A_i\boldsymbol{e}_i\,, \qquad \boldsymbol{B} = B_i\boldsymbol{e}_i\,, \qquad \bar{A}_3 = A_3\bar{s}\,, \tag{8.32}$$

and A_i, B_i, K *are constants* $(i = 1, 2, 3)$. *The constants* A_α *and* $\bar{A}_3$ *represent measures of curvature and stretch, respectively, which are given in terms of the resultants* $\mathscr{M}_\alpha^0$ *and* $\mathscr{R}_3^0$ *by the relations (8.51). Further, the coefficients* B_i *are determined in terms of* A_i *through the system of linear algebraic equations (8.48). The constant* K *is a measure of twist of the cylindrical shell, which is given by the twist-torque relation*

$$K = -\frac{\mathscr{M}_3^0}{2(1-\nu)}\left(C\,\frac{\mathscr{A}^2}{\bar{s}} + D\bar{s}\right)^{-1}, \tag{8.33}$$

while $\varphi(s)$ *is a torsion function having the expression*

$$\varphi(s) = \frac{2\mathscr{A}}{\bar{s}}\,s - \int_0^s \boldsymbol{r}\cdot\boldsymbol{n}\,\mathrm{d}s, \tag{8.34}$$

where $\mathscr{A} = \dfrac{1}{2}\displaystyle\int_0^{\bar{s}} \boldsymbol{r}\cdot\boldsymbol{n}\,\mathrm{d}s$ *is the area bounded by the closed cross-section curve* $\mathscr{C}_0$.

Proof We search for a solution $\boldsymbol{u}$ satisfying the Eqs. (8.19)–(8.22) and the conditions (8.30), such that $\frac{\partial \boldsymbol{u}}{\partial z} \in C^1(\bar{\mathscr{S}}) \cap C^2(\mathscr{S})$. According to Theorem 8.1, the field $\frac{\partial \boldsymbol{u}}{\partial z}$ verifies then the equilibrium equations (8.19)–(8.22) and the conditions (8.29) in the form

$$\mathscr{R}\left(\frac{\partial \boldsymbol{u}}{\partial z}\right) = \boldsymbol{0}, \qquad \mathscr{M}\left(\frac{\partial \boldsymbol{u}}{\partial z}\right) = \boldsymbol{0}.$$

These relations suggest to search for a solution $\boldsymbol{u}$ such that the derivative $\frac{\partial \boldsymbol{u}}{\partial z}$ is a rigid body displacement field, i.e.

$$\frac{\partial \boldsymbol{u}}{\partial z} = \tilde{\boldsymbol{c}} + \boldsymbol{d} \times \boldsymbol{r}, \tag{8.35}$$

where $\tilde{\boldsymbol{c}} = \tilde{c}_i \boldsymbol{e}_i$ and $\boldsymbol{d} = d_i \boldsymbol{e}_i$ are constant vectors. It follows by integration that

$$\boldsymbol{u} = z\,\tilde{\boldsymbol{c}} + \boldsymbol{d} \times \big(x_\alpha z\, \boldsymbol{e}_\alpha + \frac{1}{2} z^2 \boldsymbol{e}_3\big) + \tilde{\boldsymbol{w}}(s),$$

where $\tilde{\boldsymbol{w}}(s)$ is an arbitrary function of class $C^2[0, \bar{s}]$. If we introduce the notations $A_\alpha = e_{\beta\alpha} d_\beta$, $\bar{A}_3 = A_3 \bar{s} = \tilde{c}_3$, $K = d_3$, and $\boldsymbol{w}(s) = \tilde{\boldsymbol{w}}(s) + (\tilde{c}_\alpha x_\alpha)\boldsymbol{e}_3$, then we can write the last relation in the form

$$\boldsymbol{u} = -\big(\frac{1}{2} z^2 A_\alpha + K z\, e_{\alpha\beta} x_\beta\big)\boldsymbol{e}_\alpha + z(\boldsymbol{A} \cdot \hat{\boldsymbol{r}})\, \boldsymbol{e}_3 + \boldsymbol{w}(s), \tag{8.36}$$

where $\boldsymbol{A} = A_i \boldsymbol{e}_i$ and $\hat{\boldsymbol{r}} = x_\alpha \boldsymbol{e}_\alpha + \bar{s}\, \boldsymbol{e}_3$. In what follows, we determine the unknown function $\boldsymbol{w}(s) = w_i(s)\boldsymbol{e}_i$ such that the displacement field (8.36) satisfies the equilibrium equations (8.20). The strain measures (8.19) corresponding to the displacement field (8.36) are

$$\begin{aligned} &\epsilon_{ss} = \boldsymbol{w}' \cdot \boldsymbol{\tau}, \qquad \epsilon_{sz} = \epsilon_{zs} = \tfrac{1}{2}(K\boldsymbol{r} \cdot \boldsymbol{n} + w_3'), \qquad \epsilon_{zz} = \boldsymbol{A} \cdot \hat{\boldsymbol{r}}, \\ &\rho_{ss} = \boldsymbol{w}'' \cdot \boldsymbol{n}, \qquad \rho_{sz} = \rho_{zs} = -K, \qquad \rho_{zz} = -\boldsymbol{A} \cdot \boldsymbol{n}. \end{aligned} \tag{8.37}$$

where we denote by $(\cdot)' \equiv \frac{\mathrm{d}}{\mathrm{d}s}(\cdot)$ the derivative with respect to s. The stress tensors (8.22) have the form

$$\begin{aligned} &N_{ss} = C(\nu \boldsymbol{A} \cdot \hat{\boldsymbol{r}} + \boldsymbol{w}' \cdot \boldsymbol{\tau}), \quad N_{sz} = N_{zs} = \tfrac{1}{2} C(1-\nu)(K\boldsymbol{r} \cdot \boldsymbol{n} + w_3'), \\ &N_{zz} = C(\boldsymbol{A} \cdot \hat{\boldsymbol{r}} + \nu \boldsymbol{w}' \cdot \boldsymbol{\tau}), \quad M_{ss} = D(-\nu \boldsymbol{A} \cdot \boldsymbol{n} + \boldsymbol{w}'' \cdot \boldsymbol{n}), \\ &M_{sz} = M_{zs} = -D(1-\nu)K, \; M_{zz} = D(-\boldsymbol{A} \cdot \boldsymbol{n} + \nu \boldsymbol{w}'' \cdot \boldsymbol{n}) \end{aligned} \tag{8.38}$$

and the relations (8.21) yield

$$\begin{aligned} &P_{ss} = \nu \boldsymbol{A} \cdot (C\hat{\boldsymbol{r}} + D\tfrac{1}{R}\boldsymbol{n}) + C\boldsymbol{w}' \cdot \boldsymbol{\tau} - D\tfrac{1}{R}\boldsymbol{w}'' \cdot \boldsymbol{n}, \\ &P_{sz} = K(1-\nu)(\tfrac{1}{2} C\boldsymbol{r} \cdot \boldsymbol{n} + D\tfrac{1}{R}) + \tfrac{1}{2} C(1-\nu) w_3', \\ &P_{zs} = \tfrac{1}{2} C(1-\nu)(K\boldsymbol{r} \cdot \boldsymbol{n} + w_3'), \qquad P_{zz} = C(\boldsymbol{A} \cdot \hat{\boldsymbol{r}} + \nu \boldsymbol{w}' \cdot \boldsymbol{\tau}), \\ &S_s = D(\nu \boldsymbol{A} \cdot \boldsymbol{n} - \boldsymbol{w}'' \cdot \boldsymbol{n})', \qquad S_z = 0. \end{aligned} \tag{8.39}$$

We note that all these tensor components are independent of z.

Using (8.39) we deduce that the equilibrium equations $(8.20)_2$ reduces to

$$\big(K\boldsymbol{r} \cdot \boldsymbol{n} + w_3'(s)\big)' = 0.$$

Integrating this equations and imposing the continuity condition $w_3(0) = w_3(\bar{s})$ (see (8.46)), we obtain the solution

$$w_3(s) = K\Big(\frac{2\mathscr{A}}{\bar{s}}\,s - \int_0^s \boldsymbol{r}\cdot\boldsymbol{n}\,\mathrm{d}s\Big), \qquad \text{i.e.} \qquad w_3(s) = K\,\varphi(s), \tag{8.40}$$

in accordance with notation (8.34). The remaining two equilibrium equations $(8.20)_{1,3}$ can be put in the vectorial form

$$\begin{gathered}\big(-P_{ss}\boldsymbol{n} + S_s\boldsymbol{\tau}\big)' = \boldsymbol{0}, \qquad \text{i.e.}\\ -P_{ss}\boldsymbol{n} + S_s\boldsymbol{\tau} = B_\alpha\boldsymbol{e}_\alpha\,, \quad \text{or equivalently,}\\ P_{ss} = -(B_\alpha\boldsymbol{e}_\alpha)\cdot\boldsymbol{n} \qquad \text{and} \qquad S_s = (B_\alpha\boldsymbol{e}_\alpha)\cdot\boldsymbol{\tau}\,,\end{gathered} \tag{8.41}$$

where B_1, B_2 are some arbitrary constants. Using (8.39), from (8.41) we obtain

$$\begin{aligned}&C(\nu\boldsymbol{A}\cdot\hat{\boldsymbol{r}} + \boldsymbol{w}'\cdot\boldsymbol{\tau}) + D\,\tfrac{1}{R}(\nu\boldsymbol{A}\cdot\boldsymbol{n} - \boldsymbol{w}''\cdot\boldsymbol{n}) = -(B_\alpha\boldsymbol{e}_\alpha)\cdot\boldsymbol{n}\,,\\ &D(\nu\boldsymbol{A}\cdot\boldsymbol{n} - \boldsymbol{w}''\cdot\boldsymbol{n}) = B_\alpha x_\alpha + B_3\bar{s}\,,\end{aligned} \tag{8.42}$$

where B_3 is an arbitrary constant. Introducing the notation $\boldsymbol{B} = B_i\boldsymbol{e}_i$ we can write the system of Eq. (8.42) in the simpler form

$$\begin{aligned}\boldsymbol{w}'\cdot\boldsymbol{\tau} &= -\nu\boldsymbol{A}\cdot\hat{\boldsymbol{r}} - \tfrac{1}{C}\,\boldsymbol{B}\cdot\big(\boldsymbol{n} + \tfrac{1}{R}\,\hat{\boldsymbol{r}}\big),\\ \boldsymbol{w}''\cdot\boldsymbol{n} &= \nu\boldsymbol{A}\cdot\boldsymbol{n} - \tfrac{1}{D}\,\boldsymbol{B}\cdot\hat{\boldsymbol{r}}.\end{aligned} \tag{8.43}$$

Integrating the relation $(\boldsymbol{w}'\cdot\boldsymbol{n})' = \boldsymbol{w}''\cdot\boldsymbol{n} + \boldsymbol{w}'\cdot\boldsymbol{\tau}\,\frac{1}{R}$ with respect to s and using (8.43), we deduce

$$\boldsymbol{w}'\cdot\boldsymbol{n} = \nu\boldsymbol{A}\cdot\Big(e_{\alpha\beta}x_\beta\boldsymbol{e}_\alpha - \int_0^s \frac{1}{R}\,\hat{\boldsymbol{r}}\,\mathrm{d}s\Big) + \boldsymbol{B}\cdot\Big[\frac{1}{C}\,\boldsymbol{\tau} - \int_0^s\Big(\frac{1}{D} + \frac{1}{CR^2}\Big)\hat{\boldsymbol{r}}\,\mathrm{d}s\Big], \tag{8.44}$$

where we neglect some additive constants which represent rigid body displacement fields. The Eqs. $(8.43)_1$ and (8.44) determine the field $\boldsymbol{w}'(s)$. If we integrate once again with respect to s, we obtain the following solution for $\boldsymbol{w}(s)$:

$$\begin{aligned}w_\alpha(s)\,\boldsymbol{e}_\alpha = {}&\frac{1}{2}\nu(x_\gamma x_\gamma)A_\alpha\boldsymbol{e}_\alpha - \nu(\boldsymbol{A}\cdot\hat{\boldsymbol{r}})x_\alpha\boldsymbol{e}_\alpha + \frac{1}{C}\Big[s\boldsymbol{B}\times\boldsymbol{e}_3 - \int_0^s\frac{1}{R}(\boldsymbol{B}\cdot\hat{\boldsymbol{r}})\boldsymbol{\tau}\,\mathrm{d}s\Big]\\ &-\int_0^s\boldsymbol{n}\int_0^s\Big[\frac{\nu}{R}\boldsymbol{A} + \Big(\frac{1}{D} + \frac{1}{CR^2}\Big)\boldsymbol{B}\Big]\cdot\hat{\boldsymbol{r}}\,\mathrm{d}s\,\mathrm{d}s.\end{aligned} \tag{8.45}$$

Inserting the relations (8.40) and (8.45) into Eq. (8.36) we derive that the solution $\boldsymbol{u}$ has indeed the expression (8.31), which was announced in the statement of the theorem. This solution must satisfy some continuity conditions for $s = 0, \bar{s}$, namely

$$\boldsymbol{u}(0,z)=\boldsymbol{u}(\bar{s},z),\qquad \boldsymbol{u}_{,s}(0,z)=\boldsymbol{u}_{,s}(\bar{s},z),\qquad \boldsymbol{u}_{,ss}(0,z)=\boldsymbol{u}_{,ss}(\bar{s},z),\tag{8.46}$$

for all $z\in[0,\ell]$. In view of the integrations that we performed to obtain relations (8.44) and (8.45), the conditions (8.46) reduce to

$$\big(\boldsymbol{w}'\cdot\boldsymbol{n}\big)(0)=\big(\boldsymbol{w}'\cdot\boldsymbol{n}\big)(\bar{s})\qquad\text{and}\qquad w_\alpha(0)=w_\alpha(\bar{s}),\quad \alpha=1,2.\tag{8.47}$$

By virtue of (8.44) and (8.45), the last relations (8.47) can be written in the form

$$\begin{aligned}
&\nu\boldsymbol{A}\cdot\int_0^{\bar{s}}\frac{1}{R}\,\hat{\boldsymbol{r}}\,\mathrm{d}s+\boldsymbol{B}\cdot\int_0^{\bar{s}}\Big(\frac{1}{D}+\frac{1}{CR^2}\Big)\hat{\boldsymbol{r}}\,\mathrm{d}s=0,\\
&\nu\boldsymbol{A}\cdot\int_0^{\bar{s}}\frac{x_\alpha}{R}\,\hat{\boldsymbol{r}}\,\mathrm{d}s+\boldsymbol{B}\cdot\int_0^{\bar{s}}\Big[x_\alpha\Big(\frac{1}{D}+\frac{1}{CR^2}\Big)\hat{\boldsymbol{r}}+\frac{1}{C}\big(\boldsymbol{e}_\alpha+x'_\alpha\boldsymbol{\tau}\big)\Big]\,\mathrm{d}s=0.
\end{aligned}\tag{8.48}$$

From the system of three linear algebraic equations (8.48) one can determine the constants B_i in terms of the coefficients A_i $(i=1,2,3)$.

To conclude the proof, we determine the constants A_i and K in terms of the resultant loads $\mathscr{R}_3^0$ and $\mathscr{M}_i^0$. We remark first that the functionals (8.27) can be rewritten in different forms: namely, inserting (8.21) into (8.27) and making some transformations we obtain

$$\begin{aligned}
\mathscr{R}(\boldsymbol{u})&=-\int_{\mathscr{C}_0}\big[N_{sz}\boldsymbol{\tau}+N_{zz}\boldsymbol{e}_3-M_{zz,z}\boldsymbol{n}\big]\mathrm{d}s\qquad\text{and}\\
\mathscr{M}(\boldsymbol{u})&=-\int_{\mathscr{C}_0}\big[N_{zz}\boldsymbol{r}\times\boldsymbol{e}_3+M_{zz}\boldsymbol{\tau}+\big(N_{sz}\boldsymbol{r}\cdot\boldsymbol{n}-2M_{sz}+M_{zz,z}\boldsymbol{r}\cdot\boldsymbol{\tau}\big)\boldsymbol{e}_3\big]\mathrm{d}s.
\end{aligned}\tag{8.49}$$

Also, using (8.40) and (8.43) we obtain the following expressions for the stress tensors (8.38) corresponding to the solution $\boldsymbol{u}$:

$$\begin{aligned}
&N_{ss}=-\boldsymbol{B}\cdot\big(\boldsymbol{n}+\frac{1}{R}\hat{\boldsymbol{r}}\big), && N_{sz}=N_{zs}=C(1-\nu)K\,\frac{\mathscr{A}}{\bar{s}},\\
&N_{zz}=C(1-\nu^2)\boldsymbol{A}\cdot\hat{\boldsymbol{r}}-\nu\boldsymbol{B}\cdot\big(\boldsymbol{n}+\frac{1}{R}\hat{\boldsymbol{r}}\big), && M_{ss}=-\boldsymbol{B}\cdot\hat{\boldsymbol{r}},\\
&M_{zz}=-D(1-\nu^2)\boldsymbol{A}\cdot\boldsymbol{n}-\nu\boldsymbol{B}\cdot\hat{\boldsymbol{r}}, && M_{sz}=M_{zs}=-D(1-\nu)K.
\end{aligned}\tag{8.50}$$

Inserting (8.50) in (8.49), we derive after some calculations that the end edge conditions (8.30) reduce to the following equations

$$C(1-\nu^2)\boldsymbol{A}\cdot\int_0^{\bar{s}}\hat{\boldsymbol{r}}\,\mathrm{d}s-\nu\boldsymbol{B}\cdot\int_0^{\bar{s}}\frac{1}{R}\hat{\boldsymbol{r}}\,\mathrm{d}s=-\mathscr{R}_3^0,$$

$$\begin{aligned}C(1-\nu^2)\boldsymbol{A}\cdot&\int_0^{\bar{s}}\Big(x_\alpha\,\hat{\boldsymbol{r}}+\frac{D}{C}\,e_{\alpha\beta}x'_\beta\boldsymbol{n}\Big)\mathrm{d}s\\ &-\nu\boldsymbol{B}\cdot\int_0^{\bar{s}}\Big[x_\alpha\Big(\frac{1}{R}\hat{\boldsymbol{r}}+\boldsymbol{n}\Big)-e_{\alpha\beta}x'_\beta\hat{\boldsymbol{r}}\Big]\,\mathrm{d}s=e_{\alpha\beta}\mathscr{M}_\beta^0,\end{aligned}\tag{8.51}$$

together with the twist-torque relation (8.33). The Eq. (8.51) represent a linear algebraic system for the determination of the constants A_i $(i=1,2,3)$ in terms of $\mathscr{R}_3^0$, $\mathscr{M}_1^0$, $\mathscr{M}_2^0$ (since $\boldsymbol{B}$ has already been determined in (8.48)). The proof is complete. □

In Sect. 8.3 we shall present an approximation of the solution $\boldsymbol{u}$ given by Theorem 8.2 (in the thin shell limit $h\to 0$) and show the interpretation of the constants A_i as measures of stretch and curvature of the cylindrical shell considered as a beam.

8.2.2 Flexure Problem

The next result presents an exact closed-form solution to the flexure problem.

Theorem 8.3 *The flexure problem for cylindrical shells, consisting in the governing equations (8.19)–(8.22) and the end edge conditions*

$$\mathscr{R}(\boldsymbol{u})=\mathscr{R}_\alpha^0\,\boldsymbol{e}_\alpha\,,\qquad \mathscr{M}(\boldsymbol{u})=\boldsymbol{0},\tag{8.52}$$

admits the following solution

$$\begin{aligned}\hat{\boldsymbol{u}}=&-\big[(\tfrac{1}{6}z^3-\tfrac{1}{2}\nu\,z\,x_\beta x_\beta)\hat{A}_\alpha+\nu\,z\,(\hat{\boldsymbol{A}}\cdot\hat{\boldsymbol{r}})\,x_\alpha+\tilde{K}\,z\,e_{\alpha\beta}x_\beta\big]\boldsymbol{e}_\alpha\\ &+\big[\tfrac{1}{2}z^2(\hat{\boldsymbol{A}}\cdot\hat{\boldsymbol{r}})+\tilde{K}\varphi(s)+\psi(s)\big]\boldsymbol{e}_3+\frac{1}{C}\,z\Big(s\,\hat{\boldsymbol{B}}\times\boldsymbol{e}_3-\int_0^s\frac{1}{R}(\hat{\boldsymbol{B}}\cdot\hat{\boldsymbol{r}})\boldsymbol{\tau}\,\mathrm{d}s\Big)\\ &-z\int_0^s\boldsymbol{n}\int_0^s\Big[\frac{\nu}{R}\hat{\boldsymbol{A}}+\Big(\frac{1}{D}+\frac{1}{CR^2}\Big)\hat{\boldsymbol{B}}\Big]\cdot\hat{\boldsymbol{r}}\,\mathrm{d}s\,\mathrm{d}s,\end{aligned}\tag{8.53}$$

where $\hat{\boldsymbol{A}}=\hat{A}_i\boldsymbol{e}_i$, $\hat{\boldsymbol{B}}=\hat{B}_i\boldsymbol{e}_i$ are constant vectors and the constant $\tilde{K}$ is given by relation (8.64). The constant coefficients $\hat{B}_i$ are determined in terms of $\hat{A}_i$ by the system of Eq. (8.48), while the constants $\hat{A}_i$ can be calculated in terms of the resultant forces $\mathscr{R}_\alpha^0$ from the Eq. (8.55). In the above solution, $\varphi(s)$ is the torsion function (8.34) and $\psi(s)$ is a flexure function given by (8.61).

Proof To find a solution $\hat{\boldsymbol{u}}$ of the flexure problem, we employ the results of Theorems 8.1 and 8.2. We note that the solution $\boldsymbol{u}$ given in Theorem 8.2 by relation (8.31) depend on four constants A_i and K. To show this dependence explicitly, we denote the displacement field (8.31) by $\boldsymbol{u}[\boldsymbol{A},K]$.

Applying the Theorem 8.1 for the field $\hat{\boldsymbol{u}}$, we deduce that $\frac{\partial \hat{\boldsymbol{u}}}{\partial z}$ satisfies the governing equations (8.19)–(8.22) together with the end edge conditions

$$\mathscr{R}\left(\frac{\partial \hat{\boldsymbol{u}}}{\partial z}\right) = \boldsymbol{0}, \qquad \mathscr{M}\left(\frac{\partial \hat{\boldsymbol{u}}}{\partial z}\right) = e_{\alpha\beta}\mathscr{R}^0_\beta\, \boldsymbol{e}_\alpha\,.$$

Thus, $\frac{\partial \hat{\boldsymbol{u}}}{\partial z}$ is a solution of the bending problem with resultant bending moment given by $\mathscr{M}^0 = e_{\alpha\beta}\mathscr{R}^0_\beta\, \boldsymbol{e}_\alpha$. In this case, the resultant force and twisting moment are vanishing.

Combining the results of Theorems 8.1 and 8.2, we look for a solution $\hat{\boldsymbol{u}}$ such that $\frac{\partial \hat{\boldsymbol{u}}}{\partial z}$ is a solution (to the bending problem) of the form (8.31), i.e. we have

$$\frac{\partial \hat{\boldsymbol{u}}}{\partial z} = \boldsymbol{u}[\hat{\boldsymbol{A}}, \hat{K}]\,. \tag{8.54}$$

Here, the constants $\hat{A}_i$ are given by (in view of (8.51) written with $\mathscr{R}^0_3 = 0$ and bending moments $e_{\alpha\beta}\mathscr{R}^0_\beta$)

$$\begin{aligned} &C(1-\nu^2)\hat{\boldsymbol{A}}\cdot\int_0^{\bar{s}} \hat{\boldsymbol{r}}\,\mathrm{d}s - \nu\hat{\boldsymbol{B}}\cdot\int_0^{\bar{s}} \frac{1}{R}\hat{\boldsymbol{r}}\,\mathrm{d}s = 0\,,\\ &C(1-\nu^2)\hat{\boldsymbol{A}}\cdot\int_0^{\bar{s}} \Big(x_\alpha\hat{\boldsymbol{r}} + \frac{D}{C}\,e_{\alpha\beta}x'_\beta\boldsymbol{n}\Big)\mathrm{d}s\\ &\qquad -\nu\hat{\boldsymbol{B}}\cdot\int_0^{\bar{s}} \Big[x_\alpha\Big(\frac{1}{R}\hat{\boldsymbol{r}} + \boldsymbol{n}\Big) - e_{\alpha\beta}x'_\beta\hat{\boldsymbol{r}}\Big]\,\mathrm{d}s = -\mathscr{R}^0_\alpha\,, \end{aligned} \tag{8.55}$$

together with relations (8.48) (written with $\hat{\boldsymbol{A}}$, $\hat{\boldsymbol{B}}$ instead of $\boldsymbol{A}$, $\boldsymbol{B}$). By virtue of (8.33) (written with twisting moment $\mathscr{M}^0_3 = 0$), we find

$$\hat{K} = 0. \tag{8.56}$$

Taking into account that $\frac{\partial \boldsymbol{u}[\boldsymbol{A},K]}{\partial z}$ is a rigid body displacement field (see (8.35)), we obtain from the relation (8.54) and (8.56) by integration with respect to z :

$$\hat{\boldsymbol{u}} = \int_0^z \boldsymbol{u}[\hat{\boldsymbol{A}}, 0]\,\mathrm{d}z + \boldsymbol{u}[\tilde{\boldsymbol{A}}, \tilde{K}] + \hat{\boldsymbol{w}}(s), \tag{8.57}$$

where $\tilde{\boldsymbol{A}} = \tilde{A}_i\boldsymbol{e}_i$ and $\tilde{K}$ are constants, while $\hat{\boldsymbol{w}}(s) = \hat{w}_i(s)\boldsymbol{e}_i$ is an arbitrary function of class $C^2[0, \bar{s}]$. Suggested by the previous results concerning the flexure problem in three-dimensional elasticity [8, 9] and shell theory [2, 7], we search for the solution $\hat{\boldsymbol{u}}$ such that

$$\tilde{\boldsymbol{A}} = \boldsymbol{0} \quad \text{and} \quad \hat{w}_\alpha(s) = 0. \tag{8.58}$$

The choice (8.58) is made here for the sake of simplity. (Alternatively, we can work with the unknown vector $\tilde{\boldsymbol{A}}$ and unknown functions $\hat{w}_\alpha(s)$ and obtain finally from the equilibrium equations and end edge conditions that the relations (8.58) hold.) Thus, substituting (8.58) in (8.57), we search for a solution $\hat{\boldsymbol{u}}$ of the flexure problem in the form

$$\hat{\boldsymbol{u}} = \int_0^z \boldsymbol{u}[\hat{\boldsymbol{A}}, 0]\,\mathrm{d}z + \boldsymbol{u}[\boldsymbol{0}, \tilde{K}] + \psi(s)\boldsymbol{e}_3\,, \tag{8.59}$$

where we have denoted $\psi(s) = \hat{w}_3(s)$ for brevity.

In what follows, we shall determine the unknown function $\psi(s)$ from the equilibrium equations (8.20) and the constant $\tilde{K}$ from the end edge conditions (8.52). Inserting the relation (8.31) into (8.59), we deduce from the geometrical equations (8.19) the following strain measures corresponding to $\hat{\boldsymbol{u}}$:

$$\begin{aligned}
\hat{\epsilon}_{ss} &= -z\big[\nu\hat{\boldsymbol{A}}\cdot\hat{\boldsymbol{r}} + \frac{1}{C}\,\hat{\boldsymbol{B}}\cdot\big(\boldsymbol{n} + \frac{1}{R}\hat{\boldsymbol{r}}\big)\big], \qquad \hat{\epsilon}_{zz} = z(\hat{\boldsymbol{A}}\cdot\hat{\boldsymbol{r}}),\\
\hat{\epsilon}_{sz} &= \hat{\epsilon}_{zs} = \frac{1}{2}\Big\{\frac{\nu}{2}(x_\alpha x_\alpha)\hat{\boldsymbol{A}} - \nu(\hat{\boldsymbol{A}}\cdot\hat{\boldsymbol{r}})\boldsymbol{r} + \frac{1}{C}\Big(s\,\hat{\boldsymbol{B}}\times\boldsymbol{e}_3 - \int_0^s \frac{1}{R}(\hat{\boldsymbol{B}}\cdot\hat{\boldsymbol{r}})\boldsymbol{\tau}\,\mathrm{d}s\Big)\\
&\quad - \int_0^s \boldsymbol{n}\int_0^s \Big[\frac{\nu}{R}\,\hat{\boldsymbol{A}} + \Big(\frac{1}{D} + \frac{1}{CR^2}\Big)\hat{\boldsymbol{B}}\Big]\cdot\hat{\boldsymbol{r}}\,\mathrm{d}s\,\mathrm{d}s\Big\}\cdot\boldsymbol{\tau} + \tilde{K}\,\frac{\mathscr{A}}{\bar{s}} + \frac{1}{2}\,\psi'(s),\\
\hat{\rho}_{ss} &= z\big(\nu\hat{\boldsymbol{A}}\cdot\boldsymbol{n} - \frac{1}{D}\,\hat{\boldsymbol{B}}\cdot\hat{\boldsymbol{r}}\big), \qquad \hat{\rho}_{zz} = -z(\hat{\boldsymbol{A}}\cdot\boldsymbol{n}),\\
\hat{\rho}_{sz} &= \hat{\rho}_{zs} = -\tilde{K} + \nu\hat{\boldsymbol{A}}\cdot(e_{\alpha\beta}x_\beta\boldsymbol{e}_\alpha) + \frac{1}{C}\,\hat{\boldsymbol{B}}\cdot\boldsymbol{\tau} - \int_0^s \Big[\frac{\nu}{R}\hat{\boldsymbol{A}} + \Big(\frac{1}{D} + \frac{1}{CR^2}\Big)\hat{\boldsymbol{B}}\Big]\cdot\hat{\boldsymbol{r}}\,\mathrm{d}s.
\end{aligned} \tag{8.60}$$

We note that the dependence of these tensors on z is at most linear. Using the expressions (8.60) we can calculate the stress tensors according to relations (8.21) and (8.22), and then write the equilibrium equations (8.20). We obtain that the equilibrium equations $(8.20)_{1,3}$ are identically satisfied, while the equilibrium equation $(8.20)_2$ reduces to

$$C(1-\nu)\big(\hat{\epsilon}_{sz}\big)' + C(1-\nu^2)\hat{\boldsymbol{A}}\cdot\hat{\boldsymbol{r}} - \nu\hat{\boldsymbol{B}}\cdot\big(\boldsymbol{n} + \frac{1}{R}\hat{\boldsymbol{r}}\big) = 0,$$

where $\hat{\epsilon}_{sz}$ is given by $(8.60)_3$. Integrating the last equation twice with respect to s, we find the solution

$$\begin{aligned}
\psi(s) &= -\int_0^s \Big\{\boldsymbol{\tau}\cdot\Big[\frac{\nu}{2}(x_\alpha x_\alpha)\hat{\boldsymbol{A}} - \nu(\hat{\boldsymbol{A}}\cdot\hat{\boldsymbol{r}})\boldsymbol{r} + \frac{1}{C}\Big(s\hat{\boldsymbol{B}}\times\boldsymbol{e}_3 - \int_0^s \frac{1}{R}(\hat{\boldsymbol{B}}\cdot\hat{\boldsymbol{r}})\boldsymbol{\tau}\,\mathrm{d}s\Big)\\
&\quad - \int_0^s \boldsymbol{n}\int_0^s \Big[\frac{\nu}{R}\hat{\boldsymbol{A}} + \Big(\frac{1}{D} + \frac{1}{CR^2}\Big)\hat{\boldsymbol{B}}\Big]\cdot\hat{\boldsymbol{r}}\,\mathrm{d}s\,\mathrm{d}s\Big]\\
&\quad + \frac{2}{C(1-\nu)}\int_0^s \big[C(1-\nu^2)\hat{\boldsymbol{A}}\cdot\hat{\boldsymbol{r}} - \nu\hat{\boldsymbol{B}}\cdot\big(\boldsymbol{n} + \frac{1}{R}\hat{\boldsymbol{r}}\big)\big]\mathrm{d}s\Big\}\mathrm{d}s + K_0 s,
\end{aligned} \tag{8.61}$$

where the constant K_0 is determined by the continuity condition $\psi(0) = \psi(\bar{s})$ in the form

$$K_0 = \frac{1}{\bar{s}} \int_0^{\bar{s}} \Big\{ \boldsymbol{\tau} \cdot \Big[\frac{\nu}{2}(x_\alpha x_\alpha)\hat{\boldsymbol{A}} - \nu(\hat{\boldsymbol{A}} \cdot \hat{\boldsymbol{r}})\boldsymbol{r} + \frac{1}{C}\Big(s\hat{\boldsymbol{B}} \times \boldsymbol{e}_3 - \int_0^s \frac{1}{R}(\hat{\boldsymbol{B}} \cdot \hat{\boldsymbol{r}})\boldsymbol{\tau}\, \mathrm{d}s \Big)$$
$$- \int_0^s \boldsymbol{n} \int_0^s \Big[\frac{\nu}{R}\hat{\boldsymbol{A}} + \Big(\frac{1}{D} + \frac{1}{CR^2}\Big)\hat{\boldsymbol{B}}\Big] \cdot \hat{\boldsymbol{r}}\, \mathrm{d}s\, \mathrm{d}s \Big]$$
$$+ \frac{2}{C(1-\nu)} \int_0^s \big[C(1-\nu^2)\hat{\boldsymbol{A}} \cdot \hat{\boldsymbol{r}} - \nu \hat{\boldsymbol{B}} \cdot \big(\boldsymbol{n} + \frac{1}{R}\hat{\boldsymbol{r}}\big)\big] \mathrm{d}s \Big\} \mathrm{d}s. \tag{8.62}$$

We mention that the conditions $\psi'(0) = \psi'(\bar{s})$ and $\psi''(0) = \psi''(\bar{s})$ are also satisfied, by virtue of the relation $(8.55)_1$.

Finally, we impose the end edge conditions (8.52) in order to determine the constant $\tilde{K}$. In view of (8.60), (8.61) and (8.21), the stress tensors $\hat{\boldsymbol{N}}$ and $\hat{\boldsymbol{M}}$ corresponding to the solution $\hat{\boldsymbol{u}}$ have the components

$$\hat{N}_{ss} = -z\, \hat{\boldsymbol{B}} \cdot \big(\boldsymbol{n} + \frac{1}{R}\hat{\boldsymbol{r}}\big), \qquad \hat{N}_{zz} = z\big[C(1-\nu^2)\hat{\boldsymbol{A}} \cdot \hat{\boldsymbol{r}} - \nu \hat{\boldsymbol{B}} \cdot \big(\boldsymbol{n} + \frac{1}{R}\hat{\boldsymbol{r}}\big)\big],$$
$$\hat{N}_{sz} = \hat{N}_{zs} = C(1-\nu)\Big(\tilde{K}\frac{\mathscr{A}}{\bar{s}} + \frac{1}{2}K_0\Big) - \int_0^s \big[C(1-\nu^2)\hat{\boldsymbol{A}} \cdot \hat{\boldsymbol{r}} - \nu \hat{\boldsymbol{B}} \cdot \big(\boldsymbol{n} + \frac{1}{R}\hat{\boldsymbol{r}}\big)\big] \mathrm{d}s,$$
$$\hat{M}_{ss} = -z(\hat{\boldsymbol{B}} \cdot \hat{\boldsymbol{r}}), \qquad \hat{M}_{zz} = -z\big[D(1-\nu^2)\hat{\boldsymbol{A}} \cdot \boldsymbol{n} + \nu \hat{\boldsymbol{B}} \cdot \hat{\boldsymbol{r}}\big],$$
$$\hat{M}_{sz} = \hat{M}_{zs} = D(1-\nu)\Big(-\tilde{K} + \nu \hat{\boldsymbol{A}} \cdot (\boldsymbol{r} \times \boldsymbol{e}_3) + \frac{1}{C}\hat{\boldsymbol{B}} \cdot \boldsymbol{\tau}$$
$$- \int_0^s \Big[\frac{\nu}{R}\hat{\boldsymbol{A}} + \Big(\frac{1}{D} + \frac{1}{CR^2}\Big)\hat{\boldsymbol{B}}\Big] \cdot \hat{\boldsymbol{r}}\, \mathrm{d}s \Big). \tag{8.63}$$

Using the relations $(8.49)_1$ and (8.63), we remark that the end edge condition $\mathscr{R}(\hat{\boldsymbol{u}}) = \mathscr{R}^0_\alpha \boldsymbol{e}_\alpha$ is satisfied, since it reduces to the Eq. $(8.55)_2$. On the other hand, the remaining end edge condition $\mathscr{M}(\hat{\boldsymbol{u}}) = \boldsymbol{0}$ yields (with the help of $(8.49)_2$ and (8.63)) an algebraic equation for the determination of $\tilde{K}$. Thus, we find the value

$$\tilde{K}\Big[2(1-\nu)\Big(C\frac{\mathscr{A}^2}{\bar{s}} + D\bar{s}\Big)\Big] = -C(1-\nu)\mathscr{A}K_0 + 2D(1-\nu)\int_0^{\bar{s}} \Big\{ \nu \hat{\boldsymbol{A}} \cdot (\boldsymbol{r} \times \boldsymbol{e}_3)$$
$$- \int_0^s \Big[\frac{\nu}{R}\hat{\boldsymbol{A}} + \Big(\frac{1}{D} + \frac{1}{CR^2}\Big)\hat{\boldsymbol{B}}\Big] \cdot \hat{\boldsymbol{r}}\, \mathrm{d}s \Big\} \mathrm{d}s + \int_0^{\bar{s}} \Big\{ (\boldsymbol{r} \cdot \boldsymbol{\tau})\big[D(1-\nu^2)\hat{\boldsymbol{A}} \cdot \boldsymbol{n} + \nu \hat{\boldsymbol{B}} \cdot \hat{\boldsymbol{r}}\big]$$
$$+ (\boldsymbol{r} \cdot \boldsymbol{n}) \int_0^s \big[C(1-\nu^2)\hat{\boldsymbol{A}} \cdot \hat{\boldsymbol{r}} - \nu \hat{\boldsymbol{B}} \cdot \big(\boldsymbol{n} + \frac{1}{R}\hat{\boldsymbol{r}}\big)\big] \mathrm{d}s \Big\} \mathrm{d}s, \tag{8.64}$$

where the constant K_0 has been determined in (8.62). This concludes the proof. □

In the next section we present an approximated form of the exact closed-form solution to Saint-Venant's problem, which could be more useful in applications.

8.3 Simplified Solution

Let us choose the origin O of the Cartesian coordinate system in the centroid of the cross-section, i.e. we assume that

$$\int_0^{\bar{s}} x_\alpha(s)\,\mathrm{d}s = 0 \quad (\alpha = 1, 2), \quad \text{and let} \quad I_{\alpha\beta} := \int_0^{\bar{s}} x_\alpha x_\beta\,\mathrm{d}s\,. \tag{8.65}$$

We consider that the shell is very thin, i.e. $\frac{h}{R} \ll 1$. Thus, we shall neglect some small terms which are of higher order in $\frac{h}{R}$ and find a simpler form of the solution. For instance, in the solution we have the term

$$\frac{1}{D} + \frac{1}{CR^2} = \frac{1}{D}\Big(1 + \frac{D}{CR^2}\Big) \simeq \frac{1}{D}\,, \tag{8.66}$$

since $\frac{D}{CR^2} = \frac{1}{12}\big(\frac{h}{R}\big)^2 \ll 1$. Furthermore, we consider that the length $\bar{s}$ of the cross-section curve $\mathscr{C}_0$ is of the same order of magnitude as R, while the area $\mathscr{A}$ of the cross-section is of the same order as $\bar{s}^{\,2}$ (for instance, for a circular tube $\bar{s} = 2\pi R$ and $\mathscr{A} = \frac{1}{4}\bar{s}^{\,2}$). Hence, we also have $\frac{h}{\bar{s}} \ll 1$ and we can approximate the torsional rigidity in (8.33) as follows

$$C\,\frac{\mathscr{A}^2}{\bar{s}} + D\bar{s} = C\,\frac{\mathscr{A}^2}{\bar{s}}\Big(1 + \frac{D}{C}\,\frac{\bar{s}^2}{\mathscr{A}^2}\Big) \simeq C\,\frac{\mathscr{A}^2}{\bar{s}}\,,$$

since $\frac{D}{C}\,\frac{\bar{s}^2}{\mathscr{A}^2} = \frac{h^2}{12}\,\frac{\bar{s}^2}{\mathscr{A}^2} = O\big((\frac{h}{\bar{s}})^2\big) \ll 1$. Then, the twist-torque relation (8.33) simplifies to

$$K = -\frac{\mathscr{M}_3^0\,\bar{s}}{2C(1-\nu)\mathscr{A}^2} = -\frac{\mathscr{M}_3^0\,\bar{s}}{4\mu h\,\mathscr{A}^2}\,, \tag{8.67}$$

where $\mu = \frac{E}{2(1+\nu)}$ is the shear modulus (Lamé's constant). Thus, K is a global measure of twist for the cylindrical shell.

In view of (8.65) and (8.66) one can write the system of equations (8.48) in the simplified form

$$\begin{aligned} I_{\alpha\beta}\,B_\beta &= -\nu D\Big(A_\beta \int_0^{\bar{s}} \frac{x_\alpha x_\beta}{R}\,\mathrm{d}s + A_3\,\bar{s}\int_0^{\bar{s}} \frac{x_\alpha}{R}\,\mathrm{d}s\Big),\\ B_3 &= -\frac{\nu D}{\bar{s}^2}\Big(A_\alpha \int_0^{\bar{s}} \frac{x_\alpha}{R}\,\mathrm{d}s + A_3\,\bar{s}\;2\pi\Big), \end{aligned} \tag{8.68}$$

since $\int_0^{\bar{s}} \frac{1}{R}\,\mathrm{d}s = 2\pi$. From Eq. (8.68) we determine the constants B_i in terms of the coefficients A_i $(i = 1, 2, 3)$ and we observe that $\boldsymbol{B}$ has the same order of magnitude as $\frac{D}{\bar{s}}\,\boldsymbol{A}$. This remark allows us to neglect some small terms in the Eq. (8.51) and to write the system (8.51) in the approximated form

$$C(1-\nu^2)\,\boldsymbol{A}\cdot\int_0^{\bar{s}}\hat{\boldsymbol{r}}\,\mathrm{d}s=-\mathscr{R}_3^0\,,\qquad C(1-\nu^2)\,\boldsymbol{A}\cdot\int_0^{\bar{s}}x_\alpha\,\hat{\boldsymbol{r}}\,\mathrm{d}s=e_{\alpha\beta}\mathscr{M}_\beta^0\,,$$

or equivalently, in view of (8.65) and $C(1-\nu^2)=Eh$,

$$I_{\alpha\beta}\,A_\beta=\frac{e_{\alpha\beta}\mathscr{M}_\beta^0}{Eh}\quad\text{and}\quad\bar{A}_3=A_3\bar{s}=-\frac{\mathscr{R}_3^0}{\bar{s}\,Eh}\,,\tag{8.69}$$

where we denote by $\bar{A}_3:=A_3\bar{s}$. From (8.69) we can see that $\bar{A}_3$ is a global measure of axial strain and A_α are global measures of axial curvature of the cylindrical shell.

Similarly, the constants $\hat{B}_i$ can be determined in terms of $\hat{A}_i$ $(i=1,2,3)$ from the same relations (8.68) and the system of equations (8.55) reduces to

$$I_{\alpha\beta}\,\hat{A}_\beta=-\frac{\mathscr{R}_\alpha^0}{Eh}\quad\text{and}\quad\hat{A}_3=0\,.\tag{8.70}$$

Thus, the constants $\hat{A}_1$, $\hat{A}_2$ can be interpreted as global measures of strain appropriate to flexure. Due to the fact that $\boldsymbol{B}\sim\frac{D}{\bar{s}}\boldsymbol{A}$, we can neglect the terms of the form

$$\frac{1}{C}\Big(s\,\boldsymbol{B}\times\boldsymbol{e}_3-\int_0^s\frac{1}{R}(\boldsymbol{B}\cdot\hat{\boldsymbol{r}})\boldsymbol{\tau}\,\mathrm{d}s\Big)\quad\text{and}\quad\hat{\boldsymbol{B}}\cdot\Big(\boldsymbol{n}+\frac{1}{R}\hat{\boldsymbol{r}}\Big)$$

in the solutions (8.31) and (8.53), as well as in the relations (8.61)–(8.64). Hence, neglecting the small terms of order $\left(\frac{h}{\bar{s}}\right)^2\ll 1$, we find from (8.61)–(8.64) the following simplified expression of the flexure function

$$\begin{aligned}\psi(s)=K_0 s-\int_0^s\Big\{\hat{A}_\alpha\Big[\frac{\nu}{2}\,x'_\alpha(x_\beta x_\beta)-\nu x_\alpha(x_\beta x'_\beta)+2(1+\nu)\int_0^s x_\alpha\,\mathrm{d}s\Big]\\ -\boldsymbol{\tau}\cdot\int_0^s\boldsymbol{n}\int_0^s\Big(\frac{\nu}{R}\hat{\boldsymbol{A}}+\frac{1}{D}\hat{\boldsymbol{B}}\Big)\cdot\hat{\boldsymbol{r}}\,\mathrm{d}s\,\mathrm{d}s\Big\}\mathrm{d}s,\end{aligned}\tag{8.71}$$

while the relations for the constants $\tilde{K}$ and K_0 reduce to

$$\begin{aligned}\tilde{K}&=-\frac{1}{2\mathscr{A}}\int_0^{\bar{s}}\Big\{\hat{A}_\alpha\Big[\frac{\nu}{2}\,x'_\alpha(x_\beta x_\beta)-\nu x_\alpha(x_\beta x'_\beta)-\frac{(1+\nu)\bar{s}}{\mathscr{A}}\,x_\alpha\varphi(s)\Big]\\ &\qquad -\boldsymbol{\tau}\cdot\int_0^s\boldsymbol{n}\int_0^s\Big(\frac{\nu}{R}\hat{\boldsymbol{A}}+\frac{1}{D}\hat{\boldsymbol{B}}\Big)\cdot\hat{\boldsymbol{r}}\,\mathrm{d}s\,\mathrm{d}s\Big\}\mathrm{d}s,\\ K_0&=-\tilde{K}\,\frac{2\mathscr{A}}{\bar{s}}-\frac{1+\nu}{\mathscr{A}}\int_0^{\bar{s}}x_\alpha\int_0^s\boldsymbol{r}\cdot\boldsymbol{n}\,\mathrm{d}s\,\mathrm{d}s.\end{aligned}\tag{8.72}$$

Consequently, on the basis of Theorems 8.2 and 8.3 we deduce the following simplified solution to our problem.

Theorem 8.4 *Consider the relaxed Saint-Venant's problem characterized by the Eqs. (8.19)–(8.22) and the end edge conditions (8.28). Then, the approximated form of the solution for the displacement field* $\boldsymbol{u} = u_i \boldsymbol{e}_i$ *is*

$$\begin{aligned} u_\alpha &= -\hat{A}_\alpha\left(\tfrac{1}{6}\, z^3 - \tfrac{1}{2}\, \nu z\, x_\beta x_\beta\right) - \tfrac{1}{2} A_\alpha (z^2 - \nu x_\beta x_\beta) - \nu x_\alpha (z\, \hat{A}_\beta x_\beta + A_\beta x_\beta + \bar{A}_3) \\ &\quad -(K+\tilde{K}) z\, e_{\alpha\beta} x_\beta - e_{\alpha\beta} \int_0^s x'_\beta \int_0^s \Big[\frac{\nu}{R}(z\, \hat{\boldsymbol{A}} + \boldsymbol{A}) + \frac{1}{D}(z\, \hat{\boldsymbol{B}} + \boldsymbol{B})\Big] \cdot \hat{\boldsymbol{r}}\, \mathrm{d}s\, \mathrm{d}s, \\ u_3 &= \tfrac{1}{2} z^2 (\hat{A}_\alpha x_\alpha) + z(A_\alpha x_\alpha + \bar{A}_3) + (K + \tilde{K}) \varphi(s) + \psi(s). \end{aligned} \tag{8.73}$$

Here, the torsion function $\varphi(s)$ *is given by (8.34), while the flexure function* $\psi(s)$ *and the constants* K, $\tilde{K}$, $\boldsymbol{A}$, $\hat{\boldsymbol{A}}$, $\boldsymbol{B}$ *and* $\hat{\boldsymbol{B}}$ *are determined in terms of the resultant loads* $\mathscr{R}_i^0$ *and* $\mathscr{M}_i^0$ *by the relations (8.67)–(8.72).*

Remark 1 In relations (8.73) one can recognize the form of classical Saint-Venant's solution for the extension, bending, torsion, and flexure of three-dimensional cylinders, see e.g. [8, 9]. Except for the integral term in u_α, which is specific to shells, the structure of the solution (8.73) coincides with the expression of Saint-Venant's solution. However, the torsion function for shells (8.34) and the relations (8.67), (8.69), (8.70) have different forms as compared to the three-dimensional counterparts. Thus, the solutions obtained above are the analogues of the classical Saint-Venant's solutions in the classical theory of shells.

2. The stress state of the shell corresponding to the solution (8.73) is given by

$$\begin{aligned} &N_{ss} = 0, \qquad N_{zz} = C(1-\nu^2)(z\hat{\boldsymbol{A}} + \boldsymbol{A}) \cdot \hat{\boldsymbol{r}}, \\ &N_{sz} = N_{zs} = C(1-\nu) K \frac{\mathscr{A}}{\bar{s}} - C(1-\nu^2)\, \hat{\boldsymbol{A}} \cdot \Big(\int_0^s \boldsymbol{r}\, \mathrm{d}s + \frac{1}{2\mathscr{A}} \int_0^{\bar{s}} \boldsymbol{r} \int_0^s \boldsymbol{r} \cdot \boldsymbol{n}\, \mathrm{d}s\, \mathrm{d}s \Big), \\ &M_{ss} = -(z\, \hat{\boldsymbol{B}} + \boldsymbol{B}) \cdot \hat{\boldsymbol{r}}, \qquad M_{zz} = -D(1-\nu^2)(z\, \hat{\boldsymbol{A}} + \boldsymbol{A}) \cdot \boldsymbol{n} - \nu (z\, \hat{\boldsymbol{B}} + \boldsymbol{B}) \cdot \hat{\boldsymbol{r}}, \\ &M_{sz} = M_{zs} = -D(1-\nu)\Big[K + \tilde{K} - \nu e_{\alpha\beta}\, \hat{A}_\alpha\, x_\beta + \int_0^s \Big(\frac{\nu}{R}\, \hat{\boldsymbol{A}} + \frac{1}{D}\, \hat{\boldsymbol{B}}\Big) \cdot \hat{\boldsymbol{r}}\, \mathrm{d}s\Big]. \end{aligned} \tag{8.74}$$

In (8.74) some small terms of order $\left(\frac{h}{\bar{s}}\right)^2 \ll 1$, such as, e.g., $\boldsymbol{B} \cdot \left(\boldsymbol{n} + \frac{1}{R}\hat{\boldsymbol{r}}\right)$ and $\frac{1}{C}\hat{\boldsymbol{B}} \cdot \boldsymbol{\tau}$ have been neglected, see relations (8.50) and (8.63) for comparison. The dependence of the stress tensors on the axial coordinate z in (8.74) is at most linear.

3. The torque-twist relation (8.67) and the relations (8.69), which express the measures of curvature A_α and stretch $\bar{A}_3$ in terms of the resultants $\mathscr{M}_\alpha^0$ and $\mathscr{R}_3^0$, can be found in various forms in the literature, see e.g., the papers [12, 13] and the classical books [15, Sect. 47] and [16, Sects. 94, 98, 102], for some special cases.

8.4 Circular Cylindrical Tubes

The results presented above are valid for cylindrical tubes with arbitrary cross-sections. Let us specialize these results to the case of *circular* cylindrical shells and derive the solution of the relaxed Saint-Venant's problem.

The parametrization (8.15) of the midsurface has in this case the following form

$$x_1(s) = R_0 \cos\frac{s}{R_0}\,, \qquad x_2(s) = R_0 \sin\frac{s}{R_0}\,, \qquad s \in [0, \bar{s}]\,, \tag{8.75}$$

where R_0 is the radius of the circular cylindrical surface and $\bar{s} = 2\pi R_0$. From (8.75) we deduce that

$$\begin{aligned} &x'_\alpha = e_{\beta\alpha}\frac{x_\beta}{R_0}\,, \quad x''_\alpha = -\frac{x_\alpha}{R_0^2}\,, \quad \boldsymbol{\tau} = \frac{1}{R_0}\,e_{\beta\alpha}x_\beta\boldsymbol{e}_\alpha\,, \quad \boldsymbol{n} = \frac{1}{R_0}\,x_\alpha\boldsymbol{e}_\alpha\,, \\ &\int_0^{\bar{s}} x_\alpha\,\mathrm{d}s = 0, \qquad I_{\alpha\beta} = \int_0^{\bar{s}} x_\alpha x_\beta\,\mathrm{d}s = \pi\,R_0^3\,\delta_{\alpha\beta}\,, \qquad \frac{2\mathscr{A}^2}{\bar{s}} = \pi\,R_0^3\,. \end{aligned} \tag{8.76}$$

From (8.68) and (8.76) we can express the constants B_i in terms of the coefficients A_i by the relations

$$B_i = -\frac{\nu D}{R_0}\,A_i \quad \text{and} \quad \hat{B}_i = -\frac{\nu D}{R_0}\,\hat{A}_i \quad (i = 1, 2, 3). \tag{8.77}$$

Hence, we have

$$\frac{\nu}{R_0}\boldsymbol{A} + \frac{1}{D}\boldsymbol{B} = \boldsymbol{0}$$

and the integral term in the solution (8.73) vanishes.

Further, the relations (8.67), (8.69) and (8.70) reduce to

$$K = \frac{-\mathscr{M}_3^0}{2\pi\,R_0^3\mu h}, \qquad A_\alpha = \frac{e_{\alpha\beta}\mathscr{M}_\beta^0}{\pi\,R_0^3 Eh}, \qquad \bar{A}_3 = \frac{-\mathscr{R}_3^0}{2\pi\,R_0 Eh}, \qquad \hat{A}_\alpha = \frac{-\mathscr{R}_\alpha^0}{\pi\,R_0^3 Eh}. \tag{8.78}$$

In view of (8.76), the torsion function (8.34) is vanishing

$$\varphi(s) = 0$$

and from (8.71) and (8.72) we find

$$\psi(s) = \frac{4+3\nu}{2}\,R_0^2\,\hat{A}_\alpha\,x_\alpha\,, \qquad K_0 = 2(1+\nu)R_0^2\,\hat{A}_2 \quad \text{and} \quad \tilde{K} = 0. \tag{8.79}$$

Using the above results, we obtain from Theorem 8.4 the following solution.

Corollary 1 *Consider the relaxed Saint-Venant's problem (8.19)–(8.22) with boundary conditions (8.28) for circular cylindrical shells. Then, the solution for the displacement field is given by (up to a rigid body displacement field)*

$$
\begin{aligned}
u_\alpha &= -\tfrac{1}{6}\, z^3 \hat{A}_\alpha - \tfrac{1}{2}\, z^2 A_\alpha - \nu\, z\, (\hat{A}_\beta x_\beta)\, x_\alpha - \nu (A_\beta x_\beta + \bar{A}_3)\, x_\alpha - K\, z\, e_{\alpha\beta}\, x_\beta\,, \\
u_3 &= \big[\tfrac{1}{2}\, z^2 + 2(1+\nu) R_0^2\big] (\hat{A}_\alpha x_\alpha) + z\, (A_\alpha x_\alpha + \bar{A}_3),
\end{aligned}
\tag{8.80}
$$

where the constants K, A_α, $\bar{A}_3$ and $\hat{A}_\alpha$ represent measures of twist, curvature, stretch and flexure, respectively, which are given in terms of the resultant loads $\mathscr{R}_i^0$ and $\mathscr{M}_i^0$ by the relations (8.78).

These results are consistent with classical solutions for the deformation of circular cylindrical shells, see e.g. [15, 16].

Finally, we mention that the solution procedure presented in this chapter can be extended to solve the relaxed Saint-Venant's problem for orthotropic or anisotropic shells [3, 5] or to investigate thermal stresses in cylindrical shells [4, 6].

References

1. Berdichevsky, V., Armanios, E., Badir, A.: Theory of anisotropic thin-walled closed-cross-section beams. Comp. Eng. **2**, 411–432 (1992)
2. Bîrsan, M.: The solution of Saint-Venant's problem in the theory of Cosserat shells. J. Elast. **74**, 185–214 (2004)
3. Bîrsan, M.: On Saint-Venant's problem for anisotropic, inhomogeneous, cylindrical Cosserat elastic shells. Int. J. Engng. Sci. **47**, 21–38 (2009)
4. Bîrsan, M.: Thermal stresses in cylindrical Cosserat elastic shells. Eur. J. Mech. A/Solids **28**, 94–101 (2009)
5. Bîrsan, M., Altenbach, H.: Analysis of the deformation of multi-layered orthotropic cylindrical elastic shells using the direct approach. In: Altenbach, H., Eremeyev, V. (eds.) Shell-Like Structures: Non-classical Theories and Applications, pp. 29–52. Springer, Berlin Heidelberg (2011)
6. Bîrsan, M., Sadowski, T., Pietras, D.: Thermoelastic deformations of cylindrical multi-layered shells using a direct approach. J. Therm. Stresses **36**, 749–789 (2013)
7. Bîrsan, M.: Closed-form Saint-Venant solutions in the Koiter theory of shells. J. Elast. **140**, 149–169 (2020)
8. Ieşan, D.: On Saint-Venant's problem. Arch. Rat. Mech. Anal. **91**, 363–373 (1986)
9. Ieşan, D.: Saint-Venant's problem. In: Lecture Notes in Mathematics, vol. 1279. Springer, New York (1987)
10. Ladevèze, P., Sanchez, P., Simmonds, J.: Beamlike (Saint-Venant) solutions for fully anisotropic elastic tubes of arbitrary cross section. Int. J. Solids Struct. **41**, 1925–1944 (2004)
11. Lebedev, L.P., Cloud, M.J., Eremeyev, V.A.: Tensor Analysis with Applications in Mechanics. World Scientific, New Jersey (2010)
12. Reissner, E.: On torsion of thin cylindrical shells. J. Mech. Phys. Solids **7**, 157–162 (1959)
13. Reissner, E., Tsai, W.: Pure bending, stretching, and twisting of anisotropic cylindrical shells. J. Appl. Mech. **39**, 148–154 (1972)

14. de Saint-Venant, B.: Mémoire sur le calcul des pièces solides à simple ou à double courbure, et prenant simultanément en considération les divers efforts auxquelles elles peuvent être soumises dans tous les sens. C. R. Acad. Sci. Paris **17**(942–954), 1020–1031 (1843)
15. Sokolnikoff, I.: Mathematical Theory of Elasticity. McGraw-Hill, New York (1956)
16. Timoshenko, S., Goodier, J.: Theory of Elasticity. McGraw-Hill, New York (1951)

Printed in the United States
by Baker & Taylor Publisher Services